The Observatory Experiment

In this innovative history of the science of meteorology, Simon Naylor focuses our attention on the spaces in which it was pursued: meteorological observatories. During the nineteenth century, meteorologists established or converted sites where observers and their instruments could be housed, where they collected and analysed data and developed meteorological theories. He examines a number of these sites around the British Empire, along with the governmental, military and commercial networks connecting them. Taking many shapes to capture the weather in different environments, these observatories brought various social groups into contact with the practice of science, including sailors on naval surveying vessels, climbers ascending Scottish peaks and families checking their rain gauges at home. Through a study of these spaces, Naylor argues for the treatment of meteorology as an experimental observatory science, on which the development of knowledge about local, regional, national and global weather and climate relied.

Simon Naylor is Professor of Historical Geography at the University of Glasgow. He is the author of *Regionalizing Science: Placing Knowledges in Victorian England* (2010).

Science in History

Series Editors

Science in History is a major series of ambitious books on the history of the sciences from the mid-eighteenth century through the mid-twentieth century, highlighting work that interprets the sciences from perspectives drawn from across the discipline of history. The focus on the major epoch of global economic, industrial and social transformations is intended to encourage the use of sophisticated historical models to make sense of the ways in which the sciences have developed and changed. The series encourages the exploration of a wide range of scientific traditions and the interrelations between them. It particularly welcomes work that takes seriously the material practices of the sciences and is broad in geographical scope.

A full list of titles in the series can be found at: www.cambridge.org/sciencehistory

The Observatory Experiment

Meteorology in Britain and Its Empire

Simon Naylor

University of Glasgow

Shaftesbury Road, Cambridge CB2 8EA, United Kingdom

One Liberty Plaza, 20th Floor, New York, NY 10006, USA

477 Williamstown Road, Port Melbourne, VIC 3207, Australia

314–321, 3rd Floor, Plot 3, Splendor Forum, Jasola District Centre, New Delhi – 110025, India

103 Penang Road, #05–06/07, Visioncrest Commercial, Singapore 238467

Cambridge University Press is part of Cambridge University Press & Assessment, a department of the University of Cambridge.

We share the University's mission to contribute to society through the pursuit of education, learning and research at the highest international levels of excellence.

www.cambridge.org
Information on this title: www.cambridge.org/9781009207232

DOI: 10.1017/9781009207225

When citing this work, please include a reference to the DOI 10.1017/9781009207225

First published 2024

A catalogue record for this publication is available from the British Library

Library of Congress Cataloging-in-Publication data
Names: Naylor, Simon, author.
Title: The observatory experiment : meteorology in Britain and its empire / Simon Naylor.
Description: Cambridge, United Kingdom; New York, NY : Cambridge University Press, 2024. | Series: Science in history | Includes bibliographical references.
Identifiers: LCCN 2024011998 | ISBN 9781009207232 (hardback) | ISBN 9781009207218 (paperback) | ISBN 9781009207225 (ebook)
Subjects: LCSH: Meteorological stations – Great Britain –History. | Meteorology – Observations.
Classification: LCC QC875.G7 N395 2024 | DDC 551.509171/241–dc23/eng/20240509
LC record available at https://lccn.loc.gov/2024011998

ISBN 978-1-009-20723-2 Hardback

Contents

Figures

Acknowledgements

This book has taken more years to produce than I care to admit and the debts I have incurred have accumulated accordingly. It has been an immense privilege and pleasure to carry out much of the research in various libraries and archives across the UK. I would like to thank the staff at the British Library, London; the National Archives, Kew; the Royal Society Archives, London; the Royal Geographical Society, London; the National Meteorological Library and Archive, Exeter; the UK Hydrographic Office Archives, Taunton; the National Records of Scotland, Edinburgh; the National Library of Scotland, Edinburgh; and Special Collections at the University of Glasgow. Special thanks are due to Mark Beswick and Catherine Ross at the National Meteorological Archive for their continued support over many years. Thank you to the National Records of Scotland, the Royal Meteorological Society and National Library of Scotland for granting permission to reproduce images from their collections in this book. Much of the archival work was supported by grants and fellowships. Thank you to the British Academy and Leverhulme Trust for a Small Research Grant, which allowed me to conduct research into maritime and colonial meteorology; to the National Library of Scotland for awarding me the Graham Brown Research Fellowship, which allowed me to research the history of the Ben Nevis Observatory (with particular thanks to Paula Williams and Chris Taylor); and to the Leverhulme Trust for a Research Fellowship that allowed me to conduct research into the British Rainfall Organisation and to write the book's Introduction and Conclusion.

Several chapters in this book incorporate material that appeared in earlier forms in journals and edited books. Chapter 1 pulls on material from S. Naylor, 'Log Books and the Law of Storms: Maritime Meteorology and the British Admiralty in the Nineteenth Century', *Isis*, 106 (2015), 771–797; and S. Naylor, 'Weather Instruments All at Sea: Meteorology and the Royal Navy in the Nineteenth Century', in F. MacDonald and C. W. J. Withers (eds.) *Geography, Technology and Instruments of Exploration* (Aldershot: Ashgate, 2015), pp. 77–96.

Chapter 2 pulls on material from S. Naylor and M. Goodman, 'Atmospheric Empire: Historical Geographies of Meteorology at the Colonial Observatories', in M. Mahony and S. Randalls (eds.), *Weather, Climate and the Geographical Imagination: Placing Atmospheric Knowledges* (Pittsburgh: University of Pittsburgh Press, 2020), pp. 25–42. Thank you to the editors and publishers for granting permission to reprint and include this material here.

Professor Jim Secord originally spoke to me about submitting a proposal to the Science in History series during the British Society for the History of Science conference at St Andrews in 2014. Since then, Lucy Rhymer, my editor at Cambridge University Press, has been incredibly encouraging and supportive as she has helped me to bring this project to completion. Rosa Martin and Natasha Whelan, also at Cambridge University Press, have been a terrific help during the process of getting the manuscript ready for publication. Special thanks to Jim and Professor Simon Schaffer, the series' editors, and to the anonymous reviewers, for their helpful comments and suggestions on the proposal and on earlier drafts of the manuscript. I've given a lot of conference and seminar papers on aspects of this book. Thanks to the organisers of those events for the opportunity to share my research and to attentive audiences for their feedback. I am incredibly privileged to work at the interstices of the fields of historical geography and the history of science. I owe a huge debt to a great many people who also work at this junction, including Robert Mayhew, Innes Keighren, Martin Mahony, Charles Withers, David Livingstone, Diarmid Finnegan, James Ryan, Fraser MacDonald, Hayden Lorimer, Becky Higgitt, Sam Alberti, Katharine Anderson, Vladimir Jankovic, Maria Lane, Felix Driver, Michael Bravo, Kirsten Greer, Miles Ogborn, Neil Macdonald, Georgina Endfield, Peter Martin, Lawrence Dritsas, Simone Turchetti, Katrina Dean, Peter Jones, Tara Jonell and Adam Lucas. Research for this book began when I was based at the University of Exeter and ended at the University of Glasgow. Thank you to colleagues and graduate students at both those institutions for supporting me personally and for developing and sustaining collegiate environments in which humanities research can flourish. Lastly, the very biggest thank you to Larissa, Tristan and Cameron for all the support, encouragement and almond croissants over the years.

Introduction: Observatory Experiments

At the 1881 Annual General Meeting of Britain's Meteorological Society, the President, George Symons, presented a history of the Society and its predecessors. Symons began his talk with an outline of the history of the Meteorological Society of London, which held its first meeting at the London Coffee House, Ludgate Hill, in October 1823. After an uncertain start the London Society published its first volume of *Transactions* in 1839. The Society quickly got into debt and disbanded in 1843 but was replaced by the British Meteorological Society in 1850, which included members of the previous organisation and took possession of some of its library. The latter was renamed the Meteorological Society in 1866, before settling on the title Royal Meteorological Society in 1883, after Queen Victoria granted the use of the appellation. Symons pointed out the symptoms of the latter organisation's success: its increasing number of Fellows; its growing set of meteorological and climatological stations; its burgeoning list of publications; and its financial security.

In his history, Symons quoted at length from an article that featured in the first volume of the *Transactions* in 1839, written by the young John Ruskin. In the article, entitled 'The present state of Meteorological Science', Ruskin had also been keen to reinforce the value of the London Meteorological Society in particular, and those like it more generally, while clarifying why such meteorological organisations might struggle to achieve their goals:

> There is one point, it must now be observed, in which the science of meteorology differs from all others. A Galileo, or a Newton, by the unassisted workings of his solitary mind, may discover the secrets of the heavens, and form a new system of astronomy. A Davy in his lonely meditations on the crags of Cornwall, or, in his solitary laboratory, might discover the most sublime mysteries of nature, and trace out the most intricate combinations of her elements. But the meteorologist is impotent if alone; his observations are useless, for they are made upon a point, while the speculations to be derived from them must be on space. It is of no avail that he changes his position, ignorant of what is passing behind him and before; he desires to estimate the movements of space, and can only observe the dancing of

atoms; he would calculate the currents of the atmosphere of the world, while he only knows the direction of a breeze. It is perhaps for this reason that the cause of meteorology has hitherto been so slightly supported; no progress can be made by the enthusiasm of an individual; no effect can be produced by the most gigantic efforts of a solitary intellect, and the co-operation demanded was difficult to obtain, because it was necessary that the individuals should think, observe, and act simultaneously, though separated from each other, by distances, on the greatness of which depended the utility of the observations.

The Meteorological Society, therefore, has been formed, not for a city, nor for a kingdom, but for the world. It wishes to be the central point, the moving power, of a vast machine, and it feels that unless it can be this, it must be powerless; if it cannot do all, it can do nothing. It desires to have at its command, at stated periods, perfect systems of methodical, and simultaneous observations; it wishes its influence and its power to be omni-present over the globe, so that it may be able to know, at any given instant, the state of the atmosphere at every point on its surface. Let it not be supposed that this is a chimerical imagination – the vain dream of a few philosophical enthusiasts. It is co-operation which we now come forward to request, in full confidence that, if our efforts are met with a zeal worthy of the cause, our associates will be astonished, *individually*, by the result of their labours in a body. ... [E]ach, who alone would have been powerless, will find themselves a part of one Mighty Mind, – a ray of light into one vast Eye, – a member of a multitudinous Power, contributing to the knowledge, and aiding the efforts, which will be capable of solving the most deeply hidden problems of Nature, penetrating into the most occult causes, and reducing to principle and order, the vast multitude of beautiful and wonderful phenomena.[1]

The first decades of the nineteenth century were both challenging and exciting times for the science of meteorology. As Jankovic shows in his seminal study of Enlightenment weather *Reading the Skies*, meteorologists in the early years of the nineteenth century began to abandon their preoccupation with local weather and unusual atmospheric events.[2] The so-called meteoric tradition – typically undertaken by savants embedded in both local topographical and chorographical social hierarches and a more expansive republic of letters – was gradually replaced by a preoccupation with a study of atmospheric air and its planetary circulation. Observation of local weather still mattered but emphasis

[1] J. Ruskin, 'Remarks on the Present State of Meteorological Science', *Transactions of the Meteorological Society*, 1 (1839), 56–9, quoted in G. Symons, 'The History of English Meteorological Societies, 1823 to 1880', *Quarterly Journal of the Meteorological Society*, 7 (1881), 65–98, 72. For discussion of Ruskin's interests in weather and climate, see respectively, K. Anderson, 'Looking at the Sky: The Visual Context of Victorian Meteorology', *British Journal for the History of Science*, 36 (2003), 301–32; S. S. S. Cardoso, J. H. E. Cartwright and H. E. Huppert, 'Stokes, Tyndall, Ruskin and the Nineteenth-Century Beginnings of Climate Science', *Philosophical Transactions of the Royal Society A*, 378 (2020), 1–15.

[2] V. Jankovic, *Reading the Skies: A Cultural History of English Weather, 1650–1820* (Manchester: Manchester University Press, 2000).

shifted from the documentation of meteoric events to records of everyday weather. In line with, and usually following, wider shifts across the physical sciences, increasing store was placed on the principles of synchronisation, standardisation and quantification, such that data could travel freely. Scientific periodicals and newspapers continued to publish accounts of unusual weather but increasingly devoted their meteorological coverage to printing tables of mean temperature and atmospheric pressure and, later, weather maps.[3]

Ruskin's evaluation of the state of British meteorology provided a contemporary justification for this shift from 'places of life to places on the map'.[4] The crucial questions of the revivified science of meteorology could no longer rely on the isolated labour of even the most brilliant practitioner. The atomism of meteoric reportage revealed nothing more than the way the wind was blowing; what mattered was the coordination of observations across space such that they could be compared, combined together and used to track the movement of atmospheric currents. Nebeker refers to promoters of this meteorological agenda as the empiricists and argues that they dominated the science across much of the nineteenth century.[5] These actors worked within scientific societies and alongside government departments and military outfits to build regional, national and global machineries that would monitor the world's weather. Building and running observatories was a crucial part of this endeavour. Observatory personnel policed observational practices and regimens across their bailiwicks; advised on the deployment of instruments; acted as clearing houses for observations, their reduction and statistical manipulation; sustained communication with other bodies internationally; and adjudicated on the development of meteorological theories and laws. In her study of the establishment of a magnetic and meteorological observatory in Singapore in 1841, Williamson argues that the site was part of an 'observatory experiment' into global terrestrial physics.[6] Williamson notes that this mid-nineteenth-century experiment had a direct bearing on scientific activities in the second half of the century, with attempts to observe the weather across the Straits Settlements by public works officials, lighthouse staff, medical officers and naval personnel. This book argues that we should view *all* meteorological observatory projects in the

[3] V. Jankovic, 'Ideological Crests Versus Empirical Troughs: John Herschel's and William Radcliffe Birt's Research on Atmospheric Waves, 1843–50', *British Journal for the History of Science*, 31 (1998), 21–40.

[4] Jankovic, *Reading the Skies*, p. 159.

[5] F. Nebeker, *Calculating the Weather: Meteorology in the 20th Century* (New York: Academic Press, 1995), p. 1.

[6] F. Williamson, 'Weathering the Empire: Meteorological Research in the Early British Straits Settlements', *British Journal for the History of Science*, 48 (2015), 475–92, 482.

nineteenth century as observatory experiments – as investigations both into weather science and how it might be most effectively prosecuted. Gooding, Pinch and Schaffer tell us that experiments act as powerful resources for persuasion and conviction but that any consensus is always conditional and open to challenge.[7] As the book will show, observatory experiments were far from settled in the middle years of the nineteenth century – projects at the beginning of the twentieth century were still being labelled as such.[8]

Meteorological observatories continued to be treated as sites where outcomes remained uncertain because, like the steam engines that powered industrialisation in the nineteenth century, Ruskin's vast machine of meteorology often failed to run smoothly. Even observatories like Singapore, supported as they were by the British military, powerful scientific institutions and the East India Company, faced uncertain financial futures. Observatories big and small often proved to be stubbornly caught up in local landscapes and cultures. Observers produced huge quantities of printed data but they also exhibited a lack of discrimination in data collection and data treatment, and often struggled with the rigours of state-of-the-art statistical analysis. Meteorology's one vast Eye – Ruskin's metaphor for meteorology's global observational network – turned out to be less than all-seeing. Networks were hard to maintain, especially so at extremes of range but even within national boundaries, and quickly fell apart. The value of meteorological observations was debated vigorously over the course of the nineteenth century but consensus on meteorology's utility remained elusive and contested.

This book explores the complicated and contested refashioning of meteorology as an experimental observatory science over the course of the nineteenth and early twentieth centuries. In doing so it responds to Anderson's own exemplary treatment of Victorian weather study. In *Predicting the Weather*, Anderson argues that it is difficult to grasp meteorology's development when we pay attention to only disembodied debates about physical theories, or even to the more embodied theorisers. Some of the most prominent figures in nineteenth-century science were certainly involved in the development of meteorology, as were many of the largest observatories; the discipline also received significant amounts of state funding and support. But meteorology also enjoyed extraordinary levels

[7] D. Gooding, T. Pinch and S. Schaffer, *The Uses of Experiment: Studies in the Natural Sciences* (Cambridge: Cambridge University Press, 1989).

[8] Moore focuses on the same broad period and some of the same figures but argues for a 'generational experiment: a quest to prove that earth's atmosphere was not chaotic beyond comprehension'. P. Moore, *The Weather Experiment: The Pioneers Who Sought to See the Future* (London: Chatto & Windus, 2015), p. 6.

of mass participation; it was the subject of extensive coverage in general interest periodicals; and its instruments could be found in most homes. It would therefore be a mistake to read meteorology as only an elite activity pursued in rarefied settings. Rather, it brought together a diversity of social groups, ideas and locations, even if that integration could be uncomfortable or frustrating for some of those involved.[9] For Anderson, 'the historical action is in ideas about instruments or how statistics are used, about the ways that popular weather prophets communicated with the public, how networks of observers were built, and how government officials interacted with scientists'.[10] In this regard, the book asks what happens when we develop a history of meteorology from the perspective of its geographies, rather than from the viewpoint of a notable meteorological personality or theoretical position. It takes Anderson's advice and follows the historical action into the simultaneously mundane and enigmatic space of the weather observatory. The study treats observatories as fulcrums for the mixing of facts and theories; as places where local preoccupations intermingled with metropolitan concerns; where humans communicated with instruments and hoped they talked back; where controversies were staged, fought out and sometimes resolved; where notions of meteorological utility brushed up against atmospheric abstraction; where the volunteer observer could interact with the leading meteorologist; and where the local scene came into contact with the wider world.[11]

This book explores the negotiation of these relationships in a range of spaces – ships at sea, colonial buildings, huts on mountain tops and suburban back gardens. As we will see, the success or otherwise of these negotiations could either define these experimental spaces as an observatory or consign them to ruination. In choosing to focus on such a diverse range of sites, the book deliberately establishes a set of multi-centred historical geographies of weather observatories. Following Roberts, these historical geographies 'do not privilege a single location as the active centre which other locations are considered either to (passively) feed or resist'.[12] Rather, this approach treats the sites seriously on their own terms and explores how they positioned themselves in relation to other spaces of science and how they understood those interrelations. In doing

[9] M. Mahony and A. M. Caglioti, 'Relocating Meteorology', *History of Meteorology*, 8 (2017), 1–14.
[10] K. Anderson, *Predicting the Weather: Victorians and the Science of Meteorology* (Chicago: University of Chicago Press, 2005), p. 5.
[11] Anderson, *Predicting the Weather*, p. 9.
[12] L. Roberts, 'Accumulation and Management in Global Historical Perspective: An Introduction', *History of Science*, 52 (2014), 227–246, 228.

so, Raj's circulatory perspective is followed, whereby it is assumed that a variety of actors co-produced meteorological science at different scales: 'construct spaces tailored to their own activity, cultivate solutions of continuity, and function through networks'.[13] In developing these simultaneously large-scale and fine-grained historical geographies of the weather observatory, the book asks a number of specific questions: What counted as a meteorological observatory? What was the right way to observe the weather? How were observatory networks configured? How were weather data managed? And what were the ends of observatory meteorology? Contexts for each of these questions are provided in the next sections.

What Counted as a Meteorological Observatory?

Aubin argues that the nineteenth century was a period of triumph and crisis for the observatory.[14] Although not new, the number of observatories increased globally and quite suddenly in the nineteenth century. The use of the term 'observatory' in English-language literature also increased markedly. The endowment of expensive observatories became an indispensable requirement for modern states, in Europe and elsewhere, that wanted to show off their scientific enlightenment, technical modernity, political independence and international connections. Although the early observatories were dedicated to the study of the heavens, many diversified to the study of the atmosphere and the terrestrial sphere. The study of weather conditions was necessary for corrections to astronomical readings and the study of geomagnetism was important for work on navigational astronomy, but observatories were established in the nineteenth century that were devoted solely to terrestrial physics and did no astronomy at all. They supported topographical, geodetic, oceanographic and population surveys; they helped to increase the precision of metrological units; they conducted physical experiments and they acted as repositories for statistical data and their analysis. For instance, the Royal Observatory at Greenwich was established in 1675 to service navigational astronomy but incorporated meteorological and geomagnetic work in the nineteenth century. Weather records began in 1811, when John Henry Belville, a junior member of staff, started a weather journal. Belville installed a rain gauge in 1814, although his observations stopped in advance of the constitution of James Glaisher's Magnetical and Meteorological

[13] K. Raj, 'Beyond Postcolonialism … and Postpositivism: Circulation and the Global History of Science', *Isis*, 104 (2013), 337–47, 347.

[14] D. Aubin, 'A History of Observatory Sciences and Techniques', in J.-P. Lasota (ed.), *Astronomy at the Frontiers of Science* (Heidelberg: Springer, 2011), pp. 109–21.

Department in 1841.[15] On the other side of London, Kew Observatory was established to allow King George III to observe the transit of Venus in 1769 but began to collect weather observations, store instruments and provide mathematical instruction. It was adapted for geomagnetic observations in the 1840s and became a vital centre for testing instrument standards in the 1850s. Sites like Kew were increasingly referred to as 'physical observatories'.[16]

As the observatory incorporated a wider range of scientific preoccupations under its roof, the question of what actually constituted an observatory in the first place became more difficult to answer. Aubin and other historians of the observatory sciences argue that a set of techniques defined the space of the observatory and not the other way round. Observatory techniques valued the 'calibration, manipulation, and coordination of precision instruments for making observations and taking measurements'.[17] Associative techniques included 'data acquisition, reduction, tabulation, and conservation, along with complex mathematical analysis'.[18] Observatories made use of cutting-edge visualisation techniques to represent those data. Observatory techniques also incorporated the social management of personnel and networks of international collaboration. These were the techniques that made somewhere count as an observatory – they defined a common space of knowledge, whether that space was devoted to the study of astronomy, meteorology, geomagnetism or some combination of all these and other pursuits. As Schaffer notes in his analysis of the Paramatta Observatory in New South Wales, observational techniques governed every aspect of observatory life – they 'put observatories on the map and could make or destroy reputations'.[19]

Defined by its techniques, the observatory was somewhat freed from the constraints of architectural expectation – from the geographical imagination of the neoclassical monument set in tranquil gardens. An observatory could be a canvas tent as part of an expeditionary foray,

[15] W. C. Nash, 'One Hundred Years' Greenwich Rainfall, 1815–1914', *British Rainfall, 1915*, 55 (1916), 35–9; D. Belteki, 'The Spring of Order: Robert Main's Management of Astronomical Labor at the Royal Observatory, Greenwich', *History of Science*, 60 (2022), 575–93.

[16] L. T. Macdonald, *Kew Observatory and the Evolution of Victorian Science 1840–1910* (Pittsburgh: University of Pittsburgh Press, 2018).

[17] D. Aubin, C. Bigg and H. O. Sibum, 'Introduction: Observatory Techniques in Nineteenth-Century Science and Society', in D. Aubin, C. Bigg and H. O. Sibum (eds.), *The Heavens on Earth: Observatories and Astronomy in Nineteenth-Century Science and Culture* (Durham: Duke University Press, 2010), pp. 1–32, p. 6.

[18] Aubin et al., 'Introduction', p. 6.

[19] S. Schaffer, 'Keeping the Books at Paramatta Observatory' in D. Aubin, C. Bigg and H. Otto Sibum (eds.), *The Heavens on Earth: Observatories and Astronomy in Nineteenth-Century Science and Culture* (Durham: Duke University Press, 2010), pp. 118–47, p. 122.

a wooden fort hastily built by military personnel or a hut set in polar wastes. It could even be a ship at sea or a collection of instruments set out in a suburban back garden. Some constraints remained, however. The need for solid and deep foundations in astronomical observatories proved ideal for geomagnetic observatories and their sensitive magnetometers, assuming of course the building was free of iron and brick. The stately and secluded grounds of the astronomical observatories proved useful for the erection of wooden magnetic observatories where the original building was not fit for purpose, and for the placement of meteorological instruments that demanded good exposure to the weather. Many observatories also had to provide ancillary services, most obviously office space and living space for the human workforce, and storage space for the instruments, books, data, maps and other objects. As Higgitt's work on Greenwich's physical observatory reminds us, observatories were often a collection of buildings that provided space for domestic as well as scientific needs.[20]

The domestic life of the observatory highlighted a challenge to the idea and ideal of the observatory as a place set apart from the world. Observatories staked their epistemological trustworthiness and authority on their isolation, geographically and socially. Observatories were meant to collect and record information about nature in a methodical and non-intrusive way, while the world was not meant to intrude back. Sophisticated instruments, architectural design, specific building materials, secure boundaries and an appropriate location helped in this regard, but observatories remained stubbornly connected to other activities, places and personnel. The domesticity of the observatory required the presence of cooks and cleaners, builders and telegraph operators. Deliveries had to be made. Dignitaries and tourists demanded tours of the facilities. Family members lived on site and helped out while getting on with their lives. Observatories relied on and accommodated the presence of these various social groups while pretending they did not exist. If domestic lives had to work around the demands of scientific regimen, it was also the case that science became routinised and domesticated. This was especially the case where measurement schedules had to fit around volunteers' routines (discussed in Chapter 4), or where observatories were also living spaces, such as mountaintop stations or ships at sea (the focus of Chapters 3 and 1, respectively).

[20] R. Higgitt, 'A British National Observatory: The Building of the New Physical Observatory at Greenwich, 1889–1898', *British Journal for the History of Science*, 47 (2014), 609–35.

What Was the Right Way to Observe the Weather?

Arguably the most crucial of observatory techniques were practices of observation and measurement of quantities in nature, conducted by human observers interacting with instruments. For observations to count as credible facts, and ultimately as scientific knowledge, they had to conform to contemporary standards of precision, accuracy and uniformity. The pattern science for precision observation was astronomy. To ensure that astronomy was placed on a solid calculating base, great emphasis was placed on precise measurement and systematic calculation, and both were kept under constant vigilance.[21] One particular concern was the efficacy of observatory staff. Could they be guaranteed to operate in a fashion that was sufficiently precise? Astronomers had identified the worrying fact of observational idiosyncrasy in the times recorded for transits. The 'personal equation' was the term given to this, which described attempts to calibrate the disciplined performance of the observer such that they became part of the instrument to be calibrated.[22] Self-registering instruments and the mechanisation of observation promised some relief from this problem, while they too were subject to close surveillance.

Programmes dedicated to terrestrial physics – some based in the same building complexes as astronomical observatories – took astronomy as their model and also promoted cultures of precision, accuracy and uniformity in their observational activities, even if the configuration of these variables remained open to debate and required negotiation within various contexts: within the space of the observatory itself; in the pages of printed journals; and in public debates, for instance. Which instruments to use, where and when to use them, how to calibrate and deploy them, how to read them, how to detect problems, and what to do when they stopped working were critical questions across the physical sciences. How observers should interact with their instruments and the possible effects the observing body had on measurement were also issues of fierce debate. Gooday reminds us that questions about measurement were also questions about moral conduct. The morals of measurement incorporated judgements about the integrity of the measurers, the trustworthiness of measurement practices, the appropriateness of instruments, the honesty of reports and the conjectured benefits of measurement.[23] Achieving

[21] W. J. Ashworth, 'The Calculating Eye: Baily, Herschel, Babbage and the Business of Astronomy', *British Journal for the History of Science*, 27 (1994), 401–44.

[22] S. Schaffer, 'Astronomers Mark Time: Discipline and the Personal Equation', *Science in Context*, 2 (1988), 115–45, 118.

[23] G. J. N. Gooday, *The Morals of Measurement: Accuracy, Irony, and Trust in Late Victorian Electrical Practice* (Cambridge: Cambridge University Press, 2010 [2004]), p. xvi.

good degrees of accuracy was often determined by the exercise of the qualitative notion of care. Despite the increasing prevalence of a statistical discourse in the nineteenth-century physical sciences, care remained a powerful concept through which to judge the trustworthiness of measurements and measurers. Care referred to the precautions that had to be taken to eliminate errors and sources of interference prior to measurement taking place, to the performance of experiment or practice of measurement and to the treatment of numerical outcomes and their possible errors.

Like other observatory sciences, meteorology conformed to another collective noun of scientific practice: the survey sciences. The survey sciences emphasised the conduct of large-scale and yet fine-grained information collection across space, but also placed great importance on the mobility of instruments and observers and on the value of study in particular locations and environments.[24] Unlike the geographers, geologists and naturalists, meteorologists generally persisted with the principle of observation from a fixed site rather than peripatetic body. In line with Ruskin's call to 'think, observe, and act simultaneously, though separated from each other, by distances', meteorologists established stations across extensive territories in the hope of capturing the movement of weather systems overhead. As Lehmann shows in his account of weather networks across Germany's African empire, these observation sites were often nothing more than a set of instruments, appropriately exposed.[25] The impetus to coordinate dense networks of observational sites while operating in austere environments emphasised the importance of uniformity. This could precipitate a tension, where some locations in networks of observatory sites attracted particular attention – and funding – such as the tops of mountains, particular islands or places of exceptional rainfall. Charismatic sites challenged some of the principles of the observatory sciences.[26] How could stations founded on their own uniqueness replicate themselves such that their data could be shared and compared with other sites? Managing this problem could take observatories to the brink, especially those whose credibility and trustworthiness were built on the site's status as somehow remarkable. As we will learn in Chapter 3, for

[24] S. Naylor and S. Schaffer, 'Nineteenth-Century Survey Sciences: Enterprises, Expeditions and Exhibitions', *Notes and Records of the Royal Society*, 73 (2019), 135–47.

[25] P. Lehmann, 'Average Rainfall and the Play of Colours: Colonial Experience and Global Climate Data', *Studies in History and Philosophy of Science*, 70 (2018), 18–49. See also J. Vetter, *Field Life: Science in the American West during the Railroad Era* (Pittsburgh: University of Pittsburgh Press, 2016).

[26] C. Sanhueza-Cerda, 'Stabilizing Local Knowledge: The Installation of a Meridian Circle at the National Astronomical Observatory of Chile (1908–1913)', *Isis*, 113 (2022), 710–27.

instance, personnel at the Ben Nevis Observatory based their claims both to scientific authority and government funding on their exceptional location, but others saw the site as compromised and flawed as a result of that same trait.

How Were Observatory Networks Configured?

Nineteenth-century meteorology was an accessible science. Its instruments were relatively modest in terms of cost and operational requirements and were easy to obtain from metropolitan and provincial instrument makers. Observations of air pressure, temperature and precipitation could be done almost anywhere and readings taken by thermometers, rain gauges and barometers could be compared to those from locations elsewhere. Interest in the weather, unlike interest in geomagnetism, say, was widely shared and commented on in daily discourse. As we will see in Chapter 4, recruiting observers into observation networks was not hard, but assembling willing bodies was not enough to guarantee success – they had to be acting in concert and following an agreed pattern of activity. Networks functioned well only when observations of the same phenomena were taken using the same patterns of instrument at the same times of day. Perhaps more than any survey science, meteorology required extreme levels of cooperation, timeliness, reliability and sobriety, similar to what Musselman in her work on astronomy refers to as a 'kind of moral and physical purity'.[27] Developing and enforcing a culture of uniformity across a dispersed and densely populated network of observers threatened the network's commitment to precision measurement, especially in cases where labour was given voluntarily or where observational sites were at extreme distances from central sites and standard instruments. In networks where observers had to supply their own instruments, the costs had to be kept low and recommended instruments had to be robust and simple to use, while observation schedules had to fit around working and domestic lives. The precision, or degree of accuracy, of the measurements taken were often seen as the necessary casualties of the push for uniformity across observation networks. These and other challenges of operating networks at a distance were particularly, although not uniquely, apparent in military and imperial contexts.

The use of science in the service of empire was long established, whether in the form of the individual traveller-naturalist, ethnologist and surveyor – part of what Pratt famously called the 'anti-conquest' – or in institutional

[27] E. G. Musselman, 'Worlds Displaced: Projecting the Celestial Environment from the Cape Colony', *Kronos*, 29 (2003), 64–85, 77.

guise, such as a museum or a botanical garden, cultivating valuable local plants and seeds for shipment back to institutions like Kew Gardens.[28] Observatories were also critical components of imperial science, positioned and justified as of utility to imperial systems of government and enterprise. Astronomical observatories justified themselves as of use in a range of imperial ventures. They helped improve knowledge of navigation by mapping the heavens and educating naval officers, and they aided timekeeping in ports through time balls and the setting of ships' chronometers.[29] In Schaffer's evocative phrase, astronomical observatories made the empire tick.[30] The addition of physical observatories in the grounds of the astronomical observatory promised to extend science's usefulness to empire, by mapping regional magnetic variations, contributing to understanding of the local tides and improving knowledge of local and regional weather and climate. Support went both ways – many observatories and scientific expeditions would have been unrealisable without support from state and military personnel, finances and hardware. Military activity, imperial infrastructure and international trade provided opportunities to collect scientific information otherwise denied to savants. Even so, sustaining scientific activities at the ends of long networks proved extremely challenging. Colonial observatories were far from metropolitan centres of attention, were often under-resourced and 'engaged in a constant battle to remain relevant and recognized'.[31] Networks became overstretched and often broke down, observers waited for advice from metropolitan authorities that sometimes never came and instruments struggled to operate at the limits of their environmental thresholds.[32] Uniformity was often not achieved due to resistance, indifference, incompetence or other demands

[28] F. Driver and L. Martins (eds.). *Tropical Visions in an Age of Empire* (Chicago: University of Chicago Press, 2005); J. Endersby, *Imperial Nature: Joseph Hooker and the Practices of Victorian Science* (Chicago: University of Chicago Press, 2008); M. L. Pratt, *Imperial Eyes: Travel Writing and Transculturation* (London: Routledge, 1992); L. Schiebinger and C. Swan (eds.). *Colonial Botany: Science, Commerce, and Politics in the Early Modern World* (Philadelphia: University of Pennsylvania Press, 2005).

[29] J. McAleer, '"Stargazers at the World's End": Telescopes, Observatories and "Views" of Empire in the Nineteenth-Century British Empire', *British Journal for the History of Science*, 46 (2013), 389–414; P. M. P. Raposo, 'Time, Weather and Empires: The Campos Rodrigues Observatory in Lourenço Marques, Mozambique (1905–1930)', *Annals of Science*, 72 (2015), 279–305.

[30] S. Schaffer, 'Astronomy at the Imperial Meridian: The Colonial Production of Hybrid Spaces', Plenary Lecture at the International Conference of Historical Geographers, London, 2015.

[31] McAleer, 'Stargazers at the World's End', 403.

[32] S. Naylor and M. Goodman, 'Atmospheric Empire: Historical Geographies of Meteorology at the Colonial Observatories', in M. Mahony and S. Randalls (eds.), *Weather, Climate and the Geographical Imagination* (Pittsburgh: University of Pittsburgh Press, 2020), pp. 25–42.

on observers' time, while successful networks relied on finding solutions to problems of trust, authority and moral order.[33] The challenges of sustaining meteorological networks at sea and in imperial contexts are explored in Chapters 1 and 2, while the development of a volunteer network of rain observers is the subject of Chapter 4.

How Were Weather Data Managed?

Society became both numerical and statistical in the nineteenth century. Nation states began to collect more information about their subjects and their possessions, increasingly in the form of numbers. Agrarian reforms, public health initiatives, warfare, imperial expansion, factory systems and railways, amongst other nineteenth-century projects, all produced, demanded and ran on numbers. New bureaucracies were established to manage and make sense of this information – to classify and enumerate data as well as to print, publish and deploy data sets. Across Europe and its empires, the collection of population and census information intensified and complexified over the course of the nineteenth century. Ian Hacking referred to this movement as the nineteenth century's avalanche of printed numbers.[34] The sciences experienced their own nineteenth-century avalanche. Although observation and measurement were mainstays of astronomy, it was only in the nineteenth century that they were really promoted in terrestrial physics. Cannon deployed the term 'Humboldtian science' to describe the nineteenth-century commitment to numerical expression, arrangement and comparison of instrumental observations, borrowing the name of the celebrated Prussian scientific traveller Alexander von Humboldt, who contributed to the study of the world's biogeography, weather patterns and magnetic field.[35] Cannon observed that the mass of scientific data was great by the 1850s; by the 1880s, Glasgow physicist William Thomson was able to pronounce that

> when you can measure what you are speaking about and express it in numbers you know something about it; but when you cannot measure it, when you cannot express it in numbers, your knowledge is of a meagre and unsatisfactory kind: it may be the beginning of knowledge, but you have scarcely, in your thoughts, advanced to the stage of science, whatever the matter may be.[36]

[33] K. Raj, *Relocating Modern Science: Circulation and the Construction of Knowledge in South Asia and Europe, 1650–1900* (Basingstoke: Palgrave Macmillan, 2007).
[34] I. Hacking, *The Taming of Chance* (Cambridge: Cambridge University Press, 1990).
[35] M. Dettelbach, 'Humboldtian Science', in N. Jardine, J. A. Secord and E. C. Spary (eds.), *Cultures of Natural History* (Cambridge: Cambridge University Press, 1996), pp. 287–304.
[36] M. N. Wise and C. Smith, 'Work and Industry in Lord Kelvin's Britain', *Historical Studies in the Physical and Biological Sciences*, 17 (1986), 147–73.

The practice of statistics was a crucial tool in the armoury of data managers and developed in two related contexts: servicing a politico-administrative system that wanted to make sense of society through numerical descriptions; and providing mathematical tools to make sense of and bring order to seemingly diverse data sets – to act as the logic of measurement.[37] In terms of the former, the study of population statistics was driven by efforts to render enumeration scientific: creating standardised procedures that would go further than simple population counts and providing objective, verifiable and comparable descriptions of the population.[38] In terms of the latter, astronomers, long in the business of measurement, required some way of assessing their observations' accuracy. The mathematical study of probability was applied to the assessment of the accuracy of astronomical observations, known as the theory of errors or method of least squares.[39] For instance, the work of Adolphe Quetelet, astronomer, mathematician and director of the Brussels Observatory, combined statistics, the calculus of probabilities and astronomical constants based on disparate observations.[40] Quetelet promoted the movement towards measuring with numbers and was instrumental in extending statistical techniques into nascent disciplines like sociology and meteorology, helping to identify regularities in observations of seemingly random or unpredictable behaviour.[41] In fact, Donnelly argues that Quetelet's social physics acted as a model for his global physics.[42] In the social sciences, ideas of normalcy and deviations become prevalent, while in the physical sciences, observations at the ends of Gaussian curves became outliers or errors.

Data management became a crucial and pressing problem for astronomers, meteorologists, geomagneticians and others. As will be discussed in Chapter 2, observatory personnel had to become accountants to manage the books. As information from data-gathering projects flowed in and piled up, individual errors in tables of figures had to be identified and removed. Data also had to be reduced and standardised to allow direct comparison between information from different sites, especially

[37] A. Desrosieres, *The Politics of Large Numbers: A History of Statistical Reasoning* (Cambridge, MA: Harvard University Press, 1998); S. Stigler, *The History of Statistics: The Measurement of Uncertainty before 1900* (London: Belknap Press of Harvard University Press, 1986).

[38] C. von Oertzen, 'Machineries of Data Power: Manual versus Mechanical Census Compilation in Nineteenth-Century Europe', *Osiris*, 32 (2017), 129–50, 134.

[39] Stigler, *The History of Statistics*. [40] Desrosieres, *The Politics of Large Numbers*.

[41] K. Donnelly, *Adolphe Quetelet, Social Physics and the Average Man of Science 1796–1874* (London: Pickering and Chatto, 2015).

[42] K. Donnelly, 'Redeeming Belgian Science: Periodic Phenomena and Global Physics in Brussels, 1825–1870', *History of Meteorology*, 8 (2017), 54–73.

important under conditions of imperialism where observatories operated at the end of long and tenuous networks and under diverse environmental conditions. The reduction process involved the retrospective production of data stability, converting observations 'into something like the uncanny vision of a single eye'.[43] The subsequent application of statistical techniques, including the calculation of arithmetical mean values, further standardised data, averaging away contingencies, accidental readings and personal idiosyncrasies to provide regularity. Writing about Quetelet's social statistics, Stigler referred to the calculus of averages as having an alchemical quality, transforming individual data points into stable, transmissible aggregates that could be debated, assessed and acted upon. As Chapter 4 will show, what applied to Quetelet's 'average man' also applied to average rain.[44]

Statistical techniques were one important means by which observatory personnel managed information overload. Data visualisation was another. Graphs, charts, illustrations and maps found and foregrounded meaning otherwise lost in masses of data. Whether in the form of Gaussian curves or Humboldtian isoline maps, these techniques smoothed out and marginalised errors and inconsistencies while summarising and translating data for audiences elsewhere. As we will see in all of the chapters to come, images sat alongside text and tables of figures in meteorological reports, manuals, scholarly articles, guidebooks and newspaper reports. Reports that summarised data numerically, visually and textually were often positioned as the ultimate end of observatory science, to be circulated amongst other observatories and distributed to the world's public and private libraries, where they would join company with other records. State-of-the-art means of printing the observation records, which included large amounts of tabular data as well as charts and graphs, became as important to observatories as state-of-the-art instruments.[45] The printing of observatory returns secured manuscript data – the innumerable pages of raw observations mounting up in observatory store rooms – from loss and ruin, while the widespread circulation of observatory reports meant that data archives could also circulate widely and form distributed data archives far from their place of provenance. As we will see in Chapter 2, these reports could make or break reputations, putting observatories on the map or consigning them to oblivion.

[43] Schaffer, 'Keeping the Books at Paramatta Observatory', 122.

[44] Stigler, *The History of Statistics*, p. 328.

[45] J. Ratcliff, 'Travancore's Magnetic Crusade: Geomagnetism and the Geography of Scientific Production in a Princely State', *British Journal for the History of Science*, 49 (2016), 325–52.

Publication could expose incompetence or fraud. More often than not, however, vast sections of these dispersed data archives ended up simply ignored or deemed useless.

What Were the Ends of Observatory Meteorology?

Debates about the value and purpose of observatories never went away. Observatory personnel had to spend as much time justifying their work as they did taking observations. As already noted, colonial observatories positioned themselves as of value to military and expeditionary activities, including timekeeping, mapmaking, resource assessments and the servicing and storing of instruments. Campaigners for the observatory sciences, including Humboldt and the English astronomer John Herschel, also asserted the value of data collection campaigns for theory development in terrestrial physics. Herschel advocated coordinated, global observations, not for their own sake, but to facilitate the production of mathematically stated laws. Humboldt promoted a globalised measurement regimen that would reveal universal laws governing the natural world, producing innovative hemispheric isothermal maps that gave life to new ideas of climatic distribution, later globalised by the American oceanographer Matthew Maury and the Scottish meteorologist Alexander Buchan.[46] Developments in climatology were closely connected to meteorology and supported claims regarding meteorology's national value, particularly in relation to issues of medical topography, public health, sanitation, agriculture, tourism and tropical acclimatisation.[47] But Victorian climate science often failed to live up to expectations, including the claims that it could support agricultural development, aid imperial settlement schemes or solve the spread of disease.

Perhaps more than any science, meteorology was judged a success or failure in terms of its usefulness in helping society see into the future – specifically, to predict what the state of the atmosphere would be tomorrow, in two days' or a week's time. As will be discussed in Chapter 1, mid-century concerns about safety at sea led to international congresses aimed at encouraging and standardising meteorological observations on board ships. European governments set up departments dedicated to harnessing meteorology for national ends. Britain's Meteorological

[46] M. Heymann, 'The Evolution of Climate Ideas and Knowledge', *WIREs Climate Change*, 1 (2010), 581–97; M. Mahony, 'Climate and Climate Change', in M. Domosh, M. Heffernan and C. W. J. Withers (eds.), *The SAGE Handbook of Historical Geography*, Vol. 2 (London: Sage, 2020), pp. 579–601.

[47] D.N. Livingstone, 'Tropical Climate and Moral Hygiene: The Anatomy of a Victorian Debate', *British Journal for the History of Science*, 32 (1999), 93–110.

Department was established in 1854 and led by the naval officer and surveyor Robert FitzRoy. Barometers were supplied to coastal towns and ports around the British Isles to alert seafarers to changes in the weather.[48] The expanding telegraph network was used to collect and share information about the changing state of the weather, domestically and internationally. In the early 1860s, FitzRoy pressed for a system of 'cautionary signals', or storm warnings, for shipping. Although by no means a new idea, a storm-warning system – and the idea of weather prediction more generally – was nonetheless a controversial proposal, due partly to its association with astrology, weather wisdom and folk prophecy, and partly due to philosophical resistance to utilitarian values in science and science's monetisation.[49] There was great public enthusiasm for FitzRoy's weather forecasts, but many prominent men of science – Herschel and Maury amongst them – condemned the work; FitzRoy tragically committed suicide and his storm-warning system was suspended.[50] The question of whether meteorology should put its observations to work in the service of a philosophical ideal or the public good remained under debate through the rest of the nineteenth century and well into the twentieth century. As we will see in Chapter 3, Britain's only mountain weather observatory was judged a costly failure in part because its observations did not adequately service the Meteorological Office's forecasting efforts. Organisers of Britain's rain network, explored in Chapter 4, promoted the scientific value of their work but had to rely on consultancy work for water boards, parliamentary commissions and reservoir engineers to finance their observations.

The Scope of the Book

This book is organised into four chapters, each of which deals with many of the issues raised in the preceding sections of this Introduction. Chapter 1 considers the relations between science and the maritime world by examining the British Admiralty's participation in meteorological projects in the nineteenth century. It engages with debates about the ship as a site of scientific labour through an examination of the study of meteorology at sea in the period between the conclusion of the Napoleonic Wars and the Conference on Maritime Meteorology in

[48] S. Dry, 'Safety Networks: Fishery Barometers and the Outsourcing Judgement at the Early Meteorological Department', *British Journal for the History of Science*, 42 (2009), 35–56.

[49] Anderson, *Predicting the Weather*, p. 85.

[50] M. Walker, *History of the Meteorological Office* (Cambridge: Cambridge University Press, 2012).

1874. It examines the roles played by individuals and institutions, guidebooks and regulations, in promoting a culture of instrumental meteorology on board voyages of exploration, and on Royal Naval and Hydrographic Office survey ships.[51] Particular attention is paid to attempts to establish national and international standards for the study of meteorology at sea. The chapter discusses the consent given by the British Admiralty to allow its ships to be turned into floating meteorological observatories. Reflecting on this in his 1845 address to the British Association for the Advancement of Science (hereafter British Association), Herschel idealised naval ships as 'itinerant observatories', operated by obedient crew, run according to military discipline and replete with the latest instruments.[52] Sorrenson argues that ships were conceived as more than floating observatories. They were themselves instruments of geographical discovery – conferring authority on their user, leaving traces on maps and providing 'superior, self-contained, and protected views' of the landscapes viewed from them.[53]

Over the course of the nineteenth century, an informal, even idiosyncratic, culture of meteorological inquiry was gradually formalised by the British Admiralty, at a time when there remained significant ambiguities about the place of science on ships at sea. Uniform forms and meteorological instruments were introduced together with prescribed observations and practices. These, in turn, were authorised in Admiralty regulations and guidebooks. Voyages of exploration and the Hydrographic Office survey vessels were the experimental sites for this new culture, expanded to include all Royal Naval vessels, and, later, Britain's merchant marine. The Admiralty hydrographer Francis Beaufort was instrumental in negotiating a place for meteorology on naval ships, but the chapter demonstrates that he struggled to persuade the Admiralty of the value of his inductive approach to the collection of weather information, and to persuade sailors to employ a sufficiently precise language to record both normal and extreme weather at sea. It took an Army engineer, William Reid, to convince the Admiralty to adopt their hydrographer's recommendations regarding the collection of wind and weather data. Reid used ships' logbooks and the plotting of the ships' movements themselves to develop William Redfield's theory of revolving storms. His use of maps and charts imposed order on a mass of data spread over a large geographical area. In

[51] R. Dunn, '"Their Brains Over-Taxed": Ships, Instruments and Users', in D. Leggett and R. Dunn (eds.), *Re-inventing the Ship: Science, Technology and the Maritime World* (Aldershot: Ashgate, 2012), pp. 131–55.

[52] A. Winter, '"Compasses All Awry": The Iron Ship and the Ambiguities of Cultural Authority in Victorian Britain', *Victorian Studies*, 38 (1994), 69–98, 75.

[53] R. Sorrenson, 'The Ship as a Scientific Instrument in the Eighteenth Century', in H. Kuklick and R. E. Kohler (eds.), *Science in the Field* (Chicago: Chicago University Press, 1996), pp. 221–36, p. 222.

doing so he overcame the limits of language and helped to more effectively recast meteorology as a science that was useful to the maritime world. Reid's charts explained the science of storms to the naval officer in familiar terms, such that a storm's potentially catastrophic effects could be avoided. The arguments Reid, Beaufort and others made for a science of storms and study of weather at sea were later supported by a series of manuals and international congresses on maritime meteorology.

Chapter 2 examines attempts to develop a new model space in which the physical sciences could be investigated. In Britain, physical observatories were based on a blueprint provided by astronomy and propelled forward by an obsession with the mapping of the earth's magnetic field. New or repurposed observatories were established to this end in London and Dublin and across the British Empire. Although often positioned as the poor cousin to the pursuit of terrestrial magnetism, the study of meteorology was a critical component of activities at physical observatories at home and overseas and was required to conform to the same exacting requirements. The chapter is divided into two parts. The first traces the drawing up of the blueprints for the model physical observatory, the problems it would have to overcome – managing instruments and personnel, effectively reducing observations and so on – and the uses to which it should be put, scientifically *and* socially.

The second part of the chapter investigates the long and difficult history of one physical observatory in particular – the Colaba observatory in Bombay, India – and its place in Indian meteorology more generally. Although it began its life in the 1820s as an astronomical observatory with timekeeping obligations, by the end of the century it had been formally subsumed into the ambit of the India Meteorological Department. The historical geography of this particular observatory experiment allows us to trace out how a model physical observatory functioned at the imperial periphery, particularly with regard to the management of instruments and observers, its computational regimen and its contribution to phenomenal laws. Finally, the chapter considers Colaba's place within a set of imperial and climatic geographies that extended across the subcontinent.

Chapter 3 also focuses on the history of one observatory: Britain's only mountaintop weather station, on the summit of Ben Nevis in Scotland. The idea for an observatory on Britain's highest peak came from members of the Scottish Meteorological Society, who were in turn influenced by wider arguments for the benefits of high-altitude meteorology. Financing such a scheme was beyond the means of the Scottish meteorologists until the Englishman Clement Wragge offered to test the value of the summit for weather observations by walking daily to the summit from the lochside town of Fort William. Wragge conducted himself as if he were an

Alpine mountaineer or Arctic adventurer, treating Ben Nevis as a challenging environment and a demanding journey to be confronted and conquered on a daily basis. The emphasis that Wragge gave to place as a factor in his quest for credibility in claim making was shown by his articulation of a mountain sublime in his reports and popular writing. Wragge's self-denial, bodily suffering and observational labour in turn helped members of the Scottish Meteorological Society to sell the concept of a permanent summit observatory to funders and donors.

The chapter traces the establishment of a purpose-built observatory on the summit plateau, and later its brethren observatory in Fort William, four miles distant horizontally and close to sea level. Scientific and domestic life on the top of the Ben proved to be a significant challenge – the very first winter there exposed, quite literally in some cases, the observatory's many inadequacies and led to numerous adaptations to the structure and to observational practice and routine. Observers embraced a culture of physical hardship that mirrored Clement Wragge's, the mountaineers who climbed the Ben, and the polar explorers who spent time there. The presence of day trippers and overnight tourists on the summit disrupted observatory routine and undermined the station's culture of suffering, denial and isolation. Despite its celebrity, the observatory remained under scrutiny its entire existence. Its financial security was in constant doubt, which brought the observatory directors into conflict with the Meteorological Council in London. The Council's refusal to increase the observatory's annual grant was blamed for its closure in 1904, while the observatory's critics blamed the observatory's failure to make useful contributions to forecasting efforts.

Chapter 4, the final empirical chapter, examines efforts in the second half of the nineteenth century and first years of the twentieth century to better understand Britain's rain. Despite its supposed role as life-blood of the atmosphere, meteorologist and geographer Andrew J. Herbertson claimed that rain was much less well studied than temperature or pressure. Meteorologists had attempted to investigate the distribution of rain prior to the 1850s, but observation points remained few and inadequately distributed. This was especially problematic for an atmospheric phenomenon that seemed so local – where rainfall totals varied significantly over only short distances. The solution to answering questions about the geographies of the rain was the establishment of a rainfall observatory network that covered the entirety of the British Isles. The network of rainfall observing stations – small observatories – was established by George Symons and became known as the British Rainfall Organisation. It relied almost exclusively on volunteer labour – on individuals willing to purchase a rain gauge, situate it in a suitable location, monitor it on a regular basis and supply the returns to Symons in London. Three key projects defined the success of the network. The first was the assembly of a rainfall archive, where Symons collected together past records with current observations. The second was a series of rain collection experiments at various locations across England. The third project was the production of a rainfall map, through which statistical techniques were brought to bear on the question of rainfall distribution.

Across these three key projects, organisers set out to promote uniformity of approach and accuracy of measurement, while defining and mitigating against errors. What was a poor or faulty rainfall reading and how could it be known and managed? One crucial response was the pursuit of the statistical mean as an indicator of trustworthiness, while maps of rainfall averages were promoted as visual and geographical expressions of trust in numbers. The first section of the chapter details the early years of Symons's Rainfall Organisation and its key administrative features, before moving on to discussions about rain gauges and station exposures. The following section examines a series of experimental trials that ran from 1863 to 1890 and the ensuing controversy regarding the value of the experiments and of the observatory network more generally. The chapter then examines discussions about the value of various statistical treatments of rain data, before finishing with Alexander Buchan's and H. R. Mill's rainfall maps and the maps' contributions to data management and public utility.

The Conclusion discusses the key factors that determine the success, and indeed failure, of the observatory experiments considered across the book's four empirical chapters: the significance of geographical particularity in justifications of observatory operations; the sustainability of coordinated observatory networks at a distance; the ability to manage, manipulate and interpret large data sets; and the potential public value of meteorology as it was prosecuted in observatory settings. It ends with a discussion of one notable afterlife of observatory meteorology: as a component of big data histories of possible future weather and climate.

1 Meteorology All at Sea

This chapter considers the importance of maritime exploration and surveying for the development of a culture of meteorology on board naval vessels. In his study of the whaler, explorer and magnetician William Scoresby, Bravo observes that exploration in the late eighteenth and early nineteenth centuries 'had become a much more specialised set of scientific practices that required training, the provision of expensive precision measurements, and new time-intensive methods of working and record-keeping'.[1] These methods and practices were extended to the investigation of the marine world. Reidy suggests that during the nineteenth century, 'the British Admiralty, maritime community, and scientific elite collaborated to bring order to the world's seas, estuaries, and rivers'.[2] The vast emptiness of the oceans was transformed 'into an ordered and bounded grid, inscribed with isolines of all kinds – tidal, magnetic, thermal, and barometric – in areas uncharted and on coasts unseen'.[3] Or at least that was the intention. The Royal Navy played a number of specific roles in the development of science, such as training personnel in scientific techniques useful to shipbuilding; carrying out surveying and navigation; and imparting knowledge through institutions such as the Royal Naval Academy at Portsmouth, the Greenwich Hospital, the Navy's domestic and overseas dockyards and the Admiralty's Hydrographic Office. Established in 1795, the Hydrographic Office gradually increased in importance,

[1] M. Bravo, 'Geographies of Exploration and Improvement: William Scoresby and Arctic Whaling (1722–1822)', *Journal of Historical Geography*, 32 (2006), 512–38, 519.

[2] M. S. Reidy, *Tides of History: Ocean Science and Her Majesty's Navy* (Chicago: Chicago University Press, 2008), p. 6. See also M. Deacon, *Scientists and the Sea, 1650–1900: A Study of Marine Science* (London: Academic Press, 1971); S. Millar, 'Science at Sea: Soundings and Instrumental Knowledge in British Polar Expedition Narratives, c.1818–1848', *Journal of Historical Geography*, 42 (2013), 77–87.

[3] Reidy, *Tides of History*, p. 6. See also E. J. Larson, 'Public Science for a Global Empire: The British Quest for the South Magnetic Pole', *Isis*, 102 (2011), 34–59; H. Rozwadowski, *Fathoming the Ocean: The Discovery and Exploration of the Deep Sea* (Cambridge: The Belknap Press of Harvard University Press, 2005).

despite its remit being heavily circumscribed in its early years.[4] John Croker and John Barrow, the First and Second Secretaries to the Admiralty, respectively, resisted expansion of the Hydrographic Office due to financial retrenchment in the post-war years after 1815, Croker's scepticism of the value of hydrography, and the two men's commitment to the Royal Society as the Admiralty's scientific advisor. During this period, the connections between the Admiralty and the Royal Society were strong. Croker and Barrow took it in turns to sit on the Society's Council as Admiralty representatives, while the Admiralty appealed to the Royal Society for advice on its expeditions so frequently that the Society was treated almost like a standing committee.[5]

Since its establishment in 1714, the Board of Longitude had acted as a research department for the Admiralty, with a remit that extended beyond solving problems of navigation. However, in 1828, the Board of Longitude was abolished by Act of Parliament. In its place the Admiralty created an internal consultative committee, called the Resident Committee of Scientific Advice, which was made up of physicist and Egyptologist Thomas Young, Army officer and magnetician Edward Sabine and Royal Institution chemist Michael Faraday. By establishing the Committee, Barrow and Croker hoped to keep the Navy's interactions with men of science out of public view and to more effectively control whom it dealt with. The Resident Committee was criticised by Britain's scientific reform movement on the grounds of nepotism and patronage and was brought to an abrupt conclusion by the death of Young in 1829 and Sabine's posting to Ireland in 1830. Faraday was left as a sole and occasional advisor to the Admiralty, a role he fulfilled until the 1850s.[6] Croker's retirement from the post of First Secretary to the Admiralty in 1831 also created new possibilities for the pursuit of science in the Royal Navy.[7] In 1831, the Hydrographic Office became a separate department of the Admiralty, something Croker had prevented

[4] A. Webb, 'More Than Just Charts: Hydrographic Expertise within the Admiralty, 1795–1829', *Journal for Maritime Research*, 16 (2014), 43–54.

[5] M. B. Hall, 'Public Science in Britain: The Role of the Royal Society,' *Isis*, 72 (1981), 627–9; D. P. Miller, 'The Revival of the Physical Sciences in Britain, 1815–1840', *Osiris*, 2 (1986), 107–34; R. Cock, 'Scientific Servicemen in the Royal Navy and the Professionalisation of Science, 1816–55', in D. M. Knight and M. D. Eddy (eds.), *Science and Beliefs: From Natural Philosophy to Natural Science, 1700–1900* (Aldershot: Ashgate, 2005), pp. 95–112; A. Friendly, *Beaufort of the Admiralty: The Life of Sir Francis Beaufort 1774–1857* (New York: Random House, 1977), p. 247.

[6] On the scientific reform movement, see J. Morrell and A. Thackray, *Gentlemen of Science: Early Years of the British Association of the Advancement of Science* (Oxford: Clarendon Press, 1981).

[7] Friendly, *Beaufort of the Admiralty*, p. 247.

so as to limit the Office's autonomy.[8] The Admiralty Scientific Branch was also established in 1831, which was overseen by the Admiralty hydrographer and encompassed the Nautical Almanac Office, the Chronometer Office, the Astronomical Observatories at Greenwich and the Cape and the Hydrographic Office itself.

The naval administration also assisted scientific projects through the provision of indispensable resources, namely passage upon and use of a naval vessel as well as its well-equipped, disciplined and trained personnel. While financial and infrastructural resources were critical to major scientific projects, the Royal Navy's emphasis on order and discipline was arguably just as important. In theory, the daily regime on board ship lent itself well to ensuring regular and reliable scientific observations.[9] For John Herschel, the benefit of using naval ships as observational platforms was their capacity to act as 'itinerant observatories' and naval officers as ideal observers.[10] The necessity of a twenty-four-hour watch and the demands of the logbook promised to make the collection of routine and numerous observations more straightforward than in other settings. It was also assumed that naval discipline turned officers and crew into regulated instruments themselves, just like the precision devices they used daily.[11] Naval seamen were meant to be the meteorological equivalents of the 'obedient drudges' that the Astronomer Royal George Airy wanted to operate astronomical observatories like Greenwich – to treat their work and their ship in the same manner that astronomical technicians were expected to operate in an observatory setting.[12]

The Admiralty applied naval personnel to projects that utilised their familiarity with the latest mathematical, scientific and technical knowledge.[13] This was especially the case after peace with France in

[8] A. Day, *The Admiralty Hydrographic Service 1795–1919* (London: Stationery Office, 1967), p. 35.

[9] C. Ward and J. Dowdeswell, 'On the Meteorological Instruments and Observations Made during the 19th Century Exploration of the Canadian Northwest Passage', *Arctic, Antarctic, and Alpine Research*, 38 (2006), 454–64, 454.

[10] Quoted in Winter, '"Compasses All Awry"', 75.

[11] C. W. J. Withers, 'Science, Scientific Instruments and Questions of Method in Nineteenth-Century British Geography', *Transactions of the Institute of British Geographers*, 38 (2012), 167–79, 173. French commentators made similar assumptions about the capacities of their navy to collect meteorological information: F. Locher, 'The Observatory, the Land-Based Ship and the Crusades: Earth Sciences in European Context, 1830–50', *British Journal for the History of Science*, 40 (2007), 491–504, 498.

[12] Winter, '"Compasses All Awry"', 74.

[13] E. Behrisch, *Discovery, Innovation, and the Victorian Admiralty: Paper Navigators* (Switzerland: Palgrave Macmillan, 2022).

1815.[14] The prospects of peace 'presented an opportunity to those, both in the Navy and outside, who had ambitions to harness to the ends of science the resources of the new smaller, more professional and career-oriented service that developed as a result'.[15] After 1815, the Royal Navy experienced financial retrenchment and disarmament: many ships were decommissioned and thousands of enlisted men lost their jobs. Naval officers had greater political influence and so few of them were retired but perhaps 90 per cent of them found themselves without a role and on the half-pay list.[16] Croker defended the reduced Navy Estimates, while Barrow argued that the Navy's ships and personnel should be employed in global exploration, on the basis that 'exploration would increase scientific knowledge, that it would be a boon to national commerce, and above all that it would be a terrible blow to national pride if other countries should open up a globe over which Britain ruled supreme'.[17]

The Royal Navy and the Admiralty Hydrographic Office made numerous contributions to science, including geographical exploration of the Northwest Passage, the Antarctic Ocean and of Africa.[18] For naval officers interested in science, a position on one of these voyages of exploration was a choice appointment. These 'scientific servicemen' gradually took on much of the scientific work from civilians and many became Fellows of the Royal Society.[19] Naval personnel were part of the emerging division of labour in science in the nineteenth century. These scientific servicemen were most important as global data gatherers, passing information back to gentlemen savants for analysis in metropolitan centres, although some officers specialised in science and exploration themselves. They were supported by newly formed scientific societies, such as the Royal Geographical Society, which were comfortable about including military personnel in their ranks and benefited from their ability to collect data from locations around the world, pursue research programmes and to bring those programmes to the attention of government.

Evidence of the Admiralty's involvement in training sailors, supporting expeditions and collecting information on a global scale has led some historians to argue that in the first half of the nineteenth century, the Navy

[14] Reidy, *Tides of History*, p. 140.
[15] R. Cock, *Sir Francis Beaufort and the Co-ordination of British Scientific Activity, 1829–55*, Unpublished DPhil Thesis, University of Cambridge, 2003, pp. 7–8.
[16] C. Lloyd, *Mr. Barrow of the Admiralty: A Life of Sir John Barrow* (London: Collins, 1970), pp. 91–2.
[17] F. Fleming, *Barrow's Boys: A Stirring Story of Daring, Fortitude and Outright Lunacy* (London: Granta, 1999), p. 11.
[18] Friendly, *Beaufort of the Admiralty*, p. 289.
[19] Miller, 'Revival of the Physical Sciences'; R. Cock, 'Scientific Servicemen in the Royal Navy', pp. 95–112.

was the principal governmental subsidiser of science in Britain.[20] The Admiralty's support for a number of voyages of exploration was certainly justified on the grounds of national scientific prestige, but just as important were issues of commercial advantage and maritime safety. Rodger argues that the Royal Navy's growth and success were bound up with Britain's prosperity in overseas trade, while Webb suggests that matters concerning safety of life at sea were given priority over scientific interests.[21] Although self-interest is an obvious explanation for the position taken by the Admiralty in this regard, it is also likely that its thinking was shaped by the Royal Society's assumption that science should constitute a form of useful knowledge, an instrument of improvement and an aid to profitable and rational economic activity.

Cultures of Instrumentation on Voyages of Exploration

Voyages of exploration in the late 1810s and 1820s served to establish standards for the conduct of physical scientific inquiry at sea, particularly in relation to the use of philosophical instruments on board ships. The Royal Society had long offered advice to the Admiralty on the scientific aspects of its expeditions, viewed by government and the military as a 'state tool for consultation'.[22] The period from Ross's 1818 Arctic voyage to Foster's South Atlantic expedition in 1828 was a tumultuous one for the Society. Joseph Banks's reign as president of the Royal Society ended with his death in 1820. Successive presidents – Humphry Davy (1820–27) and Davies Gilbert (1827–30) – were caught up in wider contests over the character and direction of British science. Davy put the Royal Society on a course that aimed to satisfy both the remnants of Banks's 'Learned Empire' and the reformist intentions of the 'Cambridge Network'.[23] The changes experienced by the Society over this period were reflected in the composition and work of its committees. In the

[20] Friendly, *Beaufort of the Admiralty*, p. 289. For similar arguments see T. Levere, *Science and the Canadian Arctic: A Century of Exploration, 1818–1918* (Cambridge: Cambridge University Press, 1993); J. Ratcliff, *The Transit of Venus Enterprise in Victorian Britain* (London: Pickering & Chatto, 2008), p. 24.

[21] N. A. M. Rodger, 'From the "Military Revolution" to the "Fiscal-Naval" State', *Journal of Maritime Research*, 13 (2011), 119–28, 123; Webb, 'More Than Just Charts', 52.

[22] S. Waring, 'The Board of Longitude and the Funding of Scientific Work: Negotiating Authority and Expertise in the Early Nineteenth Century', *Journal for Maritime Research*, 16 (2014), 55–71, 57.

[23] D. P. Miller, 'Between Hostile Camps: Sir Humphrey Davy's Presidency of the Royal Society of London, 1820–1827', *British Journal for the History of Science*, 16 (1983), 1–47; R. M. MacLeod, 'Whigs and Savants: Reflections on the Reform Movement in the Royal Society, 1830–48', in I. Inkster and J. Morrell (eds.), *Metropolis and Province: Science in British Culture, 1780–1850* (London: Hutchinson, 1983), pp. 55–90.

early years of Davy's presidency in particular, increased use was made of scientific committees.[24] Over the course of the 1820s, scientific reformers, such as Herschel, Charles Babbage and Francis Baily, joined long-standing members like Thomas Young, Henry Kater and William Hyde Wollaston, all taking a greater role in the running of these committees. Miller notes that members of the reform group 'increasingly dominated public discussion of the most important objects of research for scientific voyages'.[25] Herschel in particular 'maintained an ambition to make the surveying voyages commissioned by Barrow on behalf of the Admiralty more "scientific"'.[26] The changes effected in this period had a direct bearing on the advice that the Royal Society provided to exploring expeditions.

During the final years of Banks's presidency, William Thomas Brande, one of the Royal Society's two secretaries, wrote to Barrow to supply the Admiralty with a list of instruments that the Society recommended for use on the two 1818 expeditions then heading for the polar regions, to be led by John Ross and David Buchan, respectively. These included compasses, barometers, magnetic instruments, bottom sampling and dredging equipment, chronometers, mercurial and sea thermometers, a Wollaston micrometer, artificial horizons, electrometers, hydrometers and apparatus 'for ascertaining the quantity of air in water'.[27] Four laboratory tents were added to protect the instruments during observations to be made onshore, along with two transit instruments, four 'Small Altitude Instruments', a water sampler and a tent for astronomical observations.[28]

In 1821, a 'Committee for suggesting Experiments and Observations to Mr Fisher, about to proceed to the Arctic Seas under the command of Capt. Parry', was established.[29] Herschel, William Hyde Wollaston and Charles Hatchett bolstered a core group made up of the president, the two secretaries – Brande and Taylor Combe – and Henry Kater and Thomas Young. The expedition astronomer George Fisher was invited to attend.[30] While the advice given to Ross in 1818 laid out in detail the instruments to be used on his expedition, that provided to Fisher was

[24] Hall, 'Public Science in Britain'. [25] Miller, 'Between Hostile Camps', 34.
[26] Waring, 'The Board of Longitude and the Funding of Scientific Work', 59.
[27] Copy of letter from W. T. Brande to J. Barrow, 29 January 1818, Archives of the Royal Society, CMB/1, p. 9.
[28] Anon, 'Committee for ascertaining the Length of the Seconds Pendulum', 26 March 1818, Archives of the Royal Society, CMB/1, p. 14.
[29] Anon, 'Hints of Experiments to be made in the Arctic Expedition ... of 1821', 12 April 1821, Archives of the Royal Society, CMB/1, pp. 26–31.
[30] G. W. Roberts, 'Magnetism and Chronometers: The Research of the Reverend George Fisher', *British Journal for the History of Science*, 42 (2009), 57–72.

more direct in the scientific agenda to be pursued, emphasising terrestrial physics. Twenty experiments were proposed. The majority focused on the effects of extreme cold on atmospheric chemistry, the behaviour of fluids (including mercury) and on humans, animals, food and different metals. Of particular interest was the freezing point of pure mercury and of different amalgams of mercury and other metals. This was significant because of its effect on the performance of the thermometer and barometer.[31] Other questions related to the operation and effects of the Aurora Borealis, and the investigation of sea temperature at different depths.

The advice supplied to Captain Henry Foster's 1828 voyage on the HMS *Chanticleer* to the South Atlantic was more comprehensive still. At this committee Davies Gilbert (now president), Herschel and Kater were joined by William Fitton, president of the Geological Society of London, Sabine and the Admiralty hydrographer Francis Beaufort. James Horsburgh, the East India Company hydrographer, and Captains Parry and Foster were present by invitation. In line with the interests of Herschel, Sabine and Beaufort, the principal objects of Foster's expedition were defined as the investigation of physical astronomy, the determination of the figure of the earth and the investigation of the law of the variation of gravity, along with inquiries into ocean currents, magnetism, the longitude of significant locations, natural history and meteorology. The committee noted that meteorological observations 'form a branch of inquiry of no small amount in this and all similar expeditions' and it recommended that 'regular observations of the Barometer, Thermometer, Hygrometer, and the direction and force of the wind should be daily made; and of the actinometer or other instruments proper for measuring the Solar and terrestrial variation, at favorable opportunities and at various levels'. The result, it was hoped, would be a better understanding of 'the probable former and future climate of different regions of the Earth[,] the permanence or variability of the Solar influence at different epochs, and the stability of the actual equilibrium of meteorological agents'.[32] In its findings, the voyage was judged a success and the results were later used by Royal Society reformers and members of the Astronomical Society to affirm the analytic importance of mathematics in accurate observation and experimental research.[33]

[31] Ward and Dowdeswell, 'On the Meteorological Instruments', 455.

[32] Anon, 'At a meeting of the Committee for considering and resolving on the most advantageous objects to be attained by Capt'n Foster in the course of his intended scientific Voyage', 28 January 1828, Archives of the Royal Society, CMB/1, p. 230.

[33] Miller, 'Revival of the Physical Sciences'.

The advice given to Foster was dwarfed, however, by that supplied to James Clark Ross for his 1839 voyage to the Antarctic Ocean as part of Britain's Magnetic Crusade (discussed further in Chapter 2). The committee convened to advise on the expedition was chaired by Herschel and included Beaufort, Sabine, John Ross, Michael Faraday, John Frederic Daniell, Peter Mark Roget, Charles Wheatstone and William Snow Harris.[34] The expedition was principally intended as an investigation into terrestrial magnetism, but other sciences were pursued, including study of the tides, the figure of the earth and meteorology. Meteorology was given greater emphasis than was necessary simply to correct the performance of the magnetic instruments.[35] The committee additionally advised on the instruments with which the naval expedition should be equipped. In terms of meteorology, these included actinometers, Lind's rain gauge, an Osler anemometer and spirit thermometers for operation in Antarctic temperatures below those at which mercury freezes and mercurial thermometers become ineffective.[36] Procedures were recommended for the verification of the instruments, especially when the expedition was far from fixed observatories on land.[37] Both of the ships – HMS *Erebus* and HMS *Terror* – were to carry standard barometers and thermometers against which others were to be compared. This was especially important when instruments were taken ashore, 'so as to detect and take into account of any change which may have occurred in the interval'.[38] The standards on one ship were to act as checks upon the other.

The passage of the *Erebus* and *Terror* from the tropics to the Antarctic presented an opportunity to investigate von Humboldt's claim that atmospheric pressure at the equator was uniformly 'less in its mean amount than that at and beyond the tropics', a phenomenon that was, in turn, believed to produce the trade winds.[39] The observation of changes in the barometer when approaching the line was therefore of great scientific

[34] Anon, Joint Committee of Physics and Meteorology, 1838–39, 19 June 1839, Archives of the Royal Society, CMB/284.

[35] Anon, *Report of the President and Council of the Royal Society on the Instructions to be Prepared for the Scientific Expedition to the Antarctic Regions* (London: Richard and John E. Taylor, 1839), p. 13; E. Gillin, 'The Instruments of Expeditionary Science and the Reworking of Nineteenth-Century Magnetic Experiment', *Notes and Records of the Royal Society*, 76 (2022), 565–92.

[36] Ward and Dowdeswell, 'On the Meteorological Instruments', 459.

[37] Joint Committee of Physics and Meteorology, 1838–39, 22 August 1839, Archives of the Royal Society, CMB/284.

[38] Anon, *Report ... on the Instructions ... for the Scientific Expedition to the Antarctic Regions*, p. 13.

[39] Anon, *Report ... on the Instructions ... for the Scientific Expedition to the Antarctic Regions*, p. 14

value, as was the observation of the local effects that continents or oceanic currents had on atmospheric pressure. Periods spent at high southern latitudes also presented opportunities to calibrate the instruments. For instance, Ross was asked to verify and to register the ships' standard thermometers at the freezing point of mercury whenever the opportunity arose. This was to be effected by placing four permanent marks on the tube of each standard thermometer, and Ross was 'requested occasionally to compare these marks with the degrees of the ivory scale'. A bottle of mercury was ordered to accompany each standard thermometer.[40] The scientific instructions presented to Ross contained, in his words, 'a detailed account of every object of inquiry which the diligence and science of the several committees of that learned body could devise'.[41] This report became a standard for subsequent scientific guides.

The deployment of philosophical instruments, and the supply of precise instructions for observations and experiments, was not alone enough, however, to guarantee reliable inscriptions. The directions provided to the captains of scientific expeditions were often aspirational in tone and susceptible to compromise when in the field. The robustness of the scientific outcomes of an expedition relied as much on 'immense chains of delegated trust and labour' as they did on detailed instructions, calibrated instruments and well-organised skeleton forms.[42] Instruments could not speak for themselves effectively. The determination of their accuracy relied on the person or persons operating them. Identifying and justifying who was to operate which instruments was a crucial matter in voyages of exploration. For John Ross's 1818 voyage to the Arctic, the Royal Society committee suggested to the Admiralty that Sabine was the 'proper person to conduct certain experiments', accompanied by a sergeant of artillery to 'take care of instruments'.[43] The committee also suggested the inclusion of Fisher – 'a Gentleman of considerable mathematical talent' – while Henry Kater reported that the naval officers John Franklin, Frederick Beechey and William Parry 'had been most assiduous in acquiring a due knowledge of the use of the Instruments to be employed in the Northern Expedition, and that he considers them fully competent to prosecute the required observations and experiments'.[44]

[40] Anon, Joint Committee of Physics and Meteorology, 1838–39, 22 August 1839, Archives of the Royal Society, CMB/284.

[41] J. C. Ross, *A Voyage of Discovery and Research in the Southern and Antarctic Regions, during the Years 1839–43*, Volume 1 (London: John Murray, 1847), p. xxvii.

[42] S. Schaffer, '"On Seeing Me Write": Inscription Devices in the South Seas', *Representations*, 97 (2007), 90–122, 113.

[43] 12 February 1818, Archives of the Royal Society, CMB/1.

[44] 26 March 1818, Archives of the Royal Society, CMB/1.

Despite the various controversies surrounding Ross's 1818 Arctic expedition, the Royal Society again recommended Sabine as a member of William Parry's 1819 Arctic voyage:

It is of the opinion of this Committee that Capt'n Sabine has shown the greatest possible diligence in making the observations which were intrusted [sic] to his care and the greatest judgement and regularity in his method of recording them. And this Committee therefore suggests the propriety of recommending Capt'n Sabine to the Admiralty in the strongest manner, both as deserving every professional encouragement, and as a proper person to be again appointed to take charge of the Observations to be made in a new Expedition.[45]

The reiteration of instrumental and observational competence was crucial. The practices employed and the vagaries of the instruments' fate 'governed the status of the data they produced and the interpretations they suggested'.[46] The reputation of the observer was intrinsically linked to the data and the instruments: 'To question or doubt results or methodology was to question the character and morality of their creator.'[47]

Reforming Meteorology

After 1820, Royal Society committee members were increasingly chosen on the basis of expertise, whether intellectual or professional. This was also true in other respects, such as over the quality and use of the Society's meteorological instruments. The committee formed in 1822 to study this matter incorporated Thomas Young, William Hyde Wollaston and Henry Kater, together with Humphry Davy, Davies Gilbert, the secretaries Brande and Combe, Babbage and Herschel, as well as Luke Howard and John Frederic Daniell, included given their standing in meteorology and related fields.[48] Amongst other recommendations, the committee ordered the construction of new instruments for the Society, including two barometers from John Newman, of Lisle Street, London; these were, subsequently, the subject of experiments at the Society in December 1822.[49] The observational regime and the siting of the Society's instruments were also reviewed. At a meeting of the Society's Meteorological Committee in 1827, James South and Francis Baily, along with Beaufort and Herschel, complained about its recording forms and the quality and situation of its meteorological instruments.

45 18 March 1819, Archives of the Royal Society, CMB/1.
46 Schaffer, '"On Seeing Me Write"', 112. 47 Waring, 'The Board of Longitude', 66.
48 Anon, Committee for examining into the state of the Meteorological Instruments belonging to the Royal Society, 10 and 12 December 1822, Archives of the Royal Society, CMB/1.
49 Anon, Committee for examining into the state of the Meteorological Instruments, Archives of the Royal Society, CMB/1.

They argued that the 'local situation' of its headquarters at Somerset House did not allow for the production of 'any series of meteorological observations of material weight and importance in the present state of the science'.[50]

For Jankovic, '[w]hether fairly or not, early nineteenth-century commentators ... erupted with criticisms of a general lethargy that supposedly prevailed in the investigation of weather-systems, of the insufficiency *and* profusion of observations, of the public uselessness of the existing stock of facts, and of the imprecision of means for standardizing and using meteorological instruments'.[51] In his *Meteorological Essays*, John Daniell, Professor of Chemistry at King's College London, pointed to the Royal Society's meteorological observations as evidence of the poor science undertaken in England. He extended his criticism to the operations of overseas observatories, where, he claimed, there had been insufficient coordination of efforts, such that their 'labour and perseverance lose more than half their value by the want of a well-digested plan of mutual co-operation'.[52] Concerns about the level of training and expertise of meteorological observers similarly preoccupied William Whewell and James Forbes, the noted Edinburgh physicist, who argued that science should centre on precision observations and be conducted by trained personnel. Forbes expressed these arguments in his report on British meteorology to the 1832 British Association meeting in Oxford.[53] For Forbes, meteorological instruments 'have been for the most part treated like toys', while few of the numerous registers 'which monthly, quarterly, and annually are thrown upon the world' could be expected to afford information useful to the development of the science.[54] The situation was, in his view, so bad as to require 'a total revision upon which meteorologists have hitherto very generally proceeded'.[55]

This troubled history of meteorology at the Royal Society is important given discussions over the deployment of meteorological instruments on Admiralty ships. The review of the Royal Society's own instrumental practices was coincident with the Society's advice to captains and scientific officers on board exploring expeditions. The composition of the

[50] Anon, Minutes of the Meteorological Committee, 2 August 1827, Meteorological Committee Minutes and Letters 1830–1837, Archives of the Royal Society, DM/3.
[51] Jankovic, 'Ideological Crests', 24.
[52] J. Daniell, *Meteorological Essays* (London: Thomas & George Underwood, 1823), p. viii.
[53] J. D. Forbes, 'Report of the First and Second Meetings of the British Association for the Advancement of Science', *British Association for the Advancement of Science Second Report* (1833), 196–258.
[54] On the use, and misuse, of meteorological instruments, see J. Golinski, *British Weather and the Climate of Enlightenment* (Chicago: University of Chicago Press, 2007).
[55] Forbes, 'Report', 196–7. See also Jankovic, 'Ideological Crests', 24.

Society's committees on these issues was almost identical. It is reasonable to assume, therefore, that the reform of meteorology at the heart of British science was part of attempts to improve the conduct of science at sea. The difficulties experienced at the Royal Society illustrate the challenges inherent in the pursuit of an exacting instrumental regime. The committees established to advise Parry, Foster and others laid down scientific agendas and observational practices on the assumption that ships were floating observatories. At the same time, criticisms of the Society's own meteorological practices illustrated the challenges of meeting such demands when on dry land. When far away from instrument makers and scientific advisors, on board a moving ship in challenging conditions, operating personnel had no choice but to 'make up and mend ways of recording and transmitting what they reckoned worth noting'.[56]

Francis Beaufort and Instrumental Cultures on Hydrographic Ships

Scientific and exploring expeditions, such as those already discussed, helped establish precedents for the collection of information about terrestrial physics on board ships. The success of these and other voyages in the first half of the nineteenth century encouraged the belief that all military vessels might be employed as floating observatories. In his work on French arctic expeditions, Locher notes that the regular maintenance of the systematic naval watch offered real advantages to science, particularly if officers could be compelled to collect data in addition to the other observations they were required to undertake.[57] Naval officers received training in mathematics, navigation and astronomy and would have been comfortable operating relatively sophisticated precision instruments. For observations to be scientifically useful, however, they had to be made regularly, specific instruments had to be employed and full details had to be supplied about their constitution and conditions of use. Particular reduction protocols and computing methods had to be followed. The situation of an instrument and the state of the atmosphere around it had to be given consideration and recorded so that measurements could be reduced to a virtual common environment.

The weather was an inescapable part of life on board ship and the Admiralty required officers to keep a record of it. The 1808 edition of the Admiralty's *Regulations and Instructions Relating to His Majesty's*

[56] Schaffer, '"On Seeing Me Write"', 106.
[57] Locher, 'The Observatory, the Land-Based Ship and the Crusades', 498.

Service at Sea required the ship's master to record in the logbook 'with very minute exactness' the state of the weather and the directions of the wind, along with other observations relating to navigation, and to the state of the ship and its provisions.[58] Any shifts in the wind that might affect the ship's course were to be recorded on the log board, while special care was to be taken during periods of fog. There was no compulsion to record meteorological observations for their own sake.[59] These were to be recorded on a pre-printed pro forma outlined in Appendix 25 of the *Regulations and Instructions*. There were, however, no explicit directions as to the manner in which the weather observations were to be recorded, and the space provided to do so was very small, especially when other essential information had to be noted. This attitude to the study of maritime weather was in contrast to the approach of other services. American navy surgeons had been keeping weather journals since 1814. The French navy had been analysing ships' logs for weather patterns to aid sailing since the 1720s, while Wilkinson claims that officers of the East India Company's ships employed a more sophisticated system of wind observations than their naval equivalents.[60]

The person who did the most to persuade the Admiralty that their ships' crews should take careful weather observations was Sir Francis Beaufort (1774–1857). Beaufort left school at fourteen to join an East India Company ship, the *Vansittart*, before transferring to the Fifth Rate Royal Naval ship *Latona* as an able seaman. During the Napoleonic Wars Beaufort served on fighting and surveying vessels and rose to the rank of captain by 1810. He gained his reputation as an excellent surveyor through his work on the Rio de la Plata and along the coast of Turkey. Beaufort replaced William Edward Parry as Admiralty hydrographer in 1829 after being overlooked for the post in 1823 by John Croker. He was appointed Knight Commander of the Bath in 1848 and eventually attained the rank of rear admiral. He held the post of Admiralty hydrographer until 1855. He died in 1857.

[58] Anon, *Regulations and Instructions Relating to His Majesty's Service at Sea* (London: Stationery Office, 1808), p. 192. In the Royal Navy, the ship's master was a naval warrant officer, trained in and responsible for the navigation of the ship. The completion of logbooks and remarks books was a legal requirement and officers were required to submit them to the Admiralty at the completion of a voyage, at which time they would be paid their salary. The logbooks and remarks books would constitute the complete record of a voyage. C. Wilkinson, 'The Non-climatic Research Potential of Ships' Logbooks and Journals', *Climatic Change*, 73 (2005), 155–67.

[59] See the section 'Lieutenant' in Admiralty, *Regulations and Instructions*, pp. 171–81.

[60] N. Courtney, *Gale Force 10: The Life and Legacy of Admiral Beaufort* (London: Review, 2002). However, see D. C. Agnew, 'Robert FitzRoy and the Myth of the "Marsden Squares": Transatlantic Rivalries in Early Marine Meteorology', *Notes and Records of the Royal Society*, 58 (2004), 21–46.

Beaufort has been credited with turning the Hydrographic Office into a world leader in maritime survey.[61] Alongside his formal responsibilities as Hydrographer, Beaufort had an informal role as liaison between British science and the state. Beaufort's interests in geophysics and exploration permitted his membership on the committees of many of Britain's most eminent scientific societies, including the Royal Society, the Royal Astronomical Society and the Royal Geographical Society. He was appointed head of the scientific branch of the Admiralty Board in 1831, a position that gave him administrative responsibility for the Greenwich and Cape Observatories and Admiralty Offices related to navigation.[62] He was close to the Cambridge reformers, with whom he found common cause in 'breaking the strangehold of heirs of that regime upon voluntary and government scientific institutions and also in the promotion of common interests in geophysical science'.[63] He worked closely with Airy in his role as Astronomer Royal, aided the work of John Lubbock and William Whewell on the tides and assisted Herschel and Sabine's campaign for magnetic observation voyages to Antarctica.[64] Beaufort also made full use of the resources and networks of the Royal Navy to supply willing volunteers spread across the world with scientific instruments and advice, and to facilitate the movement of valuable information and commodities, such as botanical specimens, back to Britain for analysis.

Beaufort's longest-held scientific interest was meteorology. He kept records of the weather in his diary as a teenager whilst serving on the frigate HMS *Aquilon* and continued to do so throughout his life. In 1806, while serving on the *Woolwich*, he laid out his own fourteen-point scale for the measurement of wind force (where 0 denoted 'calm' and 13 denoted 'storm'), as well as shorthand for the description of the weather.[65] Beaufort's early attempts at a wind scale did not eliminate the possibility that two observers could attribute different categories to the same strength of wind – how was one to distinguish between Beaufort's '4. gentle breeze' and '6. fresh breeze', for instance? His solution in the following year was to correlate wind force with the amount of sail a fully

[61] Friendly, *Beaufort*, p. 248. [62] Day, *Admiralty Hydrographic Service*, pp. 47–8.

[63] Miller, 'Revival of the Physical Sciences', 114.

[64] G. S. Ritchie, *The Admiralty Chart: British Naval Hydrography in the Nineteenth Century* (London: Hollis & Carter, 1967). Day, *Admiralty Hydrographic Service*, p. 45. Severe cuts to the Hydrographic Office in 1847 and 1853 retarded the Hydrographic Office and Beaufort's ability to assist in other scientific schemes.

[65] National Meteorological Archive, MET/2/1/2/3/540, MET2/1/2/3/541a and 541b. It is generally accepted now that Beaufort's wind scale was modelled on a system of observation developed by John Smeaton in 1759. Alexander Dalrymple, the Scottish geographer and first hydrographer of the Admiralty, is credited with passing Smeaton's ideas on to Beaufort so that they might be adapted for use at sea. See Friendly, *Beaufort*, p. 143.

rigged ship would carry.[66] The making, shortening, reefing or furling of sail were tasks crucial to the effective and safe operation of a ship prior to the age of steam. These tasks demanded cooperation amongst a large group of skilled sailors, all of whose movements were controlled through standardised instructions issued by an officer on deck.[67] The use of sail as a method of measuring the force of the wind was therefore an expedient way of turning an ingrained awareness of a subject to new ends. Turning part of the architecture of a ship into an instrument of science was also not without precedent. William Snow Harris conducted research into the effects of lightning strikes on over 200 naval ships and experimented with the use of lightning conductors, arguing that these 'may be considered as so many grand experiments on the gigantic scale of nature'.[68]

Beaufort's use of sail to measure wind speed was evident in the private diary he kept while in command of HMS *Blossom* and HMS *Frederiksteen* from 1810 to 1812. Now 'Gentle breeze' was ranked '3' and described as 'That which will impel a Man of War with all sail by the winds' at four to five knots. A 'Fresh breeze' was ranked '5', and described as 'That with which Whole S[ai]l_ royals, stays &c. may be just carried full and by'.[69] If it was challenging to differentiate the subtle differences in wind strength around the midpoint of the scale, Beaufort's nomenclature really struggled at the extremes. A storm, ranked 11 in Beaufort's 1810–12 diary, was defined as that which would blow away any sail. A hurricane, at the twelfth and final point on the scale, was defined simply as 'Hurricane!' Just as a ship's sails were unable to catch the wind in the event of a hurricane, so language seemed unable to capture a precise description of extreme weather.

As he developed his wind scale and weather notation, Beaufort agitated for better use to be made of ships' logbooks as effective textual instruments in the accumulation of knowledge about the wind and weather. Writing to his brother-in-law Richard Lovell Edgeworth in 1809, he noted:

> There are at present 1000 King's vessels employed. From each of them there are from 2 to 8 Log books deposited every year in the Navy office; those log books give the wind and weather every hour . . . spread over a great extent of ocean. What better data could a patient meteorological philosopher desire? Is not the subject,

[66] Friendly, *Beaufort*, p. 144. The idea of describing wind strength in terms of sail carried was not new, and was referred to in Daniel Defoe's 1704 account *The Storm*.

[67] Anon, *Observations and Instructions for the Use of the Commissioned, the Junior and Other Officers of the Royal Navy* (London: C. Whittingham, 1804).

[68] W. S. Harris, *Remarkable Instances of the Protection of Certain Ships of Her Majesty's Navy from the Destructive Effects of Lightning* (London: Richard Clay, 1847).

[69] National Meteorological Archive, MET/2/1/2/3/540, MET2/1/2/3/541a and 541b.

not more in a scientific than a nautical point of view, deserving laborious investigation?[70]

Beaufort's appointment as Admiralty hydrographer provided him with the ideal platform from which to effect this vision. He and his officers used the Hydrographic Office as a centre for the collection of meteorological information and its surveying ships as mobile weather stations. In 1832, Lieutenant Alexander Becher, Beaufort's naval assistant, wrote an article in the *Nautical Magazine* entitled 'The Log Book', where he argued for better methods of the recording of the weather at sea.[71] Becher complained that the log contained too little space for the recording of the state of the weather given the mass of observations that officers had to record, and he advocated the system of abbreviated annotation that Beaufort had developed.[72] Beaufort encouraged the use of his meteorological schema amongst his surveyors. Probably the first surveying ship to employ it was HMS *Beagle* on its voyage to South America in 1831, captained by Robert FitzRoy. The Admiralty instructions that FitzRoy received noted that the ship's records of the wind should use Beaufort's wind scale and weather notation, as opposed to 'ambiguous terms … in using which no two people agree'. The guidance recommended that Beaufort's scale and notation be pasted on the first page of the logbook and the officer of the watch instructed to use the same terms.[73] They also gave guidance on when and how to read the barometers and thermometers on board ship.

Responses to Beaufort's meteorological plans for the Navy were generally positive but their uptake was uneven. In 1833, Beaufort received a letter from Admiral Sir George Cockburn, the commander-in-chief of the Navy's North America and West Indies Station at Bermuda, praising his system of wind and weather recording and noting that it was in general use there.[74] Herschel wrote to Beaufort

[70] Letter dated 9 December 1809, quoted in Friendly, *Beaufort*, p. 142.

[71] The *Nautical Magazine*'s aim was to collect and disseminate navigational and hydrographic knowledge with a view to the improvement of the Royal and Merchant Navy. Becher was the *Magazine*'s founder and editor. M. Barford, 'Fugitive Hydrography: The Nautical Magazine and the Hydrographic Office of the Admiralty, c.1832–1850', *International Journal of Maritime History*, 27 (2015), 208–26.

[72] Friendly, *Beaufort*, p. 146. Officers were obliged to note down all signals that were made and received, all changes of sail, 'all strange sails that are seen', any circumstances 'which may derange the order in which the Fleet is sailing', as well as 'all shifts of wind'. Anon, *Regulations and Instructions*, p. 173.

[73] 'Admiralty Instructions for the Beagle Voyage' is included in Appendix One of C. Darwin, *Voyage of the Beagle* (London: Penguin, 1989 [1839]), p. 396. John Herschel also employed Beaufort's wind scale at the Cape Observatory. Cock, *Sir Francis Beaufort*.

[74] G. Cockburn to Beaufort, 14 September 1833, UK Hydrographic Office Archives, LP1857/C.

in December 1835, saying that while visiting the Cape, Lord Auckland had volunteered his aid in establishing a proper system of meteorological and tidal observations in India, under his stewardship as Governor-General.[75] Beaufort also found support from the Admiralty Committee established in 1836 to organise the Navy's steam department, who were interested in adopting Beaufort's system on the Navy's fleet of steam vessels. Writing to the Committee in October 1836, Beaufort congratulated the Committee for the 'character of precision and utility' of the logbooks they proposed to use. In terms of records of the force of the wind, he asserted the value of his numerical scheme, urging them not to reduce the scale to six categories, worrying that such restrictions would encourage the use of fractions. Beaufort also complimented the Committee on their proposed column for the state of the sea, the observation of which could supplement evaluations of wind force.[76]

A year later, however, Beaufort was lamenting the quality of weather observations on board naval ships. In a letter to Captain Sir James Bremer, he wrote that 'once in the watch the officer generally inserts "Moderate and cloudy" or some one or other of those proverbial phrases, the ambiguity of which is quite laughable. I have tried a dozen persons and no two of them have agreed as to the expressions they would use to describe the state of the wind and weather.'[77] Although his own schema had been 'invariably adopted' by surveying vessels, only some of the admirals in general service had taken it up.[78] Even among his own surveying vessels there were inconsistencies in approach. While surveying off the coast of Sierra Leone in 1834 the crew on HMS *Ætna* collected various meteorological observations at 8am, noon and 4pm, but did not record wind force or direction, and made weather observations of the sort that Beaufort had been complaining about to Captain Bremer. Beaufort's attempt to regulate weather observation at sea suffered from the same problem as his wind scale. The use of terms such as 'variable airs', 'passing clouds' and 'pretty clear' in

[75] J. Herschel to Beaufort, 26 December 1835, UK Hydrographic Office Archives, LP1857/H.

[76] Letter from Beaufort to T. Baldock, 8 October 1836, UK Hydrographic Office Archives, LB/8. Baldock was one of three members of the Committee. Steam vessels eventually made the old correlations between press of sail and wind speed irrelevant. Cock, *Sir Francis Beaufort.*

[77] Letter from Beaufort to J. J. Gordon Bremer, 2 November 1837, UK Hydrographic Office Archives, LB/8.

[78] Bremer was twice commander-in-chief of British forces in China and it was Bremer who took formal possession of Hong Kong for Britain in 1841. W. R. O'Byrne, *Naval Biographical Dictionary* (London: John Murray, 1849), pp. 119–20.

Ætna's meteorological log demonstrated the insufficiency of language to represent weather at the mean.[79]

On the other side of the Atlantic HMS *Jackdaw* was working around the Bahamas and its commander, Lieutenant Edward Barnett, was making meteorological observations. The observations collected were much fewer than on *Ætna*, but this time wind direction and force were recorded using Beaufort's scale, along with weather observations using Beaufort's notation. Observations were taken at 9 am, noon, 3 pm, 6 pm and 9 pm.[80] *Jackdaw* was accompanied in the Bahamas by HMS *Thunder*, another hydrographic ship, which like *Ætna* and *Jackdaw* also kept a meteorological register separate from the logbook. While *Jackdaw* collected surface water temperature readings, *Thunder* made more comprehensive use of its marine and oil barometers. Readings were also taken at different times – at 4 am, 9 am, noon, 3 pm, 8 pm and midnight.[81] When Barnett took charge of HMS *Thunder* in November 1837 for another tour of North America and the West Indies, Beaufort wrote to him with detailed instructions. As with Bremer and FitzRoy, Beaufort provided Barnett with several copies of his wind scale and weather abbreviations and suggestions as to their use. He also asked the officer to record other interesting meteorological phenomena, to document the 'periods and limits' of the trade winds, monsoons and rains as they were encountered, and to pay full attention to the barometer and thermometer. In doing so, the ship would be adding to a stock of knowledge 'for the use of future labourers whenever some accidental discovery, or the direction of some powerful mind should happily rescue that science from its present neglected state'.[82]

Beaufort clearly felt that the malaise identified by Forbes in 1832 continued to plague meteorology at sea five years later. However, despite his explicit support for meteorological reform, Beaufort struck a pragmatic tone when in correspondence with his officers. He was forced to concede to Barnett that the hours of entry of meteorological information interfered with the officers' other activities while at sea. Noting that the data's 'future utility is so uncertain', Beaufort suggested that a fuller

[79] Anon, Hygronometrical Observations made on board His Majesty's Surveying Vessel *Ætna*, communicated to the Royal Society by Captain Beaufort, 1835, Archives of the Royal Society, AP/19/1.

[80] E. Barnett, *Jackdaw*'s Meteorological Register 1st January–1st November 1834, communicated to the Royal Society by Captain Beaufort, 1835, Archives of the Royal Society, AP/19/2.

[81] Anon, Meteorological Register. HMS *Thunder*. Between January 1st & June 30th 1834, communicated to the Royal Society by Captain Beaufort, 1835, Archives of the Royal Society, AP/19/18.

[82] Surveying instructions from Beaufort to Lieutenant E. Barnett of HMS *Thunder*, 9 December 1837, UK Hydrographic Office Archives, Miscellaneous Files.

record might only be possible due to some unforeseen detention in port, 'when a system of these observations might then be advantageously undertaken'.[83]

William Reid and the Law of Storms

In the late 1830s, Beaufort and Becher continued their campaign to improve meteorological observations on board Navy ships and were joined in their work by Lieutenant-Colonel William Reid (1791–1858), a British Army engineer. Reid served in the Peninsula and the Anglo-American wars and with the Ordnance Survey in Ireland, before being sent in 1832 as resident engineer to Barbados to assist in rebuilding government buildings after the devastating hurricane of August 1831.[84] Although Reid had harboured an interest in meteorology for some time, his residence in Barbados prompted him to study tropical storms. While there he familiarised himself with the work of other meteorologists who had worked on similar topics in different parts of the world. This included the writings of Colonel James Capper, of the East India Company. Capper published several works on tropical storms, including an 1801 paper *On the Winds and Monsoons*, which was based on his studies of records of eighteenth-century hurricanes that had affected the Coromandel and Malabar coasts of India. In it he argued that 'the velocity of the wind at any point was chiefly due to the velocity of rotation of a vortex of fluid, combined probably with a progressive motion'.[85]

Even more important to Reid was the work of William C. Redfield, an American transportation engineer based in New York, who published a number of papers in the 1830s on the characteristics of Atlantic storms.[86] Redfield had been informed by Benjamin Franklin's storm observations in the north-eastern United States. While in Barbados, Reid came across an 1831 paper by Redfield in the *American Journal of Science*, in which Redfield collated more than seventy sets of observations of the hurricane of 17 August 1830, to argue that these storms were whirlwinds rotating around a centre of low pressure, which moved forwards on curved tracks.[87] Reid was particularly impressed by the chart of

[83] Beaufort, Surveying Instructions to Barnett of HMS *Thunder*, p. 27.
[84] O. M. Blouet, 'Sir William Reid, F. R. S., 1791–1858: Governor of Bermuda, Barbados and Malta', *Notes and Records of the Royal Society*, 40 (1986), 169–91, 174.
[85] J. D. Forbes, 'Supplementary Report on Meteorology', *Report of the Tenth Meeting of the British Association for the Advancement of Science* (1841), 37–156, 109.
[86] J. R. Fleming, *Meteorology in America 1800–1870* (Baltimore: Johns Hopkins University Press, 1990).
[87] Anon, 'Redfield's Law of Storms:- Notice of Col. Reid's Work on Hurricanes', *The American Journal of Science and Arts*, 35 (1839), 182; W. C. Redfield, 'Remarks on the

the storm that Redfield included in his study.[88] Convinced that Mr Redfield's views were correct, Reid set about collecting more data on the wind direction of Atlantic storms, and laying down the data on large-scale charts so as to strengthen the argument that these storms conformed to the pattern of a 'progressive whirlwind'.[89]

Upon returning to England in 1836 Reid went on the half-pay list while assembling meteorological information, including storm data from the logbooks of Admiralty ships. He also initiated correspondence with Redfield, who encouraged Reid's emphasis on direct observation.[90] At the 1838 British Association meeting in Newcastle, Reid presented his own work on the subject, notably eight charts showing the path of storms at different latitudes. Although he claimed that his object was 'not to establish or support any theory, but simply to arrange and record facts', his report came out strongly in favour of the ideas of Redfield.[91] In doing so he supported Redfield's belief that a reliable system of meteorological physics should be 'grounded in direct observations'.[92] Reid also added his own embellishments, such as that the progressive rate of storms was never greater than that of the atmospheric currents; that a hurricane's destructive power was due to its rotatory velocity; and that its path traced out a parabola.[93]

Herschel, amongst others, spoke positively of Reid's work at the meeting, commending him for his judiciousness as an observer while urging him to advance a theoretical position, if only to incite debate and encourage the 'collision of intellect'.[94] Others argued against the theory that Reid's work supported. Alexander Bache, previously Professor of Natural Philosophy at the University of Pennsylvania, and from 1843 the superintendent of the Unites States Coastal Survey, spoke out against it and in support of a rival theory of James Espy, the director of the Joint Committee on Meteorology of the American Philosophical

Prevailing Storms of the Atlantic Coast, of the North American States', *American Journal of Science and Arts*, 20 (1831), 17–51.

[88] Anon, 'On Storms', *Littell's Spirit of the Magazine and Annuals*, 2 (1838), 856–8.

[89] W. Reid, *An Attempt to Develop the Law of Storms by Means of Facts, Arranged According to Time and Place, and Hence to Point out a Cause for the Variable Winds, with the View to Practical Use in Navigation* (London: John Weale, 1838), p. 3.

[90] Fleming, *Meteorology in America*, p. 38.

[91] Anon, 'A Report Explaining the Progress Made towards Developing the Law of Storms', *The Athanaeum*, 25 (August 1838), 594–6, 594.

[92] Letter from Redfield to Reid, 26 March 1838, quoted in Fleming, *Meteorology in America*, p. 39.

[93] Reid's contributions to Redfield's theory of storms were summarised in Charles Tomlinson's essay, 'The Law of Rotatory Storms', contained in J. Greenwood, *The Sailor's Sea-Book: Rudimentary Treatise on Navigation* (London: John Weale, 1850).

[94] Herschel, quoted in 'A Report Explaining the Progress', p. 595.

Society.[95] Espy supported the centripetal theory of the German mathematician H. W. Brandes, proposing that the wind blew in all directions towards the centre of a storm, with the inward flow at the surface balanced by a corresponding outflow above a rising column of air.[96] While Espy's ideas gathered some support in America, they were largely rejected in Britain, where his theory was considered to be 'wholly contradicted by the facts'.[97]

Reid's research appeared in print in 1838, first in a long article entitled 'On Hurricanes' in the Corps of Royal Engineers' professional papers, and then in a nearly 600-page book on the subject, *An Attempt to Develop the Law of Storms*. The volume was in effect an extended demonstration of the validity of Redfield's theory. The first two chapters of the book discussed Redfield's storm observations, and then moved on in subsequent chapters to the storms that affected particular regions of the world, notably the hurricanes of the western Atlantic, the typhoons of the China Sea and the cyclones of the Indian Ocean. That a book on storms would focus on these areas was unsurprising, given the high incidence of storm events around the equator. It was also unsurprising that much of the discussion focused on islands under British control – Barbados, Mauritius, Antigua and Bermuda – or on those of significant commercial importance to the British, such as Macao. These were places where naval and merchant ships would visit on a regular basis. Weather data followed the paths of British ships and traced a geography of storms that conformed to the contours of Britain's imperial interests. Whether this was intentional or not, these were the places where a law of storms mattered most to British shipping.

Each chapter was a compilation of meteorological data from ships' logs and from observers on land, including both instrumental observations and anecdotal remarks. In the case of many of the storms, the data were traced out on large foldout charts appended at the end of the book. For instance, Reid's chart of the Great Hurricane of 10 October 1780 mapped the track of the hurricane across the Caribbean and then back across the Atlantic (Figure 1.1). Also included were the daily positions of various

[95] C. Carter, 'Magnetic Fever: Global Imperialism and Empiricism in the Nineteenth Century', *Transactions of the American Philosophical Society*, 99 (2009), i–xxvi and 1–168. For a fuller discussion of the rival storm theories of Redfield and Espy, see Fleming, *Meteorology in America*; and Jankovic, 'Ideological Crests'.

[96] J. Burton, 'Robert FitzRoy and the Early History of the Meteorological Office',*British Journal for the History of Science*, 19 (1986), 147–76, 148.

[97] H. Piddington, *The Sailor's Horn-Book for the Law of Storms: Being a Practical Exposition of the Theory of the Law of Storms, and Its Uses to Mariners of all Classes, in all Parts of the World, Shewn by Transparent Storm Cards and Useful Lessons* (London: Williams and Norgate, 1848[1860]), p. 6.

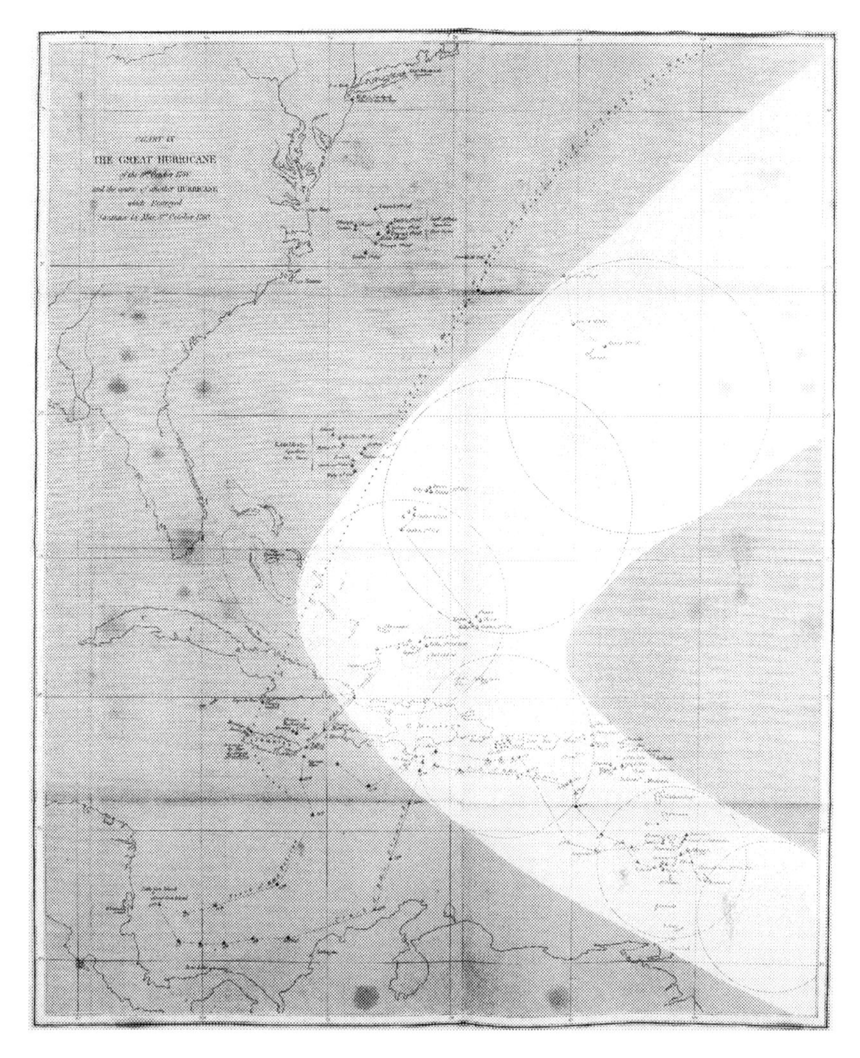

Figure 1.1 Chart of the Great Hurricane of 1780. (Source: Reid, *Attempt to Develop the Law of Storms*, Chart IV.)

ships that had supplied data, their movements between these points, and in several cases, the places where they had been wrecked or lost. Reid replicated this approach on many of his other charts. For him, the movement of those ships that encountered revolving storms was as important and useful as the information collected in their logbooks.[98]

[98] For a similar argument, see Sorrenson, 'The Ship as Scientific Instrument', p. 222.

Reid's epistemic authority was founded on the extent of the observations he had collected and their distribution across space, while that authority was bolstered by the cartographic representation of the observations. Reid argued that 'by collating a great number of reports of storms made at different places, as well at sea as on shore, the changes of wind in a separate storm are now understood'.[99] However, these charts were far from transparent representations of meteorological reality. Rather, they were opportunities to smooth out errors of observation and random fluctuations in the wind or barometric pressure. The charts revealed order that would otherwise have been concealed in tables of numbers, while retaining particular details, notably the positions and tracks of ships.[100]

The strongest endorsement for Reid's approach came from Henry Piddington in his *Sailor's Horn-Book for the Law of Storms*, published in 1848. Piddington was a merchant captain who had worked around India and China, and later became president of the Marine Court of Inquiry at Calcutta. In his *Horn-Book*, Piddington laid out the relations between scientific theory, proof and application, where the theory was 'the supposition that a thing always occurs according to certain rules, the proof or Law that it does and will always so occur, and the Application of that law to the business of common life'. According to Piddington, Reid's analysis of more than 2,000 logs and of some hundreds of storms had provided the proofs of the theory of storms developed by Redfield.[101] Reid's work certainly conformed to prevalent models for the pursuit of terrestrial physics in the first half of the nineteenth century (discussed further in Chapter 2), with its emphasis on the gathering of large amounts of global data. His graphical representations of storms were also part of a wider movement to present scientific ideas visually in the 1830s, such as by Herschel in astronomy and Sabine in studies of terrestrial magnetism. Lastly, Reid's implicit support of bold theorising and his ambition to identify the universal laws of nature meant that he was not merely a naïve military fact-gatherer – despite his own modest claims to be just that – but was supportive of a hypothetical-deductive method of the sort advocated by Herschel.[102] Indeed, Reid's

[99] W. Reid, *The Progress of the Development of the Law of Storms and of the Variable Winds, with the Practical Application of the Subject to Navigation* (London: John Weale, 1849), p. 2.

[100] T. L. Hankins, 'A "Large and Graceful Sinuosity": John Herschel's Graphical Method', *Isis*, 97 (2006), 605–33, 606; K. Anderson, 'Mapping Meteorology', in J. R. Fleming, V. Jankovic and D. R. Coen (eds.), *Intimate Universality: Local and Global Themes in the History of Weather and Climate* (Sagamore Beach: Science History, 2006), pp. 69–92.

[101] Piddington, *Sailor's Horn-Book*, p. 8.

[102] G. Good, 'A Shift of View: Meteorology in John Herschel's Terrestrial Physics', in J. R. Fleming, V. Jankovic and D. R. Coen (eds.), *Intimate Universality: Local and Global Themes in the History of Weather and Climate* (Sagamore Beach: Science History, 2006), pp. 35–68, p. 36.

charts were effective at holding these various demands in tension, while avoiding unnecessary philosophical controversy.

Piddington's claims that Reid's work added to the business of common life remind us that Reid conformed to another principle of science in the early years of the nineteenth century: that it should produce knowledge that was of use to society. Reid's charts had an obvious practical value: they spoke to non-scientific figures such as a sailor, navigator or harbour master – all of whom would be comfortable with the language of charts – as much as they did to the meteorologist. This was demonstrated in a review of Reid's book in the *London Saturday Journal*, where *The Law of Storms* was discussed alongside the 1839 edition of *Murphy's Weather Almanac.*[103] Both volumes were in the business of trying to predict atmospheric events, but while *Murphy's Almanac* was founded on 'vague, incoherent jargon' and grandiose claims, Reid's book proceeded on an altogether more cautious footing. The reviewer went as far as to say that if the theory of the circular and progressive motion of hurricanes be established as an actual fact, 'it may ultimately be turned to great "practical use in navigation"'.[104] Similarly, Piddington praised Reid for deducing the rules that would render Redfield's theory 'of practical utility'.[105] The *Edinburgh Review* predicted that 'no sailor will study these records of atmospherical convulsions, without feeling himself better armed for a professional struggle with the elements. The navigator, indeed, who may quit the shores of Europe for either Indies without Colonel Reid's book, will discover, when it is too late, that he has left behind him his best chronometer and his surest compass.'[106] Despite the full title of Reid's book, however, the amount of direct advice given to sailors who found themselves in the path of a storm was relatively slight. This was remedied in Reid's 1849 book *The Progress of the Development of the Law of Storms*, published by John Weale, which contained a chapter dedicated to heaving-to and sailing out of a revolving storm. The book's arguments were also more routinely illustrated by schematics, most notably Reid's hemispheric circles (Figure 1.2), which were designed to be cut out and placed on a marine chart so that they might 'serve to aid the memory whilst considering how the wind veers in whirlwind storms'.

[103] Patrick Murphy's *Almanac* caused a brief sensation in 1838 when it predicted successfully the coldest day of the year to be 20 January. Anderson, *Predicting the Weather.*

[104] Anon, 'Weather Almanacs and the Law of Storms', *The London Saturday Journal*, 1 (1839), 7.

[105] Piddington, *Sailor's Horn-Book*, p. 7.

[106] Anon, 'Review of Reid's *Law of Storms*, 1838, along with Redfield's articles on Atlantic storms in *Silliman's Journal, Blunt's American Coast Pilot and US Naval Magazine*', *Edinburgh Review*, 68 (1839), 406–30, 431.

LAW OF STORMS

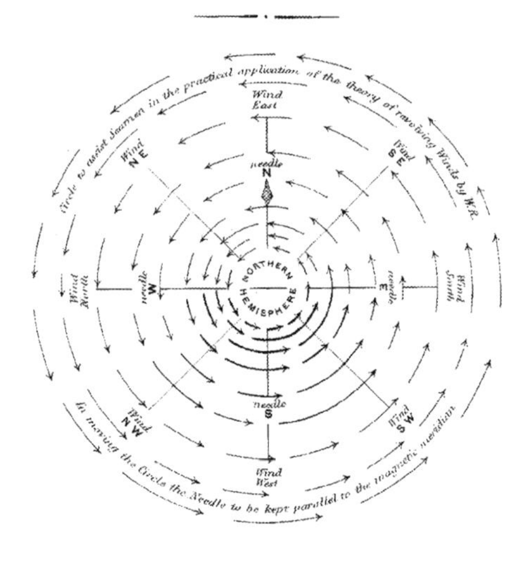

EQUATOR

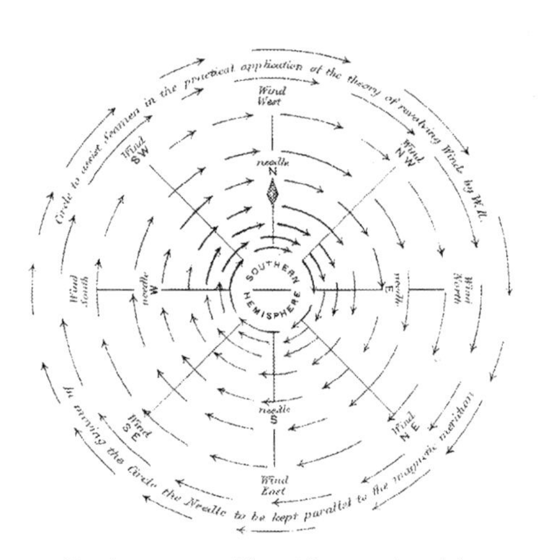

When these Circles are used they should be cut out and moved along a Marine Chart in the direction of a Storm's progress. Dipped in turpentine they will become transparent.

London, Published by John Weale, 59 High Holborn.

Figure 1.2 Hemispheric storm circles. (Source: Reid, *Progress of the Development of the Law of Storms*, facing p. 3.)

Reid, Beaufort and the Ship's Logbook

Reid's *The Law of Storms* went through several editions and was translated into various languages. In recognition of his scientific contribution Reid was made a Knight Companion of the Order of the Bath in 1838 and was elected Fellow of the Royal Society in 1839. His work and increasing prominence in the field of meteorology brought him into contact with Beaufort.[107] Reid wrote to Beaufort in May 1838 to draw the latter's attention to his work on storms. He lauded Beaufort's attempts to improve the quality of weather data collected in ships' logs and particularly his emphasis on wind speed and direction, with which a ship's position in a revolving storm could be ascertained. He also praised Beaufort's idea of inserting columns that showed a ship's track.[108]

Reid promoted Beaufort's wind force scale and weather notation in *The Law of Storms* and he petitioned the Admiralty to adopt Beaufort's meteorological additions to the logbooks of naval ships. Becher supported Reid's plan. In a memorandum of November 1838, Becher reiterated familiar criticisms of the current state of meteorological observations on board naval ships and acknowledged Reid's recommendation of Beaufort's schema, suggesting that Reid 'has evidently been enabled in his recent enquiry into Hurricanes to see the full value of it'.[109] Becher went on to argue that no seaman in command of a ship would ever go to sea without a barometer and that great advantage would arise from 'the observations of it being recorded in every weather, and in the event of storms or hurricanes occurring that the changes in its height during their progress and times of change should be carefully noted'.[110] Beaufort supported this, arguing in December 1838 that all officers who possessed a barometer 'should be permitted to observe it at least once in every watch'.[111]

Reid and Beaufort's joint campaign was a success. In December 1838, the Admiralty adopted Reid's proposal. On 3 January 1839, Reid wrote to Herschel in his role as chair of the Royal Society's Joint Committee of Physics and Meteorology, saying that 'it will be gratifying to yourself and all interested in Meteorology to know that the Lords Commissioners of

[107] Reid to Beaufort, 12 March 1838, UK Hydrographic Office Archives, LP1857/R.
[108] Reid to Beaufort, 9 May 1838, UK Hydrographic Office Archives, LP1857/R.
[109] A. Becher, Proposals for improving the Meteorological Registers in the Log Books of HM Ships, 14 November 1838, UK Hydrographic Office Archives, MB/3, pp. 91–3, on p. 91.
[110] Becher, Proposals, pp. 92–3.
[111] F. Beaufort, On inserting Meteorological Observations, according to Office Abbreviations, in the Logs of HM Ships, 5 December 1838, UK Hydrographic Office Archives, MB/3, p. 96

the Admiralty have ordered an addition to be made to the Log Books of Ships of War, to aid enquiry into the subject'.[112] Columns for the force of wind and appearance of the atmosphere were to be added to the logbook and were to be completed on an hourly basis. This instruction was reinforced in the 1844 edition of the *Admiralty Instructions*, where ships' captains were told to ensure that the logbook recorded, 'most carefully, all particulars relating to the situation of the Ship, along with the state of the weather, and the direction and force of the weather every hour'.[113] The *Admiralty Instructions* also included advice on the location and observation of the barometer on board ship, which reflected the Admiralty's decision in 1843 that all HM ships should carry one.[114] Courtney claims that it was the Royal Society's intervention that led to the formal issue of marine barometers to the naval fleet in 1843, a matter administered through the Hydrographic Office.[115] This development was reflected in the 1844 edition of the *Admiralty Instructions*, where captains were told to have the barometer 'carefully suspended in some secure and accessible part of the Ship' (and to note its location at the beginning of the logbook), and to make observations at 6 am, noon, 6 pm and midnight.[116] The British Association was quick to utilise this new development. In 1845, survey ships on the Home Station were ordered to assist the Association, which was interested in observing meteorological phenomena that affected the British Isles during the autumn. Officers were asked to keep registers of barometric observations during October and November using printed directions and blank forms issued especially, and were required to again do so in 1846.[117]

Interest in the value of meteorological instruments at sea spread beyond the survey fleet. In February 1847, Beaufort received a letter from Sir Henry John Leeke, flag captain of HMS *Queen*, a 110-gun first-rate ship of the line and the last sailing battleship to be completed before the widespread introduction of steam power.[118] Leeke wrote to Beaufort to promote the work of his Major of Marines David M. Adam, whose knowledge of the barometer and attention to changes in the weather had

[112] Letter from Reid to Herschel, 3 January 1839, Archives of the Royal Society, DM/3.

[113] Anon, *Admiralty Instructions for the Government of Her Majesty's Naval Services* (London: Stationery Office, 1844), p. 173.

[114] Day, *Admiralty Hydrographic Service*, pp. 56–7. Day notes that the barometers were supplied by the Hydrographic Office.

[115] Courtney, *Gale Force 10*.

[116] Anon, *Admiralty Instructions*, p. 173. What was meant exactly by 'secure and accessible' is not explained.

[117] Anon, 2 September 1845 and 6 October 1846, Circulars to Surveyors on the Home Station, UK Hydrographic Office Archives, LB/13 and LB/14.

[118] J. J. Colledge, *Ships of the Royal Navy: The Complete Record of all Fighting Ships of the Royal Navy from the Fifteenth Century to the Present* (London: Greenhill Books, 1970).

been of 'great use' to him on board. Adam was 'half a very clever scientific man', claimed Leeke, and he requested additional meteorological instruments to aid Adam.[119] Attached to Leeke's letter was one from Adam himself, addressed to Leeke although presumably targeted at Beaufort, forwarding his readings of the barometer, thermometer, wind direction and force, and weather while HMS *Queen* was at Plymouth Sound in November and December 1846. Adam requested a hygrometer, anemometer, rain gauge, electrometer and dipping circle, on the grounds that '[i]f there is one place where accurate knowledge of [the weather], is more useful than another, that place is a Man of War – on ship-board'. For Adam, instruments 'may give the young officers a scientific turn', and that the serious study of meteorology on a flagship could only lead to 'a more accurate knowledge of that science' throughout the fleet.[120] He promised the Admiralty Lords weekly or monthly meteorological reports in return. Columns for barometric and thermometric readings were added to the logbook, alongside those for wind and weather. The logbooks of naval ships, Reid later noted, were lodged and available for consultation at Somerset House.[121]

Interest in meteorological observation spread even further than the Navy's fighting ships. Reid appended a memorandum to his 1839 letter – written by Lord Glenelg, the Secretary of State for War and the Colonies – and addressed to all governors of British colonies on the subject of 'Keeping Journals of the Weather, and of noticing Meteorological phenomena generally'.[122] In it Glenelg also directed governors of British colonies, captains of ports, harbour masters and keepers of lighthouses to keep meteorological journals based on the principles of the logbooks of ships, and to submit them every half year to the Colonial Office, where they would be preserved in the library for future use. In the second edition of *The Law of Storms*, published in 1841, Reid drew attention to his success at persuading the Admiralty Lords and government ministers to adopt a keener interest in the weather at sea. He also noted that the inspector-general of the Coastguard had issued orders to revenue cruisers to keep hourly observations of the weather, and that the directors of the East India Company had instructed the governor-general of India to 'carry out

[119] Letter from H. Leeke to F. Beaufort, 9 February 1847, UK Hydrographic Office Archives, LP1857/L.

[120] Letter from D. Adam to H. Leeke, 30 January 1847, appended to the letter from Leeke to Beaufort, UK Hydrographic Office Archives, LP1857/L.

[121] Reid, *The Law of Storms*, p. 542.

[122] Lord Glenelg, undated, Memorandum respecting the Records to be kept of the state of the Weather, in the British Colonies, appended to letter from Reid to Herschel, 3 Jan 1839, Archives of the Royal Society, DM/3.

various suggestions on the subject of tracing the storm-tracks of the Indian seas'.[123]

For Reid, the Admiralty's willingness to support a plan that Beaufort had been promoting for some years was due to their interest in storm predictions. He noted that their lordships had ordered thirty copies of his book to be distributed among interested captains and commanders-in-chief of various stations, and that 'they would with pleasure afford any assistance in carrying on an enquiry so valuable to navigation and the interests of Humanity'.[124] Copies were also deposited in the Admiralty Library and at the Hydrographic Office. Reid's appeal for meteorological data to aid in the understanding of the behaviour of storms was clearly more persuasive than Beaufort's more general, inductive policy of meteorological observation and data gathering, and was illustrative of the priority that the Admiralty gave to fundamental matters concerning safety of life at sea over scientific interests.[125]

This sudden apparent enthusiasm for meteorology at the Admiralty and in government may also have been related to the decision in early 1839 to fund Ross's expedition to investigate terrestrial magnetism in the southern hemisphere and establish several overseas observatories, discussed earlier in the chapter and again in Chapter 2. Although meteorology and terrestrial magnetism were equally data-intensive, the coordination of magnetic research was, of the two, the 'most fully organized and most self-consciously directed toward answering questions of laws and causes'.[126] Its emphasis on theoretical explication, the use of precision philosophical instruments and the value of collective, international endeavour marginalised the individual observer in favour of a 'central scientific authority which would process all empirical information into mass data'.[127] The Magnetic Crusade set an important example for the conduct of meteorology, which was pursued during the crusade itself, while Reid's argument regarding the value of meteorological data gathering for ships caught in storms bore a close similarity to that made about the collection of magnetic data for improvements to navigation.[128] We will return to this theme in Chapter 2.

[123] Reid, *The Law of Storms*, p. 542.

[124] Reid to Herschel, 3 January 1839, Archives of the Royal Society, DM/3; Admiralty Rough Minutes, 6 January 1839, National Archives, ADM3/245.

[125] Webb, 'More than Just Charts', p. 52.

[126] G. A. Good, 'Between Data, Mathematical Analysis and Physical Theory: Research on Earth's Magnetism in the 19th Century', *Centaurus*, 50 (2008), 290–304, 301.

[127] Winter, '"Compasses All Awry"', 87; Anderson, 'Mapping Meteorology'.

[128] As noted on p. 29, a significant amount of effort was invested in the study of meteorology during the Crusade, and far more than was required to make the necessary adjustments to the magnetic and astronomical instruments. One of the five sections of the Royal

Bermuda as Island Laboratory

In late 1838, Reid was appointed governor of the Bermuda Islands, Reid surmising that Lord Glenelg had recommended him for the post on the basis of his scientific work.[129] The islands were of strategic importance to the Royal Navy and the British Empire. In 1818, the Royal Navy's North America and West Indies Station was formed when the two previous stations were combined.[130] Halifax had previously acted as headquarters of the North America Station and continued as the summer base for the new Station. Bermuda became the Station's winter headquarters and the main base of activities. It was also the site of the Royal Naval Dockyards and was well positioned to allow the Royal Navy to protect Atlantic trade routes and fisheries, to patrol for slave ships, transport troops and garrison colonial territory.[131]

As governor of Bermuda, Reid was in a position to develop his own meteorological inquiries and to assist Beaufort, and he made full use of his posting to achieve this. In doing so he positioned Bermuda as a scientific space in a number of different ways. First, Reid used the island as an archive, where he could both gather together and disseminate weather data from naval and merchant ships. Second, he treated the island as a laboratory for the testing of universal meteorological laws, a site where observational results were meant to be independent of the locale. Reid's treatment of Bermuda as an open-air weather observatory would have been reinforced by the activities of the island's temporary magnetic observatory, which was established in 1843 as part of the Hydrographic Office's contribution to the Magnetic Crusade. Reid observed the passage of a storm with Captain Barnett of HMS *Thunder* in October 1845, when Barnett was on the island to dismantle the observatory. Third, he defended his claims on the basis of an extended residence in the field, where his locatedness lent credibility to his claims. Coen has argued that a tension existed between the study of the atmosphere as laboratory and as fieldsite, but that these two approaches were in fact interdependent and impossible to isolate

Society's report was devoted entirely to meteorology. Anon, *Report ... on the Instructions ... for the Scientific Expedition to the Antarctic Regions.*

[129] Blouet, 'Sir William Reid', p. 175.

[130] Bermuda took over as the headquarters of the North America Station from Halifax, Nova Scotia. Rio had been the Navy's headquarters of the South America Station since 1808. F. Driver and L. Martin, 'Shipwreck and Salvage in the Topics: The Case of HMS *Thetis*, 1830–1854', *Journal of Historical Geography*, 32 (2006), 539–62.

[131] K. Greer, 'Zoogeography and Imperial Defence: Tracing the Contours of the Nearctic Region in the Temperate North Atlantic, 1838–1880s', *Geoforum*, 65 (2015), 454–64.

fully.[132] Reid certainly employed both types of scientific space in his work at Bermuda.[133]

Bermuda was ideally situated to act as a weather archive.[134] In October 1839, Reid wrote to Beaufort, thanking him for new charts of Bermuda and to inform him that the island's collector of customs had been distributing his 'mode of recording the winds [sic] force by symbols' and that he had asked Reid to get him a further supply of the Admiralty order. Reid also offered to distribute these among the commanders of merchant ships who regularly docked there.[135] Early the following year, he wrote to Beaufort and to the Royal Society informing them that he had arranged for the editor of the *Bermuda Royal Gazette* to be supplied with meteorological reports from the island's central signal station at Government House. This system was effective, he claimed, because the newspaper was regularly transmitted to the Colonial Office and because it was popular amongst the commanders of ships and the owners of Bermuda shipping.[136] Reid had been getting masters of vessels coming into Bermuda to supply him with information from their logbooks on their courses sailed and the direction and force of the winds they had experienced. With the help of the customs office, he was laying the information down onto a chart so that 'we can judge of the best courses to steer'.[137]

Reid used this information to investigate the storms in the region. He pulled together large numbers of extracts from ships' logs documenting the incidence of a hurricane that passed over Bermuda on 11–12 September 1839 – a storm so severe that it had 'made the people here take up the subject of storms with some earnestness'.[138] The *Bermuda Royal Gazette* published a large number of the excerpts, with an introductory commentary that located the various ships in relation to the hurricane's path. The newspaper expressed the 'hope that by continuing the enquiry, the nature of the Bermuda Hurricanes and their Courses,

[132] D. R. Coen, 'The Storm Lab: Meteorology in the Austrian Alps', *Science in Context*, 22 (2009), 463–86, 465.

[133] C. R. Weld, *A History of the Royal Society* (London: John W. Parker, 1848), p. 444; Reid, *Progress of the Development of the Law of Storms*, p. 265.

[134] Mahony has developed a similar historical geography of island meteorology in his analysis of Mauritius: M. Mahony, 'The "Genie of the Storm": Cyclonic Reasoning ad the Spaces of Weather Observation in the Southern Indian Ocean, 1851–1925', *British Journal for the History of Science*, 51 (2018), 607–33.

[135] Letter from Reid to Beaufort, 17 October 1839, UK Hydrographic Office Archives, LP1857/R.

[136] Reid to J. Russell, 8 February 1840, Archives of the Royal Society, AP/24/16.

[137] Reid to Beaufort, 30 January 1840, UK Hydrographic Office Archives, LP1857/R; Reid to Russell, 8 Feb 1840, Archives of the Royal Society, AP/24/16.

[138] Reid to Beaufort, 17 October 1839, UK Hydrographic Office Archives, LP1857/R.

may be better understood'.[139] Reid went on to use the logbooks, newspaper reports and Central Signal Station records to update the second edition of *The Law of Storms*. He used the data to prepare a chart of the course of the hurricane, which demonstrated its curved track up from the tropic to Newfoundland, and showed the positions of the various ships that encountered it on 12 September.[140] He later reworked the chart in his *Progress of the Development of the Law of Storms* to illustrate his theory of the relationship between storms and sea-swell (Figure 1.3).

The chart was accompanied by an extended commentary, making up the entirety of Chapter IX, which imposed a loosely geographical narrative upon a range of source material. The chapter began with speculation on the origin of the storm. It then traced the storm's course, introducing extracts of reports of ships, their observations of winds and weather and the crews' responses to the storm, as they encountered it. Observations at Bermuda were also included, encompassing the nature of the sea-swell, barometer readings, wind force and direction, extracts from newspapers, the Central Signal Station weather tables, the behaviour of the tide and incidents of storm damage. As the storm progressed northwards, observations from other localities, such as St Johns, Newfoundland, the Gulf of St Lawrence and New York, were incorporated.

Reid's testimonial, narrative style reflected his commitment to first-hand observation as the basis for a science of storms. Reid's locatedness also formed part of the basis of his credibility as a meteorologist. He repeatedly legitimated his claims respecting storm activity with reference to his own extended residence on an island where Atlantic hurricanes could be experienced and studied and where others' observations could be procured. Crucially though, and in conformity with Herschel's position on the matter, he positioned his own observations at Bermuda as illustrative of processes that transcended place or region.[141] For instance, in a letter to the Royal Society in 1840, Reid noted: 'Since I have been in Bermuda, I have had no reason to doubt that Great Storms of wind, (which affect the Barometer) really revolve by a fixed law; but on the contrary, I have observed much to confirm this belief.'[142] Reid used his *Progress of the Development of the Law of Storms* to make similar virtue of his time on the island. For instance, he discussed Redfield's theory that a whirlwind diminished the pressure of the atmosphere at its centre, and supported the idea with the statement: 'My observations attentively made

[139] Anon, 'The Storm', *Bermuda Royal Gazette*, 24 September 1839, included in Reid's letter to Lord Russell, 8 February 1840, Archives of the Royal Society, AP/24/16. Descriptions of many of the storms were also reported in the *Nautical Magazine*.

[140] Reid, *The Law of Storms*, facing p. 444. [141] Good, 'A Shift of View'.

[142] Reid to Russell, 8 February 1840, Archives of the Royal Society, AP/24/16.

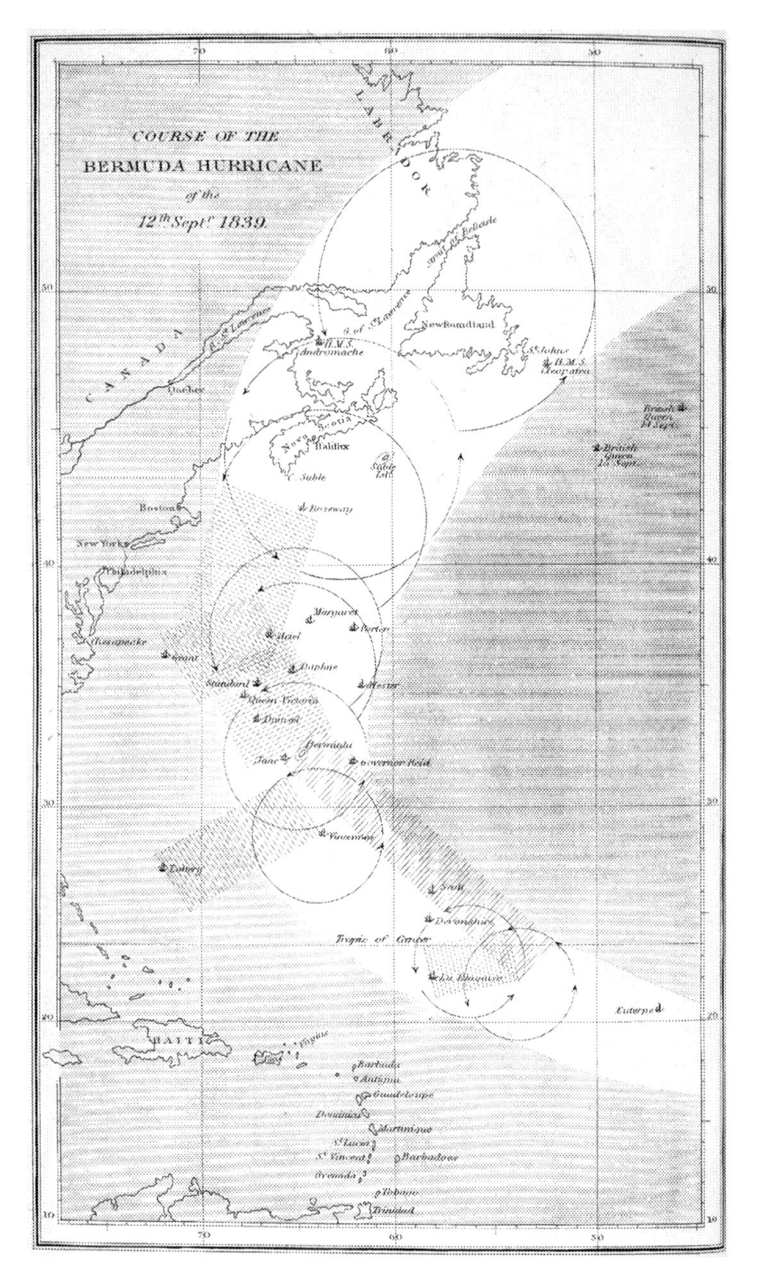

Figure 1.3 Chart of the course of the Bermuda hurricane of 1839. (Source: Reid, *Progress of the Development of the Law of Storms*, facing p. 39.)

for nearly eight years on the borders of the tropic in the Bermuda Islands, all tend to confirm the truth of this very important explanation.'[143] He defended one of his overarching suppositions on the same grounds: 'A residence of nearly eight years in the Bermudas, on the thirty-second degree of latitude, satisfied me that all the Bermuda gales, of whatever degree of force, in which the wind veers and the barometer falls, are progressive revolving gales; and I was struck when hearing the inhabitants call them "roundabouts."'[144]

Reid's regional maps of storms were used to the same end. As already noted, Reid had reworked his chart of the September 1839 Atlantic storm to illustrate his theory of the relationship between storms and sea-swell. Reid argued that 'great undulations' were raised by revolving storms along the radii of the whirlwind's circle, which then rolled straight onwards.[145] These undulations were illustrated on the chart using blocks of hatching and cross-hatching in places where sea-swells had been observed by ships. Reid also made reference to his own observations of the changing direction of the swell hitting Bermuda's shores. He came to the conclusion that '[s]ince storms obey fixed laws,' he claimed, 'and by their violence raise great undulations of the sea, these undulations probably conform to the same law'.

If Reid's researches were supported by his long residence at Bermuda, he continued to justify them on the grounds of maritime safety. Reid claimed that his extensive use of logbooks and the narratives of ships' captains helped 'seamen to study the application of the subject of revolving winds for themselves' and his advocacy of the use of barometers on board ships helped them to predict imminent changes in the weather.[146] The utility of this approach was demonstrated by several of the captains who had been caught in the 1839 hurricane. Bernard A. Ingham, the commander of the brigantine *Daphne*, experienced the storm en route from Bermuda to Halifax. He transmitted an extract of his private journal to the *Bermuda Royal Gazette* – his 'quota toward the development of the science of Storms' – that showed that he had been employing the wind and weather annotations advocated by Beaufort, and adopting Reid's advice regarding storm encounters.[147]

The utility of storm science to shipping was confirmed in Captain Robert Methven's account of a severe cyclone in March 1851 near Mauritius, which was published as the first in a putative series of

[143] Reid, *Progress of the Development*, p. 19. [144] Reid, *Progress of the Development*, p. 2.
[145] Reid, *Progress of the Development*, p. 32.
[146] Reid, *Progress of the Development*, p. 17.
[147] Letter from B. Ingham to Bermuda Royal Gazette, 7 October 1839, included in Reid's letter to Russell, 8 February 1840, Archives of the Royal Society, AP/24/16.

Narratives written by Sea Commanders – a series edited by Reid.[148] In the preface to Methven's account, Reid claimed that the captain had 'in a very striking manner applied the knowledge he had gained on the Law of Storms so as to keep his ship out of danger whilst a Hurricane was recurving South of the Island of Mauritius'.[149] Methven himself justified the study on the grounds that merchant shipping had expanded rapidly, while the pressures to move goods around the world had outstripped concerns about safety at sea.

Herschel's *Manual of Scientific Enquiry*

In the same year as Reid's *The Progress of the Development of the Law of Storms* was published by John Weale, John Murray published the *Manual of Scientific Enquiry*. The volume was commissioned by the Admiralty to provide naval officers while on foreign service with general instructions in various branches of science.[150] Beaufort was the guiding hand behind the volume while Herschel was the editor.[151] The *Manual* was part of a broader response to a growing mid-century demand from naval officers and other professional travellers for reliable guides to scientific observation.[152] Ward and Dowdeswell note that the *Manual* was effectively a reworking of the Royal Society's 1839 preparatory report for the Ross expedition to the Antarctic.[153] Like the expedition, the *Manual* positioned meteorology alongside and as an equal partner to other physical sciences, including astronomy, magnetism, tidology and statistics, as well as the natural and human sciences. The chapter on meteorology in the *Manual* was written by Herschel himself, beginning with the claim that there was 'no branch of physical science which can be advanced more materially by observations made during sea voyages than meteorology'.[154]

Alongside directions for the production of a routine meteorological log, Herschel urged naval officers to collect information on occasional atmospheric phenomena, such as squalls, storms, waterspouts, hurricanes and cyclones, even if there was no obvious place for them in the standard

[148] R. Methven, *Narratives Written by Sea Commanders, Illustrative of the Law of Storms, and of Its Practical Application to Navigation. No. 1. The Blenheim's Hurricane of 1851; with Some Observations of the Storms of the South–East Trade* (London: John Weale, 1851).
[149] W. Reid, 'Preface', no page, in Methven, *Narratives*.
[150] J. Herschel (ed.), *A Manual of Scientific Enquiry; Prepared for the Use of Officers in Her Majesty's Navy; and Travellers in General* (London: John Murray, 1849).
[151] Friendly, *Beaufort*, p. 264.
[152] See Withers, 'Science, Scientific Instruments and Questions of Method in Nineteenth-Century British Geography', for a discussion of this literature.
[153] Ward and Dowdeswell, 'On the Meteorological Instruments', 455.
[154] Herschel, 'Meteorology', in Herschel (ed.), *Manual of Scientific Enquiry*, p. 280.

entries of the register. Herschel urged officers to pay attention to these phenomena in all their phases, and to their connections to 'the state of the atmosphere preceding and subsequent, and especially every precursory appearance or fact which may have left on the observer's mind the impression of a *prognostic*'.[155] Although most familiar to medical practitioners, the application of the term prognostic to the study of meteorology was well established, used in a spate of studies in the eighteenth and nineteenth centuries.[156] The term did have some problematic connotations, given its association with folk readings of the sky and heavens, but Herschel argued that a careful study of storm prognostics, based on a large body of evidence, would in time serve to furnish the sailor with sufficient evidence and warning of an approaching hurricane. This was a position he was forced to defend publicly in 1860, after his apparent predictions of heavy floods and cold weather attracted widespread and unwanted attention. The study of past and current weather could provide indications of the future, Herschel argued, so long as the observer was credible, and their views were based on a body of trustworthy observations and on reasonable theoretical suppositions.[157] Herschel further clarified his thinking in a subsequent article in the evangelical and nonconformist periodical *Good Words*, where he dismissed folk prognostics of the weather as 'simple connotations' that displayed an ignorance of causes and modes of action, at the same time as he supported the idea of prediction of the weather a few hours into the future if based on 'an immense amount of perservering labour bestowed on daily and hourly records of the weather'.[158]

Herschel's chapter in the *Manual* also drew the reader's attention to the value of the work of Redfield, Reid and Piddington, 'which no navigator should go to sea unprovided with'.[159] These authors had shown hurricanes to be 'in the nature of vortices' which pursue a track that 'has a singular fixity of geographical situation and geometrical form'.[160] However, Herschel claimed that the habitual tracks of these storms remained 'imperfectly known', so that 'all of which tends to throw light upon this part of the subject is of the last importance to

[155] Herschel, 'Meteorology', p. 316, original emphasis.

[156] For instance, G. Adams, *A Short Dissertation on the Barometer, Thermometer, and Other Meteorological Instruments: Together with an Account of the Prognostic Signs of the Weather* (London: R, Hindmarsh, 1790); M. Waldeck, 'Natural Prognostics of the Weather', *Quarterly Journal of the Society for Literature and the Arts* (1827), 501–2; C. Clouston, *An Explanation of the Popular Weather Prognostics of Scotland on Scientific Principles* (Edinburgh: Adam & Charles Black, 1867).

[157] Anderson, *Predicting the Weather*, p. 51.

[158] J. Herschel, 'The Weather and Weather Prophets', *Good Words*, 5 (1864), 57–64, 57.

[159] Herschel, 'Meteorology', p. 320. [160] Herschel, 'Meteorology', p. 319.

navigation'.[161] Observations of the direction of the wind after the passage of a hurricane was also of interest to meteorologists, because it was still unclear whether hurricanes were constituted by the transfer of a mass of rotating air, or in the transient agitation of the air in situ.

Herschel's *Manual* became an essential component of the libraries of Royal Navy ships and an important point of reference to which potential weather observers could be referred. Beaufort routinely promoted its use and facilitated its uptake by others, on hydrographic ships and men of war, on packet and merchant vessels and at the stations of foreign consuls, where he recommended its use alongside Reid's *Law of Storms*.[162] The Admiralty's own advice was published in 1851, in the form of a pamphlet entitled *Remarks on Revolving Storms*.[163] Similar to Herschel's *Manual*, the Admiralty *Remarks* singled out Reid for special praise for collecting the facts and helping to develop the laws of storms. The *Remarks* also emphasised the value of storm prognostics alongside careful observation of the barometer, and, like Reid's *Law of Storms*, highlighted to ships' captains a geography of risk that focused on the West Indies, Madagascar and the China seas.

In spite of these guides, Beaufort remained concerned about the quality of meteorological observations at sea. In an 1852 Hydrographic Office memorandum on meteorological observation on board foreign and home men of war, Beaufort complained that '[m]uch valuable meteorologic information might undoubtedly be collected in H. M. ships if the officers could be induced with a sense of its importance – and could be induced to co-operate with zeal'.[164] Although the memo was partly addressed to the directors of the mail packet companies and to the Board of Trade and all foreign-going merchant ships, it was clear that Beaufort was frustrated at having to provide meteorological instructions and solicit information from naval ships more than thirteen years after the Admiralty had made the collection and inclusion of standardised weather data a formal component of the logbook.

Matters improved in 1853 when the Admiralty Lords decided that all HM ships should keep a meteorological journal separate from the logbook. In July that year Beaufort wrote to Sabine, the then de facto head of Britain's Magnetic Crusade, to discuss the shape of the proposed

[161] Herschel, 'Meteorology', p. 320

[162] F. Beaufort, Record of Observations by Foreign Consuls, 9 December 1851, UK Hydrographic Office Archives, MB/7.

[163] Lords Commissioners of the Admiralty, *Remarks on Revolving Storms* (London: HMSO, 1851).

[164] F. Beaufort, Meteorological Observations – General system of observing, 30 June 1852, UK Hydrographic Office Archives, MB/8.

journal, the hours at which observations would be made, the instructions that would be issued and the instruments to be used.[165] Sabine's reply urged caution due to wider developments and Beaufort quickly agreed 'on the propriety of waiting [sic] the result of Captain Beechey's mission before we decide anything'.[166] The mission to which Beaufort referred was an international maritime conference to be held in Brussels in August and September 1853 to devise a uniform system of meteorological observations at sea.

The 1853 Brussels Maritime Conference

In 1849 William Reid was appointed commanding Royal Engineer at Woolwich, having served as governor of Barbados until 1848.[167] He wrote to his former commanding officer Sir John Fox Burgoyne, Inspector General of Fortifications, to persuade him to organise meteorological observations at Royal Engineer stations overseas, the military branch responsible for operating Britain's colonial observatories. Burgoyne authorised the setting up of a network of observing stations, under the control of another Royal Engineers officer, Captain Henry E. James. James was head of the Edinburgh office of the Ordnance Survey. He had not played any significant public role in meteorology to date, but did have a professional interest in standards. In December 1851, James initiated a correspondence with Beaufort, outlining a way of calculating air pressure from wind speeds deduced from Beaufort's scale.[168] Burgoyne meanwhile wrote to the American government with a view to international cooperation on the subject of meteorological observation.[169] Matthew Fontaine Maury, director of the Naval Observatory at Washington, was one of the recipients of the proposal, who responded with the suggestion of an international conference to coordinate observations on land and sea.[170] Maury had been compiling data from ships' logbooks since the early 1840s and had produced global charts of wind and oceanic currents.[171]

[165] Letter from Beaufort to E. Sabine, 25 July 1853, UK Hydrographic Office Archives, LB/19.
[166] Beaufort to Sabine, 27 July 1853, UK Hydrographic Office Archives, LB/19.
[167] Blouet, 'Sir William Reid', p. 181. Reid later became governor of Malta, which facilitated his study of the storms of the Mediterranean.
[168] Letter from H. James to Beaufort, 24 December 1851, UK Hydrographic Office Archives, LP1857/J.
[169] Anon, *Maritime Conference Held at Brussels for Devising a Uniform System of Meteorological Observations at Sea*, MS, 1853, National Meteorological Archive, Exeter.
[170] Agnew, 'Robert FitzRoy', p. 25.
[171] Reidy, *Tides of History*, p. 287; D. G. Burnett, 'Matthew Fontaine Maury's "Sea of Fire": Hydrography, Biogeography, and Providence in the Tropics', in F. Driver and

When consulted as to Maury's plans, the Royal Society noted that different nations already had their own standards for land observations but that a conference focusing on establishing a uniform international system of meteorological observation at sea would be useful.[172] Although organised by the US government, the maritime conference was held in Brussels in 1853, and chaired by Adolphe Quetelet. Ten nations were represented at the conference: Denmark, France, Great Britain, the Netherlands, Norway, Portugal, Russia, Sweden, Belgium and the United States.[173] Captain James and Captain Frederick W. Beechey, head of the Marine Department of the Board of Trade and previously one of Beaufort's hydrographic officers, were nominated as British representatives. Delegates committed to developing a plan of uniform observation and a form of register was duly adopted, including Beaufort's nomenclature for the force of the wind. James asked Beaufort to supply him with the meteorological forms used by the British Navy and the merchant service, having also asked Maury for the American equivalents, which he planned to compare in the hope of suggesting a uniform system for the two countries.[174] He later argued, having seen the forms, that 'the proposed uniform system between the two Governments *can* be very readily effected by a little giving and taking', whilst lamenting the numerous systems employed by different bodies within and between nations.[175]

Discussion strayed inevitably onto instrumentation. Delegates recommended that ships should carry thermometers and barometers, along with 'at least one good chronometer, one good sextant, [and] two good compasses'. Beechey argued that it was impossible to recommend the adoption of any particular instruments, let alone any specific instrument makers, given that different scales and standards were in use internationally. It was feared that any standardisation of instruments would 'interfere too abruptly with long established usages and long-established records, with which the observations now to be collected would require a reduction, before they could be compared'.[176] Each nation was left to

L. Martins (eds.), *Tropical Visions in an Age of Empire* (Chicago: Chicago University Press, 2005), pp. 113–36.

[172] Agnew, 'Robert FitzRoy'.

[173] J. L. Davis, 'Weather Forecasting and the Development of Meteorological Theory at the Paris Observatory, 1853–1878', *Annals of Science*, 41 (1984), 359–82.

[174] James to Beaufort, 19 June 1852, UK Hydrographic Office Archives, LP1857/J. Worried that he would appear as an interloper in the meteorological field, James quickly appended the remark that 'Nothing can be done in this matter without the fullest concurrence of Professor Airy and yourself and I confine my ambition to the hope that I may be able to assist in promoting the object in view.'

[175] James to Beaufort, 15 July 1852, UK Hydrographic Office Archives, LP1857/J, original emphasis.

[176] Anon, *Maritime Conference held at Brussels*, p. 60.

use their own scales and standards, with the exception of the thermometer, where the centigrade scale was universally adopted (although this was not applied to the thermometer attached to the barometer), alongside any other scale currently in use. This was justified on the grounds of the possible *future* adoption of the centigrade scale – 'to accustom observers in all services to its use' – rather than its immediate use; the conference rejected the proposal that a separate centigrade column should be added to the meteorological register.[177] The conference also acknowledged the widespread use of barometers on board seagoing vessels of all types and their value as indicators of changes in relative pressure, but their use as recorders of absolute pressure was lamented: 'That an instrument so rude and so abundant in error, as is the marine barometer generally in use, should in this age of invention and improvement be found on board any ship, will doubtless be regarded hereafter with surprise.'[178]

In the aftermath of the conference, Maury acknowledged the compromises and faults inherent in the outcomes and recommendations but defended them nonetheless. In a letter to Lord Rosse, the president of the Royal Society, he explained that his support of the conference plan came:

> from the fact that with it we have in hand a grand experiment, it is an attempt to bring the sea by means of machinery already at work, regularly within the domains of systematic and scientific research, to change without cost, the common implements of navigation into philosophical instruments, and to convert the ships, for the safety of which these instruments are employed, into so many floating observatories, all cooperating together for the advancement of science, the good of mankind.[179]

There were several attempts to take forward this grand experiment in the 1860s, but these proved unsuccessful, due in part to the unstable political situation in Europe.[180] Several conferences on the topic were eventually convened in the 1870s. The first was held in Leipzig in 1872, followed by an international congress in Vienna in 1873.[181] In 1874, a private conference on maritime meteorology was held in London. The London conference set out to review participating nations' implementation of the recommendations of the 1853 Brussels conference, and to promote the recommendations of the 1872 and 1873 meetings: namely

177 Anon, *Maritime Conference held at Brussels*, p. 14.

178 Anon, *Maritime Conference held at Brussels*, p. 18. In relation to relative pressure, the conference noted the value of the aneroid barometer at sea but preferred the more delicate mercurial barometer given its ability to provide absolute results.

179 Letter from M. Maury to Rosse, 27 July 1854, National Archives, BJ 7/4.

180 Walker, *History of the Meteorological Office*.

181 R.-J. Wille, 'Colonizing the Free Atmosphere: Wladimir Köppen's "Aerology", the German Maritime Observatory, and the Emergence of a Trans-Imperial Network of Weather Balloons and Kites, 1873–1906', *History of Meteorology*, 8 (2017), 95–123.

that '[t]horough uniformity in methods and instruments should be aimed at'; 'unity of measures and scales is desirable, and to this end the introduction of millimètres for the barometer and the Centigrade scale for the thermometer should be aimed at'; and that 'the importance of the co-operation of the Navies' should be promoted.[182] In its aims the conference was part of wider movements in the 1870s to effect a system of liberal internationalism; to foster economic and social progress; to popularise a language of progress; and in particular to establish international standards, such as the international gold standard and the Treaty of the Metre.[183] As with concurrent attempts to encourage the universal adoption of the metric system, however, the implementation of a single international system of marine meteorology was stymied by national rivalries and resistance to new measures and practices on board the ships of the various European navies.

Participants' responses at the 1874 meeting revealed differences of opinion on the aims and successes of the 1853 conference and over subsequent attempts to introduce a uniform international approach to the study of meteorology at sea. Brigadier-General Myer, chief signal officer in the US Army, reported that the United States had followed the Brussels plan. J. C. de Brito-Capello, the director of the Nautical and Meteorological Observations at the Lisbon Observatory, made a similar claim on behalf of Portugal. The Danish were also supportive, although Captain Hoffmeyer of the Danish Royal Meteorological Institute conceded that some compromises had been made, such as the use of aneroid barometers on smaller vessels where a mercurial barometer 'cannot appropriately be placed'. Other nations were less positive. Professor Buys Ballot, the Dutch meteorologist and the meeting's president, argued that the Brussels conference had 'asked for too many observations' and that the hours of observation were inconvenient. The French made similar complaints. The report of Captain Rikatcheff of the Imperial Russian Navy was perhaps the most pessimistic. The thermometers used on board Russian vessels had continued to use the Reaumur scale, without the recommended addition of the centigrade scale, because of worries that 'one would often be read instead of the other'. The barometers had not been compared since 1853 and the necessary corrections not been determined, due to the want of a dedicated office to do so.

[182] Meteorological Committee, *Report of the Proceedings of the Conference on Maritime Meteorology held in London, 1874* (London: HM Stationery Office, 1875), p. 4.

[183] M. H. Geyer, 'One Language for the World: The Metric System, International Coinage, Gold Standard, and the Rise of Internationalism, 1850–1900', in M. H. Geyer and J. Paulmann (eds.), *The Mechanics of Internationalism: Culture, Society, and Politics from the 1840s to the First World War* (Oxford: Oxford University Press, 2001), pp. 55–92.

Rikatcheff complained that meteorological observations obtained at sea were not discussed or utilised in his country, or shared with other countries. For Rikatcheff, 'You ought not to be astonished, Sir, if from these answers you see that the greater part of our Maritime Meteorological Observations lie dormant till now.'[184] Discussion of the various recommendations made at the 1874 conference was wide-ranging and conflicting. Disagreements remained over which scales to use when measuring temperature; to what degree of accuracy readings should be taken; what scale should be used to record wind force; how the labour of global meteorological study should be divided; what form the meteorological register should take; and how the resultant data should be dealt with, analysed and archived. The various formal conference resolutions reflected these differences.[185]

Conclusion

> If we cannot bind [hurricanes] over to keep the peace, we may, at least, organize an efficient police to discover their ambush and watch their movements. If the bolts and bars of mechanism cannot secure our sea-borne dwellings from the angry spirit of the storm, we may at least track his course and fall into the wake of his fury.[186]

While efforts at international collaboration continued through the second half of the nineteenth century, British meteorologists sought to translate and adopt the recommendations of the conferences in a national context. Beechey reported on the outcomes of the 1853 conference to the British government and, in February 1854, the First Lord of the Admiralty, Sir James Graham, announced in the House of Commons that a new government department was to be formed, called the Meteorological Department, to be funded through the Board of Trade and the Admiralty. Its aims were to effect the recommendations of the Brussels conference; to collect and analyse meteorological observations taken at sea; to promote the observation of the weather on board ships; and, in the spirit of international cooperation, to convey reduced observations to the US Naval Observatory.[187] As discussed in the Introduction, the department's establishment was also a response to calls in the press for a force to police storms at sea. Robert FitzRoy was

[184] Meteorological Committee, *Report of the Proceedings. Appendix B*, quotes from pages, 28, 26, 31, and 32 respectively.
[185] In an attempt to encourage the more uniform pursuit of maritime meteorology, the 'Proposed English Instructions for keeping the Meteorological Log' were appended to the conference proceedings.
[186] Anon, 'Review of Reid', p. 432.
[187] Walker, *History of the Meteorological Office*, pp. 21–2.

appointed as the department's first director. FitzRoy's department began to supply instruments, instructions and registers to Royal Navy ships and British merchantmen, and to collect and compile weather logs.

It is possible to see the department as the most notable achievement of Beaufort and Reid's long campaign to promote the study of meteorology at sea, but such a reading should be treated with caution. The early, troubled, history of the department reminds us that meteorology in the nineteenth century continued to struggle to find its place in both the physical sciences and the public sphere. Perhaps the most significant criticism of the department was that levelled at FitzRoy's attempts to 'forecast' the weather (a term he coined), notably from Francis Galton, the African explorer, meteorologist and eugenicist.[188] In the aftermath of FitzRoy's death, Galton, the author of the 1866 government inquiry into the shortcomings of the department, labelled weather forecasting as unscientific, based on insufficient and poorly organised evidence, and indeed on a poor grasp of the physical laws involved. While Galton's report complimented the department on overseeing the provision of ships with instruments and registers, it was felt that too few registers had been collected and that there was insufficient global coverage.[189] FitzRoy had been distracted from his proper focus on meteorological statistics, Galton claimed, and had been diverted instead into 'the prognostication of weather'.[190] The report's criticism of FitzRoy's forecasts reflected anxieties about meteorology's supposed tendency towards the folk and the superstitious.[191] However, the deliberate comparison of Murphy's *Almanac* with Reid's *Law of Storms* in the *London Saturday Journal*, discussed earlier in the section titled 'William Reid and the Law of Storms', demonstrated that weather prediction could be countenanced under certain conditions. Herschel's discussion of storm prognostics and weather prophecy conveyed a similar sentiment and lent the approach significant intellectual weight. That Reid confined himself to the mapping of the behaviour of model storms, and did not attempt to predict the timing or location of particular storms in the future, further differentiated his work from the weather prophets.

Debates about storm prognostics were also in effect debates about meteorology's usefulness to the maritime world. Beaufort and Reid both played their roles as scientific servicemen, emphasising the public value of a maritime data-collection policy. However, it was the Army engineer's approach that proved more persuasive with the Admiralty. Beaufort argued for the construction of a large repository of basic information about the

[188] N. W. Gillham, *A Life of St Francis Galton: From African Exploration to the Birth of Eugenics* (Oxford: Oxford University Press, 2001).
[189] Burton, 'Robert FitzRoy'. [190] Walker, *History of the Meteorological Office*, p. 61.
[191] Anderson, *Predicting the Weather*, p. 124.

weather at sea, for use by a 'patient meteorological philosopher'. Reid's approach also placed great emphasis on the value of ships' logbooks for maritime meteorology, but it mirrored more closely the Royal Society's views on the usefulness of scientific knowledge, and spoke to the Admiralty's own interests in the safety of its ships. Reid's charts also successfully held in tension an inductive policy of data collection and the development of theory, while avoiding philosophical controversy.

If Beaufort and Reid differed over the ends of maritime meteorology, they also took somewhat different positions over the means by which it might be advanced. As Admiralty hydrographer, Beaufort laboured to turn naval ships into itinerant observatories and their crews into the equivalents of Airy's obedient drudges. As the chapter has demonstrated, Beaufort's maritime observatory experiment was not adopted quickly or universally across the British fleet. Sailors were not always good observers and ships often failed to conform to the model of the physical observatory. The weather routinely confounded the sailor's ability to describe it with sufficient accuracy, whether at the mean or at the extreme. Beaufort and his officers certainly aspired to bring order to the oceans, but their achievements were sporadic and geographically fragmented, while the scientific veracity of the results that were collected and preserved in the ships' logbooks was often open to question.

Reid benefited from Beaufort's campaign to turn ships' logs into weather diaries, but he went further in terms of the contribution that both ships and islands could make to the study of the atmosphere. Despite their wide geographical scope, Reid's charts were based on some very specific sites of inquiry. His work on storms relied on a handful of islands of strategic importance to Britain's Royal Navy and merchant marine, and functioned as important sites of observation and record keeping. Meanwhile, Reid treated ships not simply as floating observatories, but as meteorological instruments themselves. In his study of storm tracks, Reid was just as interested in the effects of a hurricane on the ships in and around its path as he was in the data collected on board. These ships left traces on the map, bearing mute and trustworthy witness to the actions of the atmosphere and ocean in a way that a barometer or an officer of the watch could not necessarily be trusted to do.[192] The ship produced an archive of the weather in its wake, while its paper trail could end up becalmed, languishing in Somerset House or in the offices of the Meteorological Department on Parliament Street.

[192] On similar practices of maritime surveying in Matthew Flinders' voyages of exploration, see S. Caputo, 'Exploration and Mortification: Fragile Infrastructures, Imperial Narratives, and the Self-Sufficiency of British Naval "Discovery" Vessels, 1760–1815', *History of Science*, 61 (2023), 40–59.

2 Meteorology at the Colonial Observatories

Although concerned with the development of a network of colonial observatories and specifically the establishment of a set of meteorological observatories in India, this chapter begins with a maritime voyage of exploration that was referenced in the previous chapter. In 1839, the vessels HMS *Erebus* and HMS *Terror*, commanded by Captain James Clark Ross, set off on a four-year voyage to the Antarctic Ocean. As noted in Chapter 1, the great scientific object of Ross's expedition was terrestrial magnetism – to observe the distribution of magnetic influence over the high southern latitudes along with their changes of form over time. It was hoped that the accurate mapping of such features would lead to the discovery of the causes that engendered 'the great features of the magnetic curves, and their general displacements and changes of form', while also helping the navigator at sea.[1] Ancillary scientific agendas related to the figure of the earth, to tides, meteorology, ocean currents and sea depth, and to southern astronomy and Antarctic geography, were also planned.

On board the *Erebus* and *Terror* were the staff of several fixed observatories that were to be established on various parcels of British colonial territory en route. The Royal Artillery officer John Henry Lefroy was dropped off on the island of St Helena, while another Artillery officer, Frederick Eardly-Wilmot, was left to establish a magnetic observatory in the grounds of the astronomical observatory at the Cape Colony, both men and their parties under the direction of the Master-General of Ordnance Sir Hussey Vivian. A detachment of naval personnel was left at Hobart Town, Van Diemen's Land, where a magnetic observatory was established and named Rossbank. Another observatory was established at Toronto in Canada, which Lefroy took over from 1842 after its previous superintendent, Lieutenant Charles Riddell, was invalided home. These stations were joined by several others that were operated by the British

[1] Anon, *Report ... on the Instructions ... for the Scientific Expedition to the Antarctic Regions*. See also G. A. Mawer, *South by Northwest: The Magnetic Crusade and the Contest for Antarctica* (Edinburgh: Birlinn, 2006).

East India Company, at Simla, Madras, Singapore, as well as one at Bombay. Observatories at Greenwich and Dublin acted as the standards against which the colonial stations were compared.[2]

The instantiation of this global network of fixed observatories, along with Ross's ocean-going expedition, was part of Britain's contribution to the Magnetic Crusade; what William Whewell would in 1857 claim was the greatest scientific undertaking the world had ever seen.[3] The Crusade, and in particular Britain's contribution to it, was certainly a remarkable achievement, especially when viewed from the vantage point of the late 1850s. The view was quite different during its establishment and its execution. Despite some valuable observations of the earth's magnetic field collected during various voyages of exploration to the Arctic in the 1820s, Britain lagged behind the work of its continental European counterparts. Britain's main exponents of research into geomagnetism – including Herschel, Humphrey Lloyd, George Peacock, Edward Sabine, George Airy and Whewell – recognised Christopher Hansteen in Norway, Carl Friedrich Gauss and Wilhelm Weber in Göttingen and Humboldt and François Arago in Paris as the leaders of the field.

Upon his return to Europe after his South American travels Humboldt promoted his 'cosmical tradition', where terrestrial magnetism was 'just one of a number of interconnected telluric or earth forces which were responsible for the phenomena manifest in or on the earth'.[4] He became interested in the periodic variations in the magnetic elements as well as their more abrupt changes and, with Arago at the Paris Observatory, laid out a systematic programme of research into these variations, based on observatory practice. Humboldt's Berlin Observatory served as one of the first models for instrumental standards, procedures and goals for magnetic observatories.[5] Humboldt used his widespread fame to encourage other nations, including Britain and Russia, to participate in a massive data collection endeavour to support his intention to develop a mathematically expressed universal magnetic law based on the Newtonian model.[6] It was his letter to the Royal Society (possibly

[2] See Naylor and Goodman, 'Atmospheric Empire', for a discussion of this network of colonial physical observatories.
[3] D. G. Josefowicz, 'Experience, Pedagogy, and the Study of Terrestrial Magnetism', *Perspectives on Science*, 13 (2005), 452–94.
[4] J. Cawood, 'The Magnetic Crusade: Science and Politics in Early Victorian Britain', *Isis*, 70 (1979), 492–518, 497.
[5] G. Good, 'Between Data', 294.
[6] S. Zeller, *Inventing Canada: Early Victorian Science and the Idea of a Transcontinental Nation* (Montreal: McGill-Queen's University Press, 2009), pp. 118–19.

instigated by Sabine) in 1836 that is often seen as the catalyst for Britain's own crusade.[7]

Hansteen, the author of the influential 1817 book *Magnetismus der Erde* (*The Earth's Magnetism*), proposed a four-pole theory of magnetic variation, with two areas of maximum magnetic intensity and two 'points of convergence' in both the Arctic and the Antarctic. The theory found the support of Sabine and Irish physicist Humphrey Lloyd. Meanwhile, Gauss argued that any attempt to explain terrestrial magnetic variations should be limited to the surface and interior of the planet – rather than the sun in Humboldt's view – and he regarded the earth as the source of an indefinite number of magnets. His refusal to prioritise any one area of the earth for particular study had consequences for the establishment of magnetic stations and observational work, in that it encouraged a global approach to study. Gauss and Weber also developed more precise instruments and the 'means to measure magnetic intensity in terms of length, mass and time'.[8]

Gauss's ideas were followed and developed by the British, and largely superseded those of Humboldt. That said, the British magneticians continued to promote Humboldt's 'vision of observatory-based studies of strange magnetic phenomena and allied their research proposals with the precision instruments and hard-nosed, mathematical methods of Gauss'.[9] More generally, they tended to refrain from speculation on the origins of the earth's magnetic field and placed their faith in the ability of systematic observation to reveal natural laws.[10] The globalising tendencies of the British Empire provided the ideal theatre and infrastructure in which to prosecute such a data-intensive project. The stations in Canada and Van Diemen's Land were judged as approximate to the points of the greatest intensity of the magnetic force in the northern and southern hemispheres: St Helena as approximate to the point of least intensity on the globe, and the Cape of Good Hope as a station where the changes of the magnetic elements presented features of peculiar interest.[11] The institutions of colonial administration – the various military bureaus, government offices and trading companies – were also crucial in the achievement of such an ambitious project, in terms of the supply and movement of personnel and instruments, the erection of observatories,

[7] Cawood, 'The Magnetic Crusade'; J. Cawood, 'Terrestrial Magnetism and the Development of International Collaboration in the Early Nineteenth Century', *Annals of Science*, 34 (1977), 551–87.

[8] Good, 'Between Data', 294. [9] Good, 'Between Data', 299.

[10] Cawood, 'The Magnetic Crusade', 497.

[11] R. F. Stupart, 'The Toronto Magnetic Observatory', *Journal of Geophysical Research*, 3 (1898), 145–8, 145.

the provisioning of expeditions and surveys, the communication of instructions and the computation of observations.

The Magnetic Crusade was sold to the British government in large part by Herschel, who had returned from his excursion to the Cape in 1838, where he had been completing a map of the celestial objects in the southern hemisphere. James Clark Ross and Beaufort helped to persuade the Admiralty of the scheme, while Sabine ensured the support of the Army, and Lloyd worked on the details of the fixed observatories and their cost. The newly formed British Association and the Royal Society also played key roles in the project's success. George Airy assisted with a magnetic observatory at Greenwich, even though he remained sceptical of the whole enterprise and was later outraged by Sabine's attempt to establish a geophysical observatory in London separate from Greenwich.

The crusade was meant to last the length of Ross's expedition to the Antarctic. Although the *Erebus* and *Terror* returned to Britain in 1843, funding was secured for the fixed observatories for three more years. Cawood argues that the Magnetic Crusade peaked in 1845 with an international magnetic meeting in Cambridge immediately preceding the annual British Association conference, even if only a few of the invited international luminaries actually turned up. Despite opposition from Airy and increasing scepticism from Herschel, Sabine managed to secure a further three years' funding for the observatories and for the computing staff he had employed at Woolwich, while many of the observatories managed to find funding for their activities even after the public funds dried up.

The colonial observatories that took part in the Magnetic Crusade, as well as the model observatories in Dublin, Greenwich and, later, Kew, shared as their primary concern the observation of magnetic variations, but that was by no means their only preoccupation. Some pursued astronomical research, others collected geodetic and tidal data, and the majority collected meteorological observations. These spaces of science were required to conform to the model of the physical observatory laid out by Herschel and others. In relation to the study of meteorology in particular, it was accepted that close relations existed between magnetism and meteorology: that the weather and the earth's magnetic field were subtly related and that both were under the influence of celestial forces; that both were of immediate importance to navigation; and that the atmosphere affected the performance of magnetic instruments such that information on pressure and temperature was needed to effect reductions.[12] The next section

[12] L. T. Macdonald, 'Making Kew Observatory: The Royal Society, the British Association and the Politics of Early Victorian Science', *British Journal for the History of Science*, 48 (2015), 409–33.

considers in more detail the principles and practices of observatory science. Particular attention is paid to their meteorological work.

The Eye of Reason

In his 1840 paper on 'Terrestrial Magnetism', Herschel laid out his manifesto for the physical sciences and for the observatories needed to progress them. Magnetism he argued, along with astronomy and meteorology, were sciences of observation, as opposed to sciences of experiment. Where experiments could be instantly followed up, repeated and rectified, sciences of observation relied on the assembly and comparison of large bodies of information from across the globe over long periods of time, with the constant worry of missed observations or, worse, the danger that an observation 'inaccurate or erroneously stated … poisons the stream of knowledge at its source'. He outlined what was needed:

> In order to master so large a subject multitude must be brought to contend with mass, combination and concert to predominate over extent and diffusion, and systematic registry and reduction to fix and realize the fugitive phenomena of the passing moment, and place them before the eye of reason in that orderly and methodical arrangement which brings spontaneously into notice both their correspondence and their differences.[13]

As if this was not enough, Herschel also demanded effective theory development to guide notice of natural phenomena. Theories were in this regard more important to the observatory sciences than to the experimentalists: 'to the incoherent particles of historical statement which make up the records of a science of observation, theories are as a framework which binds together what would otherwise have no unity', presenting 'fleeting impressions' as an 'existing whole'. For Herschel, theory provided hints to guide the choice of what to observe and the 'best and most available mode of making and recording our observations'.[14] These theories needed careful construction, requiring 'extraneous apparatus and shapeless masses of materials' in their early stages of growth. While for astronomy its extraneous scaffolding had since been stripped away, for other observatory sciences construction work was ongoing.

In his address to the Magnetic Conference at Cambridge in 1845, Herschel repeated a point made in his 1840 paper – that astronomy had enjoyed a disproportionate amount of public support and that it was

[13] J. Herschel, 'Terrestrial Magnetism', reproduced in J. Herschel, *Essays from the Edinburgh and Quarterly Reviews* (London: Longman, Brown, Green, Longmans and Roberts, 1857 [1840]), pp. 63–141, p. 64.

[14] Herschel, 'Terrestrial Magnetism', p. 66.

important that the observatory sciences interested in terrestrial physics should gain for themselves a greater share. The study of the 'physical constitution of this our planet – the forces which bind together its mass, and animate it with activity – the structure of its surface, – its adaptation for life, – and the history of its past changes – the nature, movements, and infinitely varied affections of the air and ocean' should be given as much importance as the study of the stars.[15] As Herschel urged astronomy to make way for terrestrial physics, he and others argued that the astronomical observatory provided the ideal model for the physical observatory – that the former should act as a school of exact practice that the latter was obliged to follow.[16] In Herschel's view, astronomers had already learnt to deal with a range of issues that would inevitably affect other fields of inquiry. These issues might be organised into two broad categories: on the one hand, there was the problem of ensuring that observers – whether under the same roof or separated by thousands of miles – were recording things in exactly the right way and in exactly the same way; and on the other, the challenge of ensuring that observations that were collected were properly processed and published in a manner that enabled their comparison.

According to Airy, too great an emphasis was placed in Britain on the value of observation and he supported instead the prevailing attitude on the continent. As he noted in his 1832 British Association report, in foreign institutions an observation was regarded as nothing more than a lump of ore, requiring only common labour for its production if the proper machinery were provided, and without value until it was smelted.[17] In other words, the crucial work of the astronomical observatory was in the act of reducing the observations rather than in the collection of the raw materials of the heavenly positions of stellar objects. Airy noted that the problem at Greenwich was not its inefficiency at making observations, but its inefficiency at reducing them. Others agreed. Herschel suggested that the reduction and printing of observations was a task just as important as their production, 'it being fully understood at present that unreduced observations are scarcely worth transmitting home', while the chemist and meteorologist John Daniell asserted that '[o]bservers would render a much greater service to science by devoting

[15] Herschel, 'Terrestrial Magnetism', pp. 68–9. Sabine later repeated Herschel's views in his own defence of the colonial observatory experiment: E. Sabine, 'On What the Colonial Magnetic Observatories Have Accomplished', *Proceedings of the Royal Society of London*, 8 (1856–1857), 395–413, 409.

[16] S. Schaffer, 'Astronomers Mark Time', 122.

[17] S. Schaffer, 'Babbage's Calculating Engines and the Factory System', *Réseaux: The French Journal of Communication*, 4 (1996), 271–98.

less of their time to the actual inspection of their instruments, and more to applying the proper corrections'.[18]

Ashworth argues that astronomy's own responses to these challenges owed much to its embrace of early nineteenth-century economic and business practices (while the principles of astronomy in turn bled back into social and economic life). At the most general level, astronomy was to treat the heavens as a stellar economy that was ordered, law-like and therefore predictable, while astronomical calculation provided the means to monitor and control the accumulation and free flow of facts – astronomy's own form of intellectual capital.[19] Viewed as capital in the economy of knowledge, scientific knowledge could accumulate interest, but could also lose it due to inferior equipment, weak productivity, poor investments and the importation of foreign goods, in this case astronomical tables and superior instruments. The accumulation and loss of capital was particularly acute for scientific projects funded by the state. For those agitating for scientific reform in the 1820s and 1830s, it was dismaying to be aware of the inadequacies of the Board of Longitude's *Nautical Almanac*, for instance, and of the poor performance of Britain's observatories. It was indefensible that publically-funded science should be providing such a slim return to the state that supported it.

The logical consequence of the economisation of scientific knowledge was the adoption of the principles of business and industry in the management of observatory staff. Observatories were deemed not dissimilar to factories, large offices or military establishments, with a clear division of labour and hierarchy, strict discipline, close supervision and an emphasis on punctuality. According to the so-called business astronomers, scientific discovery 'was the preserve of the proficient', with only severely limited tasks for devotees.[20] It was the job of the elite astronomer to foster a culture of calculated precision and to keep assistants under constant supervision for signs of deviation.[21] As Charles Babbage had expressed in *The Economy of Machinery and Manufactures* in 1832, the division of labour was just as important in mental as in physical labour so that the correct quantity of skill and knowledge could be applied to each activity.

Trustworthy observatories, like trustworthy businesses, needed to record, reduce and publish their accounts for all to see. Transparency

[18] Herschel, quoted in Schaffer, 'Keeping the Books at Paramatta Observatory', p. 122; J. Daniell, *Meteorological Essays and Observations* (London: Thomas and George Underwood, 1863 [1823]), p. 370.

[19] W. J. Ashworth, 'The Calculating Eye: Baily, Herschel, Babbage and the Business of Astronomy', *British Journal for the History of Science*, 27 (1994), 401–44, 410.

[20] Schaffer, 'Astronomers Mark Time', 130.

[21] Ashworth, 'The Calculating Eye', 427.

was crucial because the publication of accurate, standardised tables of reduced results revealed and protected the entire system from corruption, while the openness reflected in this act represented for Airy 'the exemplar of efficient organization and accountability'.[22] It should come as no surprise then that at Greenwich the computing room was deemed to be at the epicentre of the observatory's activities, the place where the performances of observers and instruments were applied to lists of necessary corrections and tabulated for publication.[23] In this new regime of astronomical bookkeeping, the astronomer assumed the role of accountant.[24]

Training in Exact Practice

In the spring of 1837, Humphrey Lloyd persuaded the University of Dublin to establish a magnetic and meteorological observatory in its grounds. Observations began in November 1838, in time for the commencement of Britain's crusade. Although Greenwich established its own magnetic observatory at around the same time, Lloyd's observatory at Dublin established its place at the epicentre of Britain's network of observatories. He used it as a model for other physical observatories within its ambit, which he set out in his *Account of the Magnetical Observatory of Dublin* in 1842.[25] In it he laid out a detailed plan of the work and ends of his own observatory and those like it. His thoughts echoed Forbes on the necessary functions of the ideal meteorological observatory, which Forbes laid out in his *Report* to the British Association in 1840, and those of Herschel on the ideal physical observatory. Herschel had produced several unpublished observatory manifestos that he had shared with key correspondents, and contributed to various published accounts, including the guidance prepared for Ross's voyage.

First, Dublin and its subordinates were to operate and keep in constant repair a set of standard philosophical instruments for which constant errors had been determined and to which local instrument makers,

[22] Ashworth, 'The Calculating Eye', 436 and 437.

[23] See D. Belteki, 'The Winter of Raw Computers: The History of the Lunar and Planetary Reductions of the Royal Observatory, Greenwich', *British Journal for the History of Science*, 56 (2023), 65–82.

[24] For studies of the computational regimes at observatories in different national contexts, see for instance A. J. Wraight and E. B. Roberts (eds.), *The Coast and Geodetic Survey 1807–1957* (Washington, DC: US Government Printing Office, 1957); D. Aubin, 'The Fading Star of the Paris Observatory in the Nineteenth Century: Astronomers' Urban Culture of Circulation and Observation', *Osiris*, 18 (2003), 79–100; S. D. Risio, *Making Space in the Capital: Scientific Knowledge in Edinburgh's Calton Hill Observatory c. 1811–1888*, unpublished PhD thesis, University of Edinburgh, 2024.

[25] H. Lloyd, *Account of the Magnetical Observatory of Dublin and of the Instruments and Methods of Observation Employed There* (Dublin: University Press, 1842).

observers and travellers could compare their own devices. Second, the observatories were to contribute to the establishment of laws of phenomena in nature. In a letter to Beaufort in 1835, Herschel claimed that observatories should provide 'regular observations of local phenomena of a variable or temporary nature' as well as 'the deduction and establishment of their laws of periodicity and local coefficients'. Chief amongst these would be the laws of magnetic intensity and direction, those of meteorology 'in all its extents', and of the tides.[26] Third, the observatories should help to fix secular constants, such as mean annual temperature or mean annual pressure at sea level. Herschel demanded accurate determinations of local data that were 'invariable, or subject only to very slow secular variation', such as magnetic force, air temperature, atmospheric pressure and sea level.[27]

This final requirement bound all three functions together. The determination of such 'normal data' with sufficient precision, Forbes argued, was 'incompatible with any but an official system of registration, which shall be conducted for very many years on exactly the *same system*, with instruments of the same kind, with unremitting attention not only to the *fidelity of the observations*, but to the *perfect repair and comparability of the instruments*'.[28] Finding laws and fixing constants required this exacting regimen to be applied in multiple localities, 'by observations strictly simultaneous, made according to the same instrumental methods, and with the same instrumental means'.[29] One solution to the problem of geographical coverage was the use of volunteer observers. It was a solution used widely by the British, who promoted a philosophy of induction and who turned a willingness to provide unpaid labour to the state and to science into a virtue. Imperial administrators, military personnel, traders, missionaries, settlers and travellers all became potential observers.[30] In his attempt to improve the extent of meteorological observation in the southern hemisphere, Herschel called on correspondents 'who may have leisure and inclination' to collect weather data and referred to them as 'lovers of knowledge'.[31] However, this approach had its limits. For many, volunteer observers presented as many problems as

[26] J. Herschel to F. Beaufort, 11 October 1835, Archives of the Royal Society, HS/3, p. 6.
[27] Herschel to Beaufort, Archives of the Royal Society, HS/3, p. 5.
[28] Forbes, 'Supplementary Report on Meteorology', 145 and 147, original emphases.
[29] Lloyd, *Account of the Magnetical Observatory*, p. 6. To ensure simultaneous observations were taken, all observatories were meant to use Göttingen time.
[30] W. J. Ashworth, 'John Herschel, George Airy, and the Roaming Eye of the State', *History of Science*, 36 (1998), 151–78, 157.
[31] J. Herschel, *Instructions for Making and Registering Meteorological Observations in Southern Africa, and Other Countries in the South Seas, and also at Sea* (London: Bradbury and Evans, 1835), p. 1.

solutions. In his British Association *Report* on the state of British meteorology in 1840, Forbes argued that while the fixing of the laws of phenomena in the physical sciences demanded extended and detailed observations, it was beyond the capacity of private research to provide this.[32] Lloyd, writing about Ireland's weather, conceded that individual observers might

> investigate successfully certain detached meteorological problems, such as the laws of the diurnal and annual changes of temperature, pressure, and humidity, at a given place; but little progress can be made in Climatology, or in the knowledge of the greater movements of the atmosphere, and their relation to the non-periodic variations of temperature and pressure, without the co-operation of many observers distributed over a large area, and acting upon a common plan.[33]

The only sure way of developing a global terrestrial physics was to provide well-funded stations at home that could verify instruments and train observers prior to despatch to state-funded colonial observatories. For Eardley-Wilmot, Lefroy and other colonial observatory superintendents, Dublin operated as a school of exact practice, where they were sent to receive instructions in observatory techniques. Lloyd taught through doing. Take his advice on the management of the observatory's barometers: Lloyd explained that the height of any observatory above sea level needed to be calculated, and noted that the height of the sill of Dublin's northern window relative to low water spring tides had been obtained using the chain of levels of the Ordnance Survey, such that suitable corrections could be made to the barometers. The observatory's Newman barometer was similar to the Royal Society's standard, against which it was compared, just as the barometers used in Canada or St Helena might in turn have been compared to those in Dublin. Lloyd used the two instruments' barometric and attached thermometric readings to explain the process and challenges of correcting for local temperature differences and capillarity. The observatory also operated a syphon form barometer by the German instrument maker Carl Pistor, for which Lloyd provided instructions as to the use of a compound microscope, a vernier and a moveable brass plate, which when placed behind the tube excluded false light and illuminated the mercurial column. The correct combination of these three features promised observations to the accuracy of 1/1,000th of an inch. Instructions were also provided for checking the perfection of the vacuum and correcting for the plasticity of the air included if not. Similar instructions were provided for

[32] Forbes, 'Supplementary Report on Meteorology', 145.

[33] H. Lloyd, 'Notes on the Meteorology of Ireland, Deduced from the Observations Made in the Year 1851, under the Direction of the Royal Irish Academy', *Transactions of the Royal Irish Academy*, 22 (1849), 411–98, 411.

the positioning and use of the other meteorological instruments, including several thermometers additional to those attached to the barometers, two hygrometers, an anemometer and a rain gauge.

Lloyd continued to advise the superintendents once in post through a steady flow of correspondence. For others less fortunate, and for times when letters were not forthcoming, instruction manuals had to suffice. Detailed instructions and correspondence could legitimately train observers on the fly. Josefowicz has noted the high degree of faith that the British – particularly Herschel – placed on written guidance, certainly when compared to Gauss, who maintained a belief in teacherly advice and direct supervision of observation and reduction. Herschel, the editor of the Admiralty's *Manual of Scientific Enquiry* – discussed in Chapter 1 – put great faith in clear and efficient communication, and the ability of guides, questionnaires and skeleton forms to 'regularize and coordinate data collection and communication'.[34] The important factor was clear and efficient communication – the philosopher's ability to 'describe, order and formalize'.[35]

Herschel's advice to meteorological travellers and enthusiasts in his 1835 *Instructions for Making and Registering Meteorological Observations in Southern Africa* demonstrates the value he placed on written instruction as the least-cost path to data accumulation. He spent a great deal of time laying out the basic principles of common observational practice, including the best way to transport instruments, the use of stated and regular hours of observation and registry, the importance of the continuity of the observations, the value of the careful training and supervision of a family member to ensure this, the use of the instruments' own scales and so on. Elsewhere, he provided advice on the proper use of skeleton forms, and the compilation of day and term-day books, a monthly abstract book and a book of curves, reinforcing the view that station superintendents were archivists as much as they were observers. The production of copies of the daily registers and more occasional term-books was encouraged, whereby the duplicate was compared with originals by two people, one reading aloud from the original, the other checking the copy, and then vice versa, whence one of the two was to be placed as a 'dead letter' in the archives of the local literary and philosophical society, where it would be 'rendered available towards the improvement of Meteorological knowledge, to the full extent of their scientific value'.[36] Monthly abstract books and curve

[34] Josefowicz, 'Experience, Pedagogy, and the Study of Terrestrial Magnetism', 466.
[35] Josefowicz, 'Experience, Pedagogy, and the Study of Terrestrial Magnetism', 466.
[36] Herschel, *Instructions for Making and Registering*, p. 4. In legal terminology a 'dead letter' was a law that had not been repealed but was no longer in use.

books facilitated the comparability of the results and helped them to circulate, while an annual report created a narrative of the preceding year and allowed the superintendent to draw out wider conclusions from the month-by-month results.[37]

If observers were archivists, so too were they accountants. If anything, *more* emphasis was placed on reduction techniques than it was on observation and archival practice. For both Gauss and Herschel the personal equation of individual observer error could be mitigated by good instruction and mechanical means – especially by self-recording instruments, where human intervention was minimised – but the British philosopher put greater faith in the observer as data reducer. Worries about the quality of the observations and their correct reduction could be set against the sheer volume of data to be collected. With enough data and with statistical techniques like Gauss's own method of least squares (discussed in more detail in Chapter 4), the detection of erroneously taken and reduced observations was a problem that could be dealt with statistically on paper rather than a priori in front of an instrument.[38]

Herschel provided very careful instructions in the reduction of various observations. For instance, there were three alterations to be made to the barometric observations. The first was for 'zero correction'. The Royal Society's Committee of Physics and Meteorology noted in 1840 that 'it is impossible to pay too much attention to the zero points of the instruments, especially the barometer'.[39] A barometer's zero point error was in turn caused by three subordinate factors: the possible faulty placing of the scale against the column of mercury; the capillary depression of the mercury in the glass tube; and any depression from unwelcome air or vapour in the upper part of the tube. The zero correction of a barometer could easily be determined by comparison with a standard instrument, or with an instrument that had been compared with a standard. 'By this means', Herschel argued, 'the zero of one standard may be transported all over the world'.[40] Even if this was not possible, he sought to reassure his observers that valuable meteorological conclusions could still be produced from a properly established barometer with an unknown zero error.

The second and most important correction was due to the temperature of the mercury in the barometer's tube at the time of observation. The

[37] Anon, *Report of the Committee of Physics, including Meteorology, on the Objects of Scientific Inquiry in those Sciences* (London: Richard and John E. Taylor, 1840), p. 117. Herschel was the Chairman of this Committee and could reasonably be viewed as the *Report's* author.

[38] Josefowicz, 'Experience, Pedagogy, and the Study of Terrestrial Magnetism'.

[39] Anon, *Report of the Committee of Physics*, p. 44.

[40] Herschel, *Instructions for Making and Registering*, p. 8.

information needed to effect this was supplied by the attached thermometer, which was to be read in advance of the barometric reading to avoid the error of breathing on it, while this thermometer was to be compared with a standard. The third and final correction dealt with the change of level of the mercurial surface in the barometer's cistern. This had a disproportionate effect on barometers with small cisterns and observers needed to be aware of the dimensions of the tube and the surface of the cistern, information that a good instrument maker should have supplied. None of these corrections were to be applied prior to registry, warned Herschel. He went so far as to suggest that their effects were 'safely applicable to mean results, and to the conclusions therefrom deduced, and a world of troublesome and often mistaken calculations may be saved by so applying them'.[41]

The proper application of corrections allowed data to travel and be compared with data from elsewhere much more efficiently, in similar fashion to Herschel's travelling standards. Meanwhile, poor reductions were worse than no reductions at all. Indeed, unreduced observations tethered to knowledge of the relevant errors were a much preferred outcome. One way or another, the data *had* to be able to move. As a science that studied processes occurring over large distances, meteorology had to spread its observations over 'the greatest extent of territory, and the greatest variety of local and geographical position'.[42] The reward was the establishment of secular constants and various laws governing weather and climate. For instance, careful analysis of the barometers' results at different observatories and during exploring expeditions promised to reveal more about the laws governing the depression of barometric pressure towards the tropics and its effects on the trade winds. In turn, the systematic development of laws to explain periodic changes in the atmosphere promised probable conjecture as to the general course of seasons to come and better preparation for singular events such as gales and droughts. This dual emphasis on data analysis and on the drawing of relations between observed phenomena and physical causes were, Good notes, two of the biggest preoccupations in the physical sciences during this period.

Colonial Science

These were the demands and the challenges bequeathed to those charged with establishing a network of magnetic and meteorological observatories domestically and across the British Empire. For the likes of Riddell,

[41] Herschel, *Instructions for Making and Registering*, p. 9.
[42] Herschel, *Instructions for Making and Registering*, p. 2.

Eardley-Wilmot and Lefroy, the prospect of setting up a magnetic and meteorological observatory in a colonial setting was daunting, although not without precedent or indeed infrastructural support. Writing to Sir George Gipps regarding the Paramatta Observatory in New South Wales, Herschel argued that such sites would be important geographical centres – 'zero points' of standard reference for local and national surveys of terrain and populations.[43] Writing to Beaufort to detail his views on observatory requirements, Herschel suggested that they could contribute to the dissemination of a global standard of weight and measure by holding a 'perfectly authentic standard Yard and Pound with local or colonial weights and measures'. Indeed, Herschel referred to observatories as 'propagating a knowledge of our standards'.[44] Sabine, in his discussion of the colonial observation 'experiment' in the late 1850s, promised that the 'investigation of physical laws, and the determination of exact data' would lead to 'great physical theories', and so to national glory and utility. New practices would follow and ancient methods would be abandoned as '*inefficient* and *uneconomical*'.[45] Colonial observatories fitted neatly into a system of imperial management that owed more to the counting house than to the arts of war.[46]

Observatories, along with botanical gardens, libraries and churches, promised to act as centres of improvement and civilisation which would encourage the development and refinement of rough-and-ready settler society at the imperial frontier.[47] 'Rapidly progressive as our colonies are, and emulous of the civilization of the mother country', opined Herschel on the colonial observatory, 'it seems not too much to hope from them, that they should take upon themselves, each according to its means, the establishment and maintenance of such institutions both for their own advantage and improvement, and as their contributions to the science of the world'.[48] In a compelling analysis of Herschel's own scientific labours

[43] J. Herschel to G. Gipps, 26 December 1837, Archives of the Royal Society, HS/19, p. 3.

[44] J. Herschel to F. Beaufort, 11 October 1835, Archives of the Royal Society, HS/3, p. 5 and p. 6, respectively. On oriental metrology, see S. Schaffer, 'Oriental Meteorology and the Politics of Antiquity in Nineteenth-Century Survey Sciences', *Science in Context*, 30 (2017), 173–212.

[45] Sabine, 'On What the Colonial Magnetic Observatories Have Accomplished', 408–9, original emphases.

[46] Ashworth, 'John Herschel, George Airy, and the Roaming Eye of the State', 152.

[47] S. Sangwan, 'From Gentlemen Amateurs to Professionals: Reassessing the Natural Science Tradition in Colonial India 1780–1840' in R. Grove, V. Damodaran and S. Sangwan (eds.), *Nature and the Orient: The Environmental History of South and Southeast Asia* (Oxford: Oxford University Press, 1998), pp. 210–36.

[48] J. Herschel, 'An Address to the British Association for the Advancement of Science at the Opening of their Meeting at Cambridge, June 19th, 1845', reproduced in J. Herschel, *Essays from the Edinburgh and Quarterly Reviews* (London: Longman, Brown, Green, Longmans and & Roberts, 1857), pp. 634–82, on p. 653.

at the Cape Observatory, Musselman argues that his transition from astronomical hunting and gathering (observing) to rational harvesting (reducing) paralleled his views of the necessary stages through which the Cape Colony needed to progress. Without the laborious task of reducing his data, Herschel's hunt through the southern skies was meaningless, while the same was true for the imperial project – civilised husbandry had to follow settlement.[49] Sabine agreed with Herschel, arguing that the observatories would help to 'arouse, to nourish' natural knowledge in the colonies by helping to form 'good observers' out of local residents and providing a haven for the travelling observer.[50]

In Herschel's 1835 letter to Beaufort, he ended his description of the ideal features of the physical observatory with the proviso that '[p]erhaps all this is dreaming. However it is at least a harmless dream.'[51] It has already been noted that many agitating for scientific reform in the 1820s and 1830s had pointed to the poor performance of Britain's observatories at home and abroad. The physical observatories established during the Magnetic Crusade at the end of the 1830s were dogged by problems of various stripes, to the extent that Herschel's melancholic musings in the mid-1830s came close to sounding prophetic. The chapter turns now to a consideration of the historical geography of one of these colonial observatories – the Colaba Observatory in Bombay – with a particular focus on the meteorological work that was undertaken there. It considers the way in which observatory staff attempted to fulfil the requirements of the model observatory, in terms of the physical establishment of the site, the management of both staff and instruments, the taking of meteorological observations and the reduction and printing of data. The chapter investigates to what extent the Colaba Observatory's staff were successful at overcoming these demands and challenges such that they could set standards, establish constants and discover laws that could travel elsewhere and so inform understandings of Indian weather and climate more widely.

Terrestrial Physics at the Colaba Observatory

Along with Madras and Calcutta, Bombay was one of the three main outposts of the East India Company in the nineteenth century. In the early 1820s, Bombay had a population of around 200,000 people. It had

[49] E. G. Musselman, 'Swords into Ploughshares: John Herschel's Progressive View of Astronomical and Imperial Governance', *British Journal for the History of Science*, 31 (1998), 419–35.

[50] Sabine, 'On What the Colonial Magnetic Observatories Have Accomplished', 411.

[51] Herschel to Beaufort, Archives of the Royal Society, HS/3, p. 8.

the only dry docks in India and was the first extra-European construction site for British naval vessels.[52] The East India Company, with its institutional reach and self-interest in matters of navigation and trade, saw itself as a patron of science.[53] As well as encouraging an interest in tropical meteorology and natural history amongst its staff, it funded the establishment of a number of astronomical observatories on its territories.[54] These sites – such as the Company's Madras Observatory (established in 1786), and those at Calcutta, Simla and Singapore – placed great emphasis on support for maritime navigation and terrestrial surveying.[55] The observatories were also part of a larger suite of scientific institutions that were modelled on European examples and imposed on the landscape, in the hopes of encouraging intellectual, moral and economic improvements.[56]

The first astronomical observatory to be established in Bombay was sited in the Company's Marine Yard, in apartments above the mould loft (Figure 2.1). Around 1815, the Literary Society of Bombay began to lobby the government of Bombay for funding for a larger observatory that would reflect the importance of the city as a centre of international trade and empire, and act as a centre of civic, imperial and scientific improvement.[57] The Literary Society received a land grant from the Company and John Curnin was chosen as the Company's official astronomer, a man apparently well versed in astronomy and natural philosophy.[58] The site at the Marine Yard was abandoned and the small island of Colaba, to the south of Bombay Island, was chosen by Curnin as the location of the new observatory, despite difficult access, the proximity of a light house and poor infrastructure. Work began in 1824. Curnin, with the support of Mountstewart Elphinstone, the governor of Bombay

[52] J. Sutton, *Lords of the East: The East India Company and Its Ships (1600–1874)* (London: Conway Maritime Press, 2000).

[53] J. Ratcliff, 'The East India Company, the Company's Museum, and the Political Economy of Natural History in the Early Nineteenth Century', *Isis*, 107 (2016), 459–517; V. Damodaran, A. Winterbottom and A. Lester (eds.), *The East India Company and the Natural World* (Basingstoke: Palgrave Macmillan, 2015).

[54] R. Grove, 'The East India Company, the Raj and the El Niño: The Critical Role Played by Colonial Scientists in Establishing the Mechanisms of Global Climate Teleconnections 1770–1930' in R. Grove, V. Damodaran and S. Sangwan (eds.), *Nature and the Orient: The Environmental History of South and Southeast Asia* (Oxford: Oxford University Press, 1998), pp. 301–23.

[55] Williamson, 'Weathering the Empire'.

[56] P. Chakrabarti, *Western Science in Modern India: Metropolitan Methods, Colonial Practices* (Delhi: Permanent Black, 2004); M. Edney, *Mapping an Empire: The Geographical Construction of British India, 1765–1843* (Chicago: Chicago University Press, 1997).

[57] J. McAleer, '" Stargazers at the World's End": Telescopes, Observatories and "views" of Empire in the Nineteenth-Century British Empire', *British Journal for the History of Science*, 46 (2013), 389–414, 409.

[58] S. Schaffer, 'The Bombay Case: Astronomers, Instrument Makers and the East India Company', *Journal of the History of Astronomy*, 43 (2012), 151–80, 152.

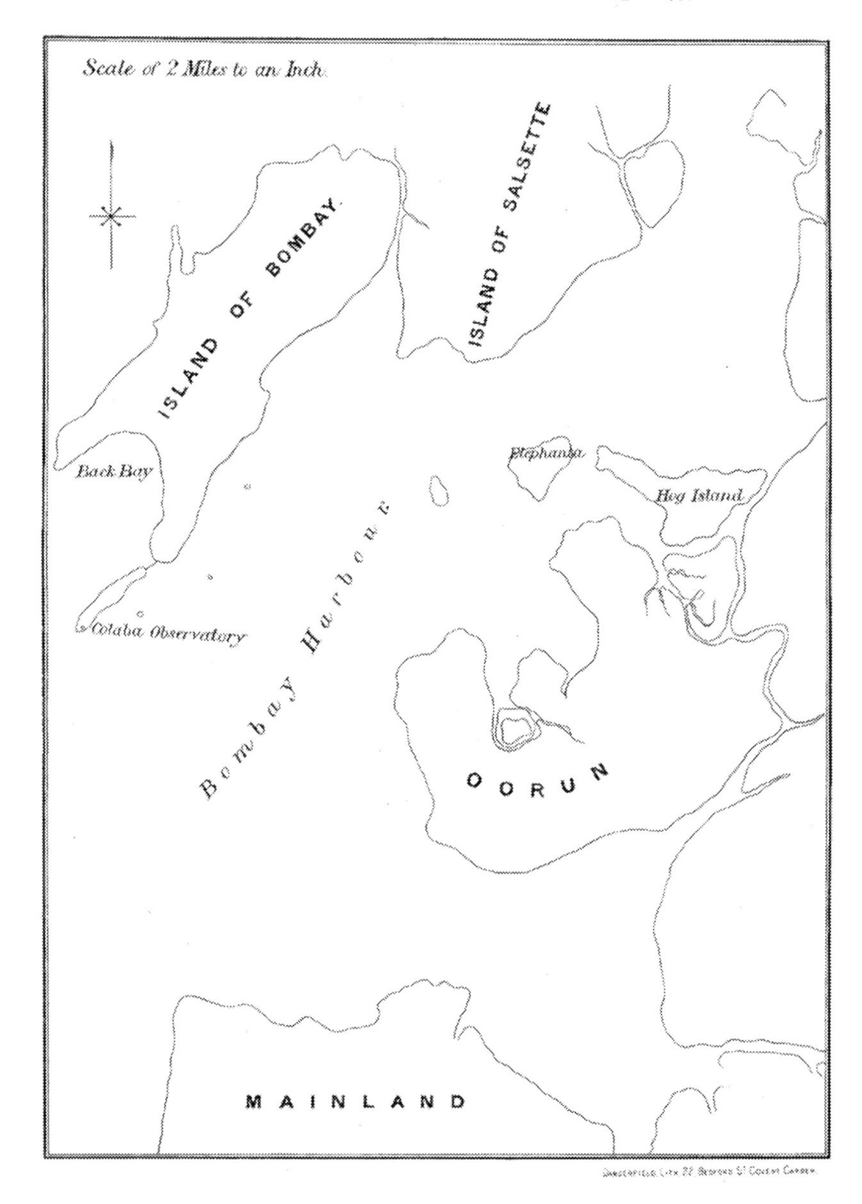

Figure 2.1 Outlines of Bombay island and harbour. (Source: C. Chambers, *Meteorology of the Bombay Presidency* (London: Her Majesty's Stationery Office, 1878).)

from 1819 to 1827, mapped out a set of aims for the observatory that would serve British mercantile and military interests: assistance in time signalling and chronometer rating, the storage of maritime surveys, the training of mariners in nautical astronomy and surveyors in astronomy and geodesy, regional surveying, as well as the study of the tides and weather.[59] In fact Elphinstone, president of the Bombay Literary Society and founding member of the Royal Geographical Society, had a personal interest in tropical weather and kept a note of it in his personal diary.[60]

Curnin began his work in March 1825, while the observatory itself was declared operational in 1826. However, it was quickly mired in controversy. From the outset Curnin was unhappy at having to use the East India Company's official instrument suppliers and reported that their instruments were faulty upon receipt of them in 1827. He also complained about the design of the building, the lack of suitable furniture and the poor quality of the road from the observer's house to the observatory itself.[61] Curnin sent the instruments back to London, where a Committee of Inquiry was established, containing Beaufort, Baily, South and James Ross, amongst others. The Committee decided in 1828 that Curnin had been wrong in his actions and he was sacked as Company astronomer, leaving India in 1829. This incident, Schaffer notes, 'became notorious as an episode of costly controversy about the construction of an observatory and its instruments in the early nineteenth-century colonial world and the epoch of political and administrative reform'.[62]

After a period in abeyance, one of the rooms of the observatory was reoccupied by the Geographical Society of Bombay in 1832. In 1834, Arthur Bedford Orlebar was appointed superintendent of the observatory, a post he held in conjunction with his Professorship of Mathematics and Astronomy at Elphinstone College, which was in turn located a few miles north of Colaba island in the main part of the city. A time ball was installed, although it was difficult to see from the harbour. Meteorological and magnetic observations began on 1 December 1841. Upon the recommendation of the Royal Society, the Bombay Observatory was one of four Company observatories – the others being Madras, Simla and Singapore – that contributed to the Magnetic Crusade, even if the

[59] J. Sen, *Astronomy in India, 1784–1876* (London: Pickering & Chatto, 2014), p. 49.

[60] G. C. D. Adamson, '"The Languor of the Hot Weather": Everyday Perspectives on Weather and Climate in Colonial Bombay, 1819–1828', *Journal of Historical Geography*, 38 (2012), 143–54.

[61] J. Curnin to Company Astronomer, 30 March 1825, British Library, IOR/P/346/16/1–3.

[62] Schaffer, 'The Bombay Case', p. 151. See also McAleer, 'Stargazers at the World's End', 400–1; Naylor and Goodman, 'Atmospheric Empire'.

Figure 2.2 The Colaba Observatory, Bombay. (Source: G. Buist, *Provisional Report on the Meteorological Observations Made at Colaba, Bombay, for the Year 1844* (Cupar: G. S. Tuller, 1845), title page.)

instruments that ended up in Bombay had originally been intended for an observatory at Aden.[63] The Company's observatories were accompanied in their work for the Crusade by the observatories of the princely states of Travancore and Lucknow.[64] A building to house the magnetic and meteorological instruments was hastily erected in the grounds at Colaba, constructed under the plan supplied by Humphrey Lloyd (Figure 2.2).[65] Orlebar expressed his dismay at being 'without guidance as to the nature of the building, but also as to the parts and objects of many of the instruments', but he received advice from John Caldecott, the superintendent at the Travancore Observatory.[66] Orlebar was assisted by three non-commissioned sappers and miners.

Although the economic, political and cultural contexts of the Bombay and Travancore observatories were quite different, they both experienced a number of similar issues and problems with regard to staffing, observational regimes, data management and publication. John Allan Broun had

[63] C. Chambers to J. F. Gassiot, Chair of the Kew Committee of the British Association for the Advancement of Science, National Archives, BJ 1/2.

[64] Ratcliff, 'Travancore's Magnetic Crusade'. [65] Buist, *Provisional Report*.

[66] W. C. Barker, *Report of the Committee of Inquiry on the Colaba Observatory*, Selections from the Records of the Bombay Government, 1865, National Archives, BJ 1/2, p. 12.

replaced Caldecott at the Travancore Observatory upon the latter's death in 1849, having previously worked as director of Sir Thomas Brisbane's physical observatory at Makerstoun in Scotland. The Sircar, or royal government, of Travancore understood that the culmination of their observatory's work had to be the reduction and publication of a set of observations that would be of general use to science. Broun decided that these tasks could most effectively be achieved in Europe, where 'machines (steam presses), the division of labour (ready-trained printers and computers) and private capital (a profit-oriented press) combined to give Europe an economic edge in transforming the raw material (data) into a finished product (the report)'.[67] Broun returned to Britain in 1865, while the first volume of magnetic results was eventually published in 1874.

The Colaba Observatory also struggled with demands on staff time and the economics of scientific publishing. When Orlebar was called away to inspect schools and seminaries in late 1841, he left the day-to-day running of the observatory to his three military assistants. Two of the men, Corporal Heywood and Corporal Saul, neglected their duties and were dismissed. Observatory work was continued by just one man, Sergeant J. H. Dunn. When Orlebar retired to Europe during a period of illness in 1842, he was replaced by George Buist. Buist was a newspaper editor who had begun his career on various papers in Scotland before becoming the editor of the *Bombay Times* in 1839. As well as his honorary role at the observatory, he was also the secretary of the Bombay Geographical Society and of the Agri-Horticultural Society of Western India.[68] In August 1842, the government sanctioned the appointment of three 'native assistants' to aid Buist, including Keru Luximon, 'a talented and accomplished Brahmin' who had been trained in observatory techniques by Orlebar.[69] Observatory staff registered horary readings of the barometer, thermometer, wet and dry bulb and the wind direction. Buist introduced thermometers for recording solar and terrestrial radiation and made accounts of remarkable events, including meteor showers and hail storms. Buist used his own newspaper to publish meteorological information. He also forwarded accounts to the *Journal of the Bombay Geographical Society*, the *Journal of the Bombay Branch of the Asiatic Society* and the *Times of India*. Buist sent his 1843 observations to Edward Sabine, who used them as the basis of his paper on the diurnal

[67] Ratcliff, 'Travancore's Magnetic Crusade', 341.

[68] K. Prior, 'Buist, George (1804–1860)', *Oxford Dictionary of National Biography* (2004), www.oxforddnb.com/view/article/3892, accessed 18 August 2016.

[69] Buist, *Provisional Report*, p. 1. Sen, in *Astronomy in India*, provides a detailed account of race and the place of Indian assistants at the Bombay Observatory.

variation of barometric pressure, in which he praised Buist and his work at Colaba as of 'particular value'.[70] Clements Markham, the geographer and explorer, also praised the work at Colaba, highlighting Buist's attempt to make the study of meteorology interesting by excavating what Buist called 'picturesque and descriptive meteorology' from under an avalanche of 'minute instrumental details'.[71]

Buist's own position at the observatory came to an unexpected end in 1845, after Orlebar returned from his convalescence in Europe and recommenced his observatory directorship. It was Buist's turn to retire to Europe, taking the observatory papers with him. Like Broun, his intention was to work up the observatory data for publication, which he justified on the grounds that the copperplate press in Bombay – the only one in the Presidency – had become unserviceable. Although Buist complained about the quality of several of the observatory's meteorological instruments, he noted that in one year alone, 1844, nearly 200,000 meteorological and magnetic observations had been read, corrected, registered and prepared for publication, and 'not one single observation bungled or lost', despite the fact that he had been occupied in preparing and delivering a course of lectures to the officers of the Indian Navy.[72] However, contrary to Broun, Buist claimed that the cost of printing the observations in Britain were 'very great' and the scheme was abandoned. The only copies of the whole report ended up with the Court of Directors of the East India Company and the observatory itself. Buist did provide a provisional report on the meteorological observations for the year 1844, fifty copies of which were published by St Andrews University Press. As well as an account of his time at the observatory, his report provided monthly accounts of the weather over the year in narrative, tabular and graphical form.

Sen argues that the problems of institutionalising European astronomy, meteorology and magnetic research in Bombay demonstrated that 'practical contingencies could be quite different at the peripheries of empire, even if the patronage of the metropole was a crucial factor'; that the running of the observatory owed more to the Literary Society of Bombay and Elphinstone College than to the Court of Directors in London.[73] The Geographical Society of Bombay also played an important role in promoting scientific work in the city and wider region,

[70] E. Sabine, 'On Some Points in the Meteorology of Bombay', *Report of the Fifteenth Meeting of the British Association for the Advancement of Science* (1846), 73–82, 75.

[71] C. R. Markham, *A Memoir on the Indian Surveys* (Cambridge: Cambridge University Press, 2014 [1871]), p. 216.

[72] Buist, Provisional Report, p. 3. These lectures began in October 1844 in a newly built lecture room. Sen, *Astronomy in India*, p. 65.

[73] Sen, *Astronomy in India*, p. 55; Grove, 'The East India Company, the Raj and the El Niño', pp. 303–4.

especially with regard to the study of the weather. In 1845, the Geographical Society recommended to the Company that meteorological and tide observations should be taken on the Indian west coast and at Aden so as to aid sanitary science and storm forecasting. The Society offered to take on much of the labour as to the procurement of instruments, forms and instructions for their use. The government and Court of Directors of the East India Company supported the plan, with the Geographical Society persuading the Admiralty, in the form of Beaufort, to supply £350 of equipment. The Society was largely preoccupied with meteorological work over the 1840s and 1850s, carrying out tests of aneroid barometers for instance, collating observations from across the subcontinent, discussing current work in the field and displaying new instruments at their meetings.[74] In parallel developments, in 1851 members of the Royal Engineers stationed overseas were ordered to collect meteorological observations, in conformity with the instructions laid out by Sir John Burgoyne.[75] These stations were instructed to collect meteorological observations every day at 9.30 am and 3.30 pm, and to collect twenty-four hourly observations for one day in every month. The Company responded by ordering its senior medical officers to establish twenty weather stations across India, including five in the Bombay Presidency (Poona, Belgaum, Deesa, Kurrachee and Bombay). The reports from these stations were collected together at the observatories at Madras, Calcutta and Colaba for examination and compiled before being forwarded to Britain.[76]

The influence of Bombay's scientific societies and educational institutions waned in the late 1840s, when the post of observatory superintendent became allied to the office of the hydrographer. William Montriou, a commander in the Indian Navy, replaced Orlebar in 1847. Montriou was in turn replaced by Lieutenant Edward Fergusson in 1851 and he held the post until 1863.[77] Both men visited Kew Observatory for training. The observatory itself, and Colaba more generally, became increasingly militarised over the period. The observatory was used for copying charts and for storing naval surveying and mathematical equipment,

[74] Anon, 'Proceedings of the Society', *Transactions of the Bombay Geographical Society, September 1850 to June 1852*, 10 (1852), i–cxi.

[75] J. Burgoyne, *Instructions for taking Meteorological Observations at the Principal Foreign Stations of the Royal Engineers* (London: John Murray, 1851).

[76] Letter from C. Chambers to the Secretary of the Marine Department, Bombay Government, 30 December 1868, National Archives, BJ 1/2.

[77] Arnold reminds us that Company officers and administrators rarely stayed in one posting for very long before promotion, illness or some other factor necessitated their move to another station. D. Arnold, *The Tropics and the Travelling Gaze: India, Landscape, and Science, 1800–1856* (Seattle: University of Washington Press, 2006), p. 15.

while the non-military occupants of Colaba were brought out by the government so as to form a complete cantonment for the entire Bombay garrison.[78] However, the work of the observatory was considered unsatisfactory and its instruments of a poor quality. Concerns about the operations of the Colaba Observatory eventually induced the Bombay government to form a Committee of Inquiry to investigate its performance, which reported its findings in January 1865. The Committee's principal aims were to establish the 'purposes intended by the present operations of the Observatory', the 'degree of accuracy with which the present purposes are fulfilled' and the improvements that were needed and desirable.[79]

The Committee of Inquiry

With regard to the meteorological operations of the observatory, the Committee began its assessment by itemising the instruments currently in use and in store. An Osler's self-registering anemometer and pluviometer, a pair of Newman wet and dry bulb thermometers, a Newman's standard barometer and a Regnault's hygrometer were set up in the astronomical room, and in the magnetic observatory were two Newman's standard barometers. Three pairs of dry and moist bulb thermometers were kept outside in a thermometer stand; at the tide-gauge house there was a self-registering anemometer; in a shed was a standard Newman thermometer, five thermometers for underground temperature and a moist bulb thermometer; and there were two rain gauges in the observatory compound. Numerous other instruments were held in the store, many of which were listed as broken or out of order.

The Committee itemised the problems with these instruments, their maintenance and use. The wind gauge was reported to have 'never been well suited for registering sudden changes of wind'[80] with a poor record of force. The vane itself was also sheltered from the wind on one side by the time-ball tower and it was concluded that all of the observations collected since the tower was erected were worthless. This instrument also affected the results of the rain gauge, having been placed too near to its vane so that it suffered from eddies of wind that carried light rain away. If that was not enough, the rain gauge was made of metal and on a blackened terrace, which was judged to 'throw off light rains in the form of vapour'.[81]

[78] Sen, *Astronomy in India*, p. 68.
[79] W. C. Barker, *Report of the Committee of Inquiry on the Colaba Observatory*, Selections from the Records of the Bombay Government, 1865, National Archives, BJ 1/2, p. 1.
[80] Barker, *Report of the Committee of Inquiry*, p. 9.
[81] Barker, *Report of the Committee of Inquiry*, p. 9.

Meanwhile the wind gauge at the tide house was judged too low to produce meaningful results and its clockwork corroded and unserviceable from damp and rust. The readings from the barometers were pronounced untrustworthy due to 'very abundant oxidation' of the mercury in the cisterns as well as the accumulation of dust and cobwebs. Those from the moist bulb thermometers were suspect because of their exposure to variations of the wind, while the cloth covering the bulbs was sometimes covered in salt, which it was assumed would influence the evaporation and 'affect the accuracy of the indications to a very considerable extent'.[82] The dry bulb thermometers were judged to be in serviceable condition, but crucially 'there does not appear to be any system of comparison and verification practised, and, as the zero point has a tendency to shift, there is no security for the correctness of the observations of absolute temperature made at the Observatory'.[83] It was concluded that 'continued registration of instruments in such a condition, it need scarcely be added, is worse than useless, it is deceptive'.[84]

Assuming that all of the instruments had begun their life in good order, the Committee concluded that poor maintenance standards had been kept and at times there seemed to have been no maintenance at all. The Committee went on to note that keeping instruments in repair, comparable and properly observed was one thing, the verification, correction and reduction of observations quite another. In both meteorology and magnetism, Barker reminded his readers, 'the condition of the instruments requires constant attention and perpetual revision under skilful superintendence in order to preserve them absolutely as well as relatively correct'. However, no verification seems to have taken place at Colaba, 'nor could anyone be trusted to verify even a thermometer', stating baldly that '[i]n the publication of the numerous large volumes of observations made at Colaba too much appears to have been trusted to mere routine – it seems to have been taken for granted that the observations should be "all right", and that nothing more could be required: no independent mind has been exercised in verifying and co-ordinating the observations'.[85] In particular, the observatory had adopted a problematic observation regime which went counter to that of all other British colonial observatories: Colaba had adopted the Göttingen day for its magnetic observations and the local day

[82] Barker, *Report of the Committee of Inquiry*, p. 11.

[83] Barker, *Report of the Committee of Inquiry*, pp. 10–11. The hygrometer, which had been at the Observatory since 1856, was the only instrument judged to be in good order.

[84] Barker, *Report of the Committee of Inquiry*, p. 10. The other departments of the Observatory – its astronomical, time-keeping, magnetic and tides work – were similarly judged to be highly defective.

[85] Barker, *Report of the Committee of Inquiry*, p. 14 and p. 15.

for the meteorological observations, a practice that seems to have begun under Buist's superintendence. This meant that magnetic observations stopped at 4 pm on Saturdays, whilst the meteorological observations stopped at midnight. It was therefore difficult to relate one to the other. It also meant that there was no record of terrestrial magnetism on Saturday night and that the Sunday observations were as good as useless because no other observatory collected information on that day. On the advice of Broun of the Trevandrum Observatory it was also presumed by the Committee that confusion over Göttingen and Bombay time had led Professor Orlebar erroneously to assume that the dry bulb thermometer readings were faulty. He had therefore decided not to publish them and so the data were lost.

Not that the Committee was complaining about the lack of observations – quite the opposite. Huge numbers of observations were published, down to the horary observations, when they could have easily been condensed. However, where mean values had been provided, 'the mere arithmetical operations of addition, subtraction, and division have been but carelessly performed, – many of the quantities being in error'.[86] In similar fashion, there had been virtually no attempt to analyse the data. It was reported: 'Graphical projections have indeed been occasionally employed to illustrate the fluctuations of the physical elements of observation, but not with the effect and success this mode of analysing results is capable of. In no case have mathematical formulæ of reduction been applied to any of the sets of observations for the discovery of the laws to which they conform.' In relation to this criticism the Committee concluded: 'what is good, bad, or indifferent among the observations is as yet unknown and cannot well be ascertained till some care has been bestowed on the observations and on the instruments with which they have been made'.

The final criticism the Committee levelled at the Colaba Observatory's meteorological work concerned its problematic locality. Given the observatory's initial maritime purposes, it had been sited close to the harbour and away from the city. This presented a number of problems. First, two 18-pounder iron guns that sat 220 yards from the observatory and fired shots to announce the arrival of ships had noticeable effects on the observatory's various delicate instruments. Although some efforts had been made to reduce this impact – using guns of smaller calibre, placing them on a patch of sand and using less powder – the installation of heavy Armstrong guns as part of the harbour battery promised even more disruption than the 18-pounders. As if that was not bad enough, the observatory was felt to be 'isolated' from the island of Bombay so that it

[86] Barker, *Report of the Committee of Inquiry*, p. 17.

was 'by no means adapted for directly ascertaining its general meteorological phenomena'.[87] It was reported for instance that rainfall differed markedly over the island, as did tension of vapour, temperature and wind. How this was known by the Committee, given the failure of the observatory to successfully measure the weather, was not explained in the report. As a result, the report recommended that the observatory should be moved to a more central and more representative locality in the city, even though such a move would interrupt the magnetic observations. This, presumably, would have gone some way to correcting the damning conclusion of the report: that 'after collecting observations for nearly a quarter of a century we do not yet know the normal data of the climatology of Bombay'.[88] It is worth noting that criticism of the operations of the Colaba Observatory was not an isolated case; others were highlighting the lack of organisation of meteorological observation across India. In 1855, Colonel Richard Strachey drew the attention of the Asiatic Society of Bengal – the journal of which routinely published meteorological reports – to the poor knowledge of India's weather and argued for all observatories in India to work together towards an improvement in this state of affairs.[89]

Re-assembling the Observatory

One of the immediate actions of the Committee was to appoint a new superintendent at Colaba, a person with 'not only perseverance but considerable scientific intelligence', and ideally 'a European of scientific education'.[90] Charles Chambers was chosen for the role, having previously been trained as an assistant at the Kew Observatory and then employed by the Indian Telegraph Service.[91] He took up a temporary position as director at Colaba in September 1865. He immediately set about reviewing the state of the observatory's instruments and practices, such that he might 'secure efficiency and completeness in the prosecution of the objects to which the institution was at the time ostensibly devoting itself'.[92] In

[87] Barker, *Report of the Committee of Inquiry*, p. 21.
[88] Barker, *Report of the Committee of Inquiry*, p. 17.
[89] D. R. Sikka, 'The Role of the India Meteorological Department, 1875–1947', in U. D. Gupta (ed.), *Science and Modern India: An Institutional History, c.1784–1947* (Delhi: Pearson, 2011), pp. 381–426, p. 387.
[90] Barker, *Report of the Committee of Inquiry*, p. 11.
[91] R. H. Scott, 'The History of the Kew Observatory', *Proceedings of the Royal Society of London*, 39 (1885), 37–86, 57 and 58.
[92] C. Chambers, *Report on the Condition and Proceedings of the Government Observatory, Colaba, for the Period from September 1865 to December 31st 1867* (Bombay: Military Secretariat Press, 1868), p. 10.

public he was careful to maintain a benign tone in his public thoughts on the observatory, noting for instance in 1866 that the observatory's failure to 'take high rank among similar institutions' was due 'not so much to its labours having been directed into wrong channels as to the injudicious manner in which those labours have been conducted'.[93] In private he was more bullish, condemning the state of the observatory's instruments and culture in letters to Stewart at Kew and John Gassiot, Chair of the Kew Committee of the British Association.[94]

Chambers's headline recommendation was that the Bombay Observatory should relinquish its claim to be a centre of astronomy and focus its efforts on magnetism, meteorology and tidal studies, retaining only enough equipment and expertise for it to determine and to signal true local time. There was a twenty-year run of magnetic and meteorological observations to work with, he argued, from which there was 'good reason to believe that by a proper, though generally laborious and lengthy, treatment much valuable information which at present lies hid in the observations may be extracted'. There was no systematic astronomical equivalent on record. Chambers went so far as to suggest that Colaba become the Central Physical Observatory of India and for Madras to devote itself to astronomy.[95] A key strategy for Chambers in this regard was to reorient Bombay's principal allegiance away from Greenwich and towards Kew. He wrote letters to both Stewart and Gassiot to ask for help in superintending the selection, purchase and despatch of new instruments for Colaba, reminding Gassiot in October 1865 of the strong support supplied by the Kew Committee to foreign and colonial governments in their scientific endeavours.[96] In a letter to Captain Young, the superintendent of marine for the Bombay government, he went further, pointing out that no other foreign or colonial observatory was at present working in concert with Greenwich, while Kew's 'complete system of observation being carried out there, as well as the form of instruments used, has been adopted in succession by no less than eight other Observatories distributed widely over the Globe'.[97] All parties acceded to Chamber's request, including Airy, although the Colaba director was nonetheless required to submit a half-yearly report on the operations of

[93] C. Chambers, 'Report on the Instrumental and Other Requirements of the Government Observatory, Colaba, for Increasing Its Efficiency and Extending the Field of Its Operations', included in *Report of the Superintendent of the Government Observatory, Colaba* (Bombay: Military Secretariat Press, 1866), p. 29.

[94] Letters and memos relating to Colaba Observatory, Bombay, National Archives, BJ 1/16.

[95] Chambers, 'Report on the Instrumental and Other Requirements of the Government Observatory', p. 25; Sen, *Astronomy in India*, p. 173.

[96] Chambers to Gassiot, 28 October 1865, National Archives, BJ 1/16.

[97] Chambers to Young, 30 January 1867, National Archives, BJ 1/16.

the observatory to the Astronomer Royal.[98] The abandonment of astronomical observation and prioritisation of meteorology and magnetic research seemed counter to Airy's scientific priorities, but his support for the observatory's shift in emphases reflected his more general views that colonial observatories should help to transform and improve the environments in which they were embedded.[99]

Given the conclusions of the Committee of Inquiry, Chambers was keen to assess the condition of the observatory's instrument collection and to effect repairs and to seek new apparatus where appropriate. He provided detailed lists of all of the observatory's instruments and their makers, where they were placed, how they were used and their present condition. The two standard barometers by Newman were observed hourly. They were located in the magnetic observatory, were in serviceable order but with corrosion to the surface of the mercury in the cistern, and had soiled glass cylinders. Chambers documented the use of a range of different thermometers by a variety of makers – Newman; Negretti and Zambra; Murray and Heath; and Barrow. They were either kept in the observatory's thermometer shed or in a revolving stand. The errors of all but the standard Newman were recorded as unknown. An Osler's self-registering wind and rain gauge was positioned on the astronomical observatory; the vane, pressure plate and rain vessel on the roof, while the recording apparatus was in the room below. The apparatus was in working order but was insufficiently sensitive: 'The force pencil rarely moves at all, while the direction pencil moves by fits.'[100] A Newman rain gauge was considered to be in good working order although it was insensitive to small amounts of precipitation and was too close to the magnetical observatory. A Daniell hygrometer and a Regnault's hygrometer were kept in the astronomical observatory and observed with the doors open. Lists of the instruments held in store were also provided – those that were serviceable (twenty-five thermometers, four barometers, three telescopes, a dip circle, four sextants, a standard brass yard and so on), those that were unrepairable and finally those 'of a useless and imperfect character that can never be turned to any account'.[101] Chambers turned to the Kew Committee for help in selecting, purchasing, testing and despatching new instruments to Bombay, so as 'to remove the more glaring defects in

[98] Deputy Secretary to Bombay Government to Chambers, 23 August 1867, National Archives, BJ 1/2.

[99] D. Belteki, '"The Grand Strategy of an Observatory': George Airy's Vision for the Division of Astronomical Labour among Observatories during the Nineteenth Century', *Notes and Records of the Royal Society*, 77 (2023), 135–51.

[100] Chambers, *Report on the Condition and Proceedings of the Government Observatory*, (1865), p. 5.

[101] Chambers, *Report on the Condition and Proceedings of the Government Observatory*, (1865), p. 6.

the present instrumental equipment of the Observatory'.[102] He requested a Barrow dip circle, a portable absolute declinometer and horizontal intensity apparatus, a Robinson's anemometer (along with 400 sheets of paper for registrations), a Kew standard thermometer and a standard barometer.[103] He managed to persuade the India Office's director general of stores to cover the costs of shipping them from London to Bombay.

Alongside his instrument inventory Chambers carried out a review of the system of observation and computation at Colaba. Hourly observations were taken of the magnetic instruments, the standard barometers, air, wet bulb and soil thermometers and rain gauges, as well as the state of the weather, clouds and wind direction. These observations began at 12 pm on Sunday and concluded at 11 pm on Saturday Bombay Civil Time (although the reading of the rain gauge was not taken until midnight on Saturday). Meteorological observations were also taken twice daily at 9.30 am and 3.30 pm so as to conform to the request of the Royal Engineers. The anemometers received new paper on a daily basis. At 10 am the meteorological observations were supplied to the Colaba telegraph office and weekly returns of rainfall to the curator of the Government Central Museum. Monthly tables of daily averages and maxima and minima of the meteorological elements were supplied to the surgeon in charge of mortuary returns.[104]

The three observers in the magnetical department were required to take and then enter the hourly magnetic and meteorological observations into the registration forms. They worked in rotation in three periods of four hours' duration every half day, and during the intervals between the hourly observations they were required, with the aid of suitable tables, to correct the barometer readings and find the tension of vapour, humidity of the air, and absolute declination, and to copy the observations of each element into pro formas, and finally to compute the daily mean values and the monthly mean values for each hour. The computer aided the observers in these operations and 'generally in average-taking, copying, and easy calculations'.[105] The first assistant was responsible for maintaining immediate oversight of both observers and instruments, and keeping general control of the department. The first assistant was

[102] Chambers, *Report of the Superintendent of the Government Observatory, Colaba* (Bombay: Military Secretariat Press, 1866), p. 3.

[103] Chambers to B. Stewart, 28 October 1865, National Archives, BJ 1/16.

[104] Chambers, *Report on the Condition and Proceedings of the Government Observatory*, (1865), p. 37.

[105] Chambers, *Report on the Condition and Proceedings of the Government Observatory*, (1865), p. 9.

also the keeper of the manuscript records, consisting of the registers of the hourly observations; monthly abstracts; the tables of results; the registration papers of the wind, tide and rain gauges; the record books of term days; and the printer's proofs, while the second assistant was responsible for the preparation of the manuscript of the observations and results for the press. There were two assistants in the separate astronomical department and four local porters. The observatory superintendent oversaw all of the staff.

In his assessment of Colaba's performance over his first few years in charge, Chambers despaired of the observatory's ad hoc and inefficient computational and archival practices. He noted that the records of observatory data were 'piled loose on the shelves of large teakwood cupboards in the Store Room . . . without any careful arrangement or any catalogue for facilitating reference or collation'.[106] He pointed to the frequent and critical errors that crept in during the acts of copying and averaging data and suggested that prior to 1866 no systematic check had been applied to computing operations. To mitigate against these issues and to shore up the observatory's reputation as a magnetic and meteorological counting house, he implemented a number of amendments to the management and surveillance of observatory staff. Instead of three observers doing two four-hour shifts every twenty-four hours, Chambers divided each twenty-four hour cycle into four periods, and assigned one to each of the three observers and the second assistant. Observations were to be reduced immediately upon entry into the register. During their terms of duty observers were banned from leaving the observatory except at fixed times for meals. Private reading could only take place between the hours of 4.50 pm and 9.20 am. All observers were required to report for computing duty from 11 am to 2 pm, aided by the first assistant and computer. Concentrating computing into a three-hour period rather than spreading it across the entire day allowed a more effectual control to be exercised over the computers, with all calculations 'systematically checked by repetition of the operation by another computer'.[107] The previous day's reductions were also checked.

Chambers set about reviewing other controversial aspects of the observatory's infrastructure and operations. He organised the removal of the time ball to the fort. He examined the exact effects of firing the Colaba battery on the instruments so that observations might be corrected

[106] Chambers, *Report on the Condition and Proceedings of the Government Observatory*, (1865), p. 10.

[107] C. Chambers, *The Meteorology of the Bombay Presidency* (London: Her Majesty's Stationery Office, 1878), p. 7.

accordingly. He reorganised and updated the observatory's library. He considered the impact of the observatory's location on the general applicability of its meteorological observations, arguing that Colaba island was as good a location as anywhere for the observatory given that 'no single station could be assumed to represent Meteorologically every part of the island'.[108] He introduced a full suite of self-recording instruments to aid in both meteorological and magnetic work. To Chambers's mind, the introduction of self-recording barographs, thermographs, anemographs and magnetographs marked a new era in the observatory's history. Replacing 'personal agency' with 'automatic mechanical agency' changed the observatory's routines and eased the burden of work for the observers, especially at night. At the same time Chambers judged the self-recording instruments to increase the workload of the superintendent, in that the 'native observers' were deemed 'quite unfit and exhibit no ambition to be otherwise regarded – to be trusted with operations which require a judgement to be formed on the particular circumstances of each case, and require sometimes also some skill in manipulating somewhat intricate mechanisms'. Native automatons – 'admirable workers in a groove to which they have been thoroughly and patiently drilled' – were deemed to be incapable of interacting with their mechanical equivalents.[109] Chambers's views on the superior capabilities of the European to manage complex instruments should, in part, be read as self-serving in nature. Chambers spent his first years as superintendent demonstrating his own indispensability because he was on a temporary contract and keen to see it made permanent. He pointed out that the Committee of Inquiry had argued for 'the need of continuous and effectual supervision in the production of the Observatory records and of intelligence and experience in making use of them when obtained', and that 'with the exception of two or three officers in the Survey Department, no persons in India, much less in Bombay, have had experience in the use of these instruments which are of comparatively modern design'.[110] Reinforcing the implication that he was the obvious man for the job, Chambers highlighted his own long training at Kew, claiming that it was now regarded as Europe's model observatory.[111]

[108] Chambers, 'Report on the Instrumental and Other Requirements of the Government Observatory', p. 27.

[109] C. Chambers, *Report on the Condition and Proceedings of the Government Observatory, Colaba, for the Year which Ended with the 30th June 1875* (Bombay: Government Central Press, 1876), p. 5. On Indian natives' 'mechanical disposition', see Schaffer, 'Oriental Metrology', 194.

[110] Chambers, *Report of the Superintendent of the Government Observatory*, p. 5 and p. 6.

[111] Sen, *Astronomy in India*, p. 179.

The Meteorology of the Bombay Presidency

In the early years of his time at the observatory Chambers prioritised the continuation of the routine of observation and then gradually introduced improved methods of observation and registration. Once he was satisfied that these were in place, he turned to his longer-term aim of producing a comprehensive study of the observations collected at Colaba and, in relation to the meteorological data at least, situating them within a wider study of the weather and climate of the Bombay Presidency. Chambers's *The Meteorology of the Bombay Presidency* was eventually published in 1878. This work made use of the observatory's long series of observations going back to 1841, even though Chambers noted gaps in the series and the variable state of the records that remained, with some original manuscript registers missing and some registers unreduced and means uncalculated. Despite Chambers's concerns about the quality of the computations prior to 1866, he admitted that they had been adopted as they appeared in the abstract sheets: that the 'enormous amount of labour of recomputing these for a period of twenty years would, if it had been undertaken, have rendered it quite impracticable to obtain any general conclusions in a reasonable time'.[112]

The first part of the book was devoted to the presentation and analysis of the Colaba observations. This part, which ran to over 100 pages, was in turn divided into sections that discussed the results of each instrument: the barometer, the dry and wet bulb thermometers, eye observations of degrees of cloudiness, ground thermometers, rain gauges and the anemometers. For each, he described the instrument in detail: its placement in the observatory; the calculation of the instrument's errors and the corrections that had been applied; the local history of observational methods with the instrument; the diurnal, monthly and annual mean values and variations; and calculations of probable error. The text was routinely broken up by tabulated and graphed data. The completion of this summary and analysis of the meteorological observations collected at Colaba marked for Chambers 'the closing of a phase in the Observatory's history – a phase in the course of which the old eye observations (ending in 1872) will have been utilized for the determination of the various meteorological and magnetic elements and of the laws of their periodic variations' for the city.[113] The release of the observations from the confines of the stores and into general view also shored up the observatory's

[112] Chambers, *Meteorology of the Bombay Presidency*, p. 7.
[113] Chambers, *Report on the Condition and Proceedings of the Government Observatory*, (1875). p. 4.

claim to government funding and defended it from accusations of abuse of public support.

The second half of the *Meteorology* was, in the most general terms, a mapping exercise which sought to pull in meteorological data from a variety of stations so that the Presidency's climatic geography could be revealed. In Part II, Chambers applied the same form of discussion and analysis used in Part I to the Presidency's four provincial observatories at the military stations run by the Royal Engineers at Kurrachee, Deesa, Poona and Belgaum. Part III considered the results collected at civil and military hospitals and dispensaries, and the rainfall returns from revenue officers; and Part IV brought the whole body of results together to construct the climate of the Presidency. The order was not insignificant. It traced in Chambers's view a geography of accuracy and trustworthiness, where 'a high order of efficiency and value is followed by gradations which reach the lower limit of all but absolute worthlessness'.[114] Despite his views on the relative value of provincial meteorological observation, Chambers *had* expressed the hope that 'the combined labours of private observers' would aid in building a wider and more detailed picture of India's weather and climate. He went as far as to propose that the observatory might, for a small fee, issue and verify instruments, inspect amateur stations and check their observations.[115]

For the military stations Chambers reported that the long series covered a seventeen-year period from 1856 to 1872, although significant gaps in the records remained for all four sites. He described the stations themselves and the situation of the instruments, but then largely presented the results by meteorological phenomenon and instrument. Although these stations were much more modest than the colonial observatories established as part of the Magnetic Crusade – with the weather only being observed twice a day – their value lay in their national extent. Chambers noted elsewhere: 'So far as I am aware this is the only system of scientific observation extending over all India that has been attempted.'[116] In Part III, Chambers pulled on the observations made at fifty-two sites across the region to give a general account of the distribution and annual variation of temperature; the average monthly and annual fall of rain across 282 stations, with particular attention to the monsoon; and a description of the

[114] Chambers, *Meteorology of the Bombay Presidency*, p. 4.
[115] Chambers, *Report of the Superintendent of the Government Observatory*, p. 27.
[116] C. Chambers, 'Report of the Superintendent of the Government Observatory, Colaba, for the period from June 27th to December 29th, 1868', attached to a letter from Chambers to Secretary of Government of the Marine Department, 30 December 1868, National Archives, BJ 1/2.

general characteristics and variations of the wind system by district and season.

Although Chambers was aware of the questionable character of much of the data upon which the final part of the book – 'The Climate of the Presidency' – rested, they also presented an opportunity for him to expand the ambitions of the study and to make a case for the wider value of Colaba's work to the science of meteorology and to Britain's governance of India. In the opening pages of the *Meteorology* Chambers argued that the observatory sciences were largely constrained to the accumulation of well-ascertained observed phenomenal laws: that 'the observer must be content to take natural phenomena as they come before him, having no power to vary the complication of circumstances which surround them, but relying upon such variations as differences of time and place afford, to give him a possible clue to efficient physical causes'.[117] The model 'professional observer' and 'high authority' for Chambers was Edward Sabine, who had demonstrated the value of amassing a sufficient body of phenomenal laws before any generalisation could be countenanced.[118] Parts I–III had followed Sabine's principle of establishing phenomenal laws through analysis of observations. Part IV was an opportunity to bring some of these laws together and to sketch out 'the theoretical relations of the distribution and variations of temperature, of the distribution and variations of rainfall, and of the wind system of the Bombay Presidency'.[119] On the one hand, Chambers applied these situated phenomenal laws to wider physical theories, such as Herschel's theory of the causes and operations of the trade winds. On the other hand, he demonstrated the wider utility and 'practical usefulness' of their application to particular locations and schemes.[120] For instance, the application of the law of rainfall distribution to the Presidency promised to aid irrigational and engineering schemes, a point Chambers illustrated through a discussion of the effects of rainfall on river basin discharge and the damage done to some of the Lower Sind Railway bridges.[121] Chambers also highlighted the usefulness of observatory science in the colonies by listing the various government and other offices to which he had supplied meteorological information and examined instruments, including military surveyors and surgeons, cavalry units, telegraph

[117] Chambers, *Meteorology of the Bombay Presidency*, p. 1.

[118] Chambers, *Meteorology of the Bombay Presidency*, p. 2.

[119] Chambers, *Meteorology of the Bombay Presidency*, p. 249.

[120] Chambers, *Report on the Condition and Proceedings of the Government Observatory*, (1875), p. 4.

[121] C. Chambers, *Report on the Condition and Proceedings of the Government Observatory, Colaba, for the Year which Ended with the 30th June 1876* (Bombay: Government Central Press, 1877), p. 7.

companies, various newspapers, sanitary and health officers, port and harbour officials and private companies. He proposed the extension of telegraphic communications from Colaba to the Koromandel and Kutch coast so that the observatory might be able to give prior warnings of impending cyclones and storms. Chambers's suggestions were put into practice. By the 1880s the observatory received daily telegraphic weather reports from ten stations along the western Indian coast, and issued warnings to six, using a simplified version of FitzRoy's drum and cone signals.[122]

Centralising Indian Meteorology

Despite the gradually extending reach of the Bombay Observatory, Chambers lamented the poverty of a national meteorological culture in India compared to that in Europe, which he noted had an area smaller than the subcontinent, but which had a much greater number of well-funded meteorological stations.[123] This criticism was answered in January 1875 when the government of India established the India Meteorological Department with the appointment of Henry F. Blanford as imperial meteorological reporter. The failure of the monsoon rains in 1866 and 1871 and the effects of a devastating tropical cyclone that hit Calcutta in October 1864 are often cited as motivations for this post, with increasing demands for a better storm warning system and improved understanding of India's weather and climate.[124] Anderson argues that the department also reflected an increasing degree of political centralisation and an active engagement with India from Britain in the years following the Indian Rebellion in 1857.[125] This reflects Bayly's more general argument that the Rebellion decisively reshaped the information order of colonial India in the second half of the nineteenth century, including the 'spread of scientific modernism'.[126]

Born in London in 1834, Blanford's training was in geology rather than meteorology. In 1855, he came to India to work on the Geological Survey of India. Ill health forced him to retire in 1862 and he took up a position as a professor at Presidency College, Calcutta, giving lectures on physics and chemistry. As a secretary of the Asiatic Society of Bengal he developed an

[122] Chambers, *Meteorology of the Bombay Presidency*, p. 255; Walker, *History of the Meteorological Office*, p. 43.
[123] Chambers to Young, 30 January 1867, National Archives, BJ 1/16.
[124] D. Arnold, *Science, Technology and Medicine in Colonial India* (Cambridge: Cambridge University Press, 2000), p. 134.
[125] Anderson, *Predicting the Weather*, p. 250.
[126] C. A. Bayly, *Empire and Information: Intelligence Gathering and Social Communication in India, 1780–1870* (Cambridge: Cambridge University Press, 1999), p. 338.

interest in meteorology, particularly in the cyclones that regularly affected Calcutta. In 1867, he was appointed meteorological reporter to the government of Bengal, with a view to developing a storm warning system for the city, before being appointed head of the Meteorological Department of India.[127] In his role as imperial meteorologist he was to rework the failed provincial model of meteorological observation, drawing India's disparate observatories together into a single network that would improve knowledge of the weather of the subcontinent and connect it to climate research developing elsewhere around the world.[128] The observatory at Alipore, Calcutta, was established as India's Central Meteorological Observatory, where standard instruments were kept and comparisons and verifications were performed.[129] Annual reports of meteorological data were published from there and it acted as the principal site for communication with neighbouring countries and with Europe.

Blanford published many papers on Indian meteorology, as well as several books, including his detailed report on the 1864 cyclone, a handbook – or *Vade-Mecum* – for meteorological observers in India, a monumental study of monsoons and droughts, and a popular guide to the climate and weather affecting the countries surrounding the Bay of Bengal.[130] Blanford's *Vade-Mecum* was divided into two parts, the first of which provided a guide to observers keeping a meteorological register. Although it was apparently based on a short guide for observers in Bengal that had been published in 1868 – since gone out of print – the section provided a generic account of the correct use of various meteorological instruments; guidance on eye observations; and instructions on when to observe, how to reduce observations and how to establish and run a successful observatory. There was no reference to the geographical context in which the observations were to take place. Rather, the volume's tone informed its readers they were operating in a universal scientific culture; that meteorology was 'a branch of physics, and requires therefore the same care, and the same precautions, that are demanded in all

[127] Anon, 'News and Views: Henry Francis Blanford, F.R.S. (1834–93)', *Nature*, (2 June 1934), 824.

[128] R. Axelby and S. P. Nair, *Science and the Changing Environment in India*, 1780–1920 (London: British Library, 2010), p. 133.

[129] Sikka, 'The Role of the India Meteorological Department', p. 388.

[130] J. E. Gastrell and H. F. Blanford, *Report on the Calcutta Cyclone of the 5th October 1864* (Calcutta: O.T. Cutter, 1866); H. F. Blanford, *The Rainfall of India, Volume III of the Indian Meteorological Memoirs* (Calcutta: Superintendent Government Printing, 1886); H. F. Blanford, *The Indian Meteorologist's Vade-Mecum* (Calcutta: Thacker, Spink, 1877); H. F. Blanford, *A Practical Guide to the Climates and Weather of India, Ceylon and Burmah and the Storms of Indian Seas Based Chiefly on the Publications of the Indian Meteorological Department* (London: Macmillan, 1889).

experiments and observations undertaken for the discovery of physical laws'.[131]

Part II did address the characteristics of Indian meteorology directly, with particular attention to India's physical geography, its hot and rainy seasons, winds and storms. But much remained to be done. Converting his scientific interests into metaphor, Blanford suggested that Indian meteorologists had only 'gained possession of the lowest outlying spurs of a stupendous range' of meteorological knowledge, 'in the recesses of which lie the materials of untold wealth'. Published a year ahead of Chambers's account of the meteorology of the Bombay Presidency, Blanford complained in particular of 'extremely scanty' knowledge of southern and western India.[132] Just as with its inception, the department continued to be motivated by the desire to map and anticipate meteorological features operating over subcontinental scales, notably the cyclones, droughts and monsoons that caused such great disruption.[133] Writing in the early 1880s, Strachey, now inspector of railway stores at the India Office, a member of the Council of India and president of the Famine Commission (established in the aftermath of the severe drought of 1877–9), emphasised the importance of reliable weather data from across the territory, claiming that 'at present no power exists for foreseeing the atmospheric changes effective in producing the rainfall or of determining beforehand its probable amount in any season, such as it would admit of timely precautions being taken against impending drought'.[134] Blanford's *Rainfall of India* responded to Strachey's call by composing and analysing the occurrence of monsoon and drought over the 1870s. He later commenced the issuance of the long-range seasonal forecast of rains over India and developed theories of the monsoon's onset which challenged Norman Lockyer's claims regarding the importance of sunspot activity.[135]

In 1889, Blanford published his *Practical Guide to the Climates and Weather of India, Ceylon and Burmah*, by which time Calcutta's climate

[131] Blanford, *Vade-Mecum*, p. 1.

[132] Blanford, *Vade-Mecum*, p. 260. Blanford *was* aware of a paper Chambers had communicated recently to the Royal Society on the subject of the meteorology of the Presidency but knew nothing of its content. It was not until 1899 that the Colaba Observatory came under the direct control of the India Meteorological Department.

[133] V. Damodaran, 'The East India Company, Famine and Ecological Conditions in Eighteenth-Century Bengal' in V. Damodaran, A. Winterbottom and A. Lester (eds.), *The East India Company and the Natural World* (Basingstoke: Palgrave Macmillan, 2015), pp. 80–101.

[134] Quoted in Sikka, 'The Role of the India Meteorological Department', p. 389–90; Grove, 'The East India Company, the Raj and the El Niño', p. 304.

[135] M. Davis, *Late Victorian Holocausts: El Niño Famines and the Making of the Third World* (London: Verso, 2002).

had gotten the better of him: he had retired to Kent in 1888. Writing not for the dedicated weather observer but for 'those others of the general public to whom the weather and the climates of India and of its seas are practical and not scientific objects of interest', Blanford struck both a simpler and more positive tone than he had in his *Vade-Mecum*, arguing: 'we now possess a far better knowledge of the weather and climate of India than of those of any other tropical country, and, in some respects, better than of those of many parts of Europe; thanks to the greater simplicity of the processes concerned, and to the prominence and regular recurrence of the more striking phases of the seasons'.[136] The regularity of the tropical climate, the size and discrete nature of the landmass – bounded by mountains and ocean on all sides – as well as the frequent cyclones, turned India into a natural laboratory for the study of atmospheric physics. As a discrete imperial space, India presented the potential for the effective combination of science and utility on a grand scale as well as the taming of unruly natural *and* political forces.[137]

While India's size provided meteorologists with the opportunity to study the atmosphere over a large, coherent territory, its physical diversity challenged attempts to make general claims: 'the term *Indian climate* means little more than that, on the general average of the year, the sun is higher in the heavens, and the temperature some degrees greater, than in Europe or other lands of the temperate zone. In all else we must be prepared to find the greatest variation, and even the greatest contrasts.'[138] In his *Guide* Blanford exploited India's burgeoning and reorganised network of observatories to provide a simultaneously national and regional picture of the weather and climate. For instance, discussion of the diurnal variation and range of air temperature utilised diverse sources of data, from detailed accounts of the daily averages experienced over two months for Calcutta to wide-ranging comparisons of monthly average diurnal ranges for stations across India. Mean data by month over the course of a year for different parts of the country were often put together and compiled into tabular form. Isoline charts were also used to show weather conditions at varying scales, including several for the whole of India and the Bay of Bengal at fixed points in time and as monthly and yearly averages (Figures 2.3 and 2.4).

Bombay featured routinely in the text as a source of weather information and Chambers was cited as a reliable observer and authority. However, as with the other larger and older observatories – with the notable exception of Blanford's own observatory at Alipore – Bombay was positioned in the text

[136] Blanford, *Practical Guide*, p. viii and p. vii, respectively.
[137] Anderson, *Predicting the Weather*, pp. 258–60 and p. 283.
[138] Blanford, *Practical Guide*, p. 96, original emphasis.

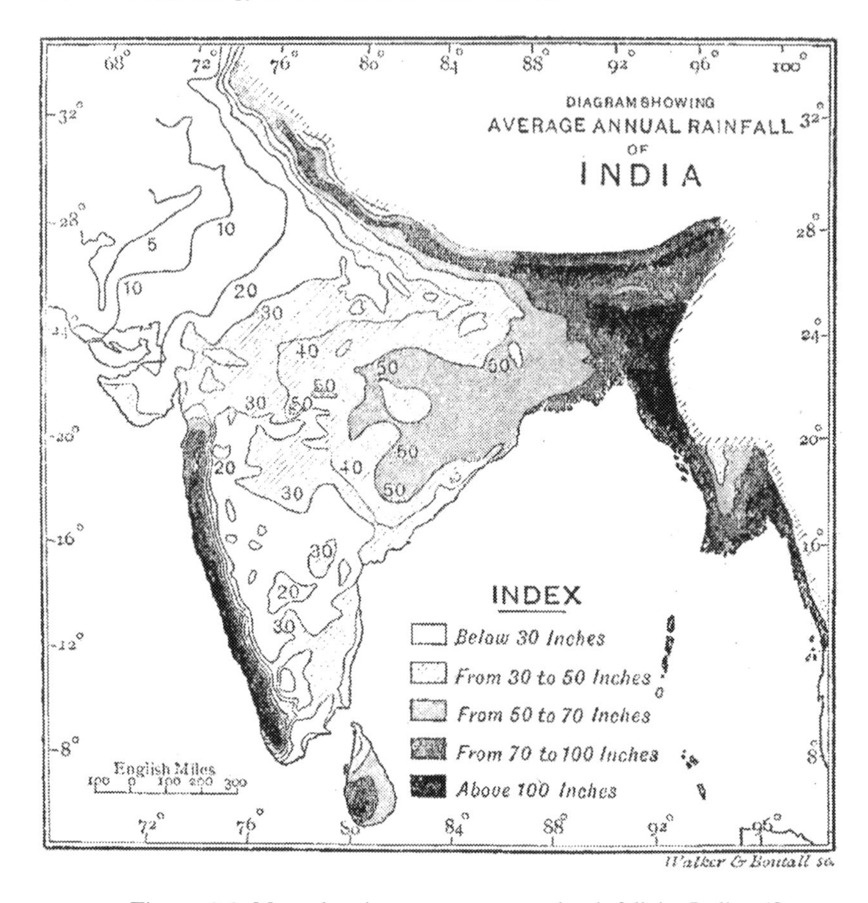

Figure 2.3 Map showing average annual rainfall in India. (Source: Blanford, *Practical Guide to the Climates and Weather of India, Ceylon and Burmah*, p. 66.)

as of no more significance than a military station at Quetta on the Afghan border or a sanatorium in the Kumaon Himalaya. Blanford put great emphasis on increasing the number of stations reporting to the department. He set up new stations in Burma, Rajputana, central India, the Bombay Presidency and in difficult-to-reach locations like Leh in the Himalayas and equipped them with standard instruments. He improved rainfall registration across India and promoted the establishment of observatories in the princely states.[139] This smoothing out of the hierarchy of observation

[139] Sikka, 'The Role of the India Meteorological Department', p. 388.

Figure 2.4 Average barometric and wind chart for May. (Source: Blanford, *Practical Guide to the Climates and Weather of India, Ceylon and Burmah*, p. 206.)

across a single imperial landscape of knowledge was exemplified in the featureless maps of barometric averages and wind direction, which 'summarized multiple observations, but subordinated them to broad patterns, using methods that could reveal how both the science and the phenomena transcended a local scale'.[140] Blanford's data empire was considered trustworthy enough to be used by others as a baseline against which to judge the veracity of particular results. When mathematician and meteorologist Douglas Archibald set out to investigate Broun's meteorological data for Travencore, he relied on the means of the pressure and temperature

[140] Anderson, 'Mapping Meteorology', p. 86.

oscillations provided by Blanford's stations, noting that '[a]s these stations represent every part of the country, the results afford a basis for deduction of sufficient extent to be reliable'.[141]

Attempts to develop a single national meteorological culture in India – one that placed heavy emphasis on meteorology's public value – reached its apotheosis with the appointment of Sir John Eliot as Blanford's successor in 1889. Eliot's career had followed Blanford's closely – he also taught physical sciences at Presidency College and acted as meteorological reporter to the government of Bengal. His research efforts also developed Blanford's work, focusing on tropical cyclones, monsoons and large-scale pressure oscillations over the tropical belt. As head of the department, Eliot instituted a system of telegraphic flood warnings to engineers on large construction works, and to those in charge of railways and bridges. He furthered Blanford's work on drought and famine warnings.[142] He was also recognised as a committed synoptic meteorologist who emphasised the value of high-quality surface meteorological data.[143] In addition to being made meteorological reporter for India, he was appointed director-general of Indian Observatories in 1899, when the observatories at Colaba and Madras came under formal control of the India Meteorological Department. By 1901 India's surface observatory network was comprised of 230 stations. Eliot also further improved the diffusion of weather information with the issuance of frequent weather reports at various locations.

The handbook Eliot produced to supersede Blanford's *Vade-Mecum* epitomised a further shift in meteorological culture in India. Eliot noted in the introduction to his *Instructions to Observers* that observers in India 'now merely take the readings of certain meteorological instruments'.[144] The guidance provided only a description of the instruments provided, instructions for taking readings and recording the observations, and advice on instrument maintenance and restoration. No explanation was given as to how the readings were to be corrected and reduced. Instead, observations were to be forwarded by telegram or post on suitable forms to the Imperial Meteorological Office at Calcutta, where the reduction and preparation of the data for subsequent use would be performed. Despite the removal of this problematic function from the observer's round of tasks – and regardless of Eliot's choice of adverb above – observers were to

[141] E. D. Archibald, 'Barometric Pressure and Temperature in India', *Nature*, 20 (1879), 54–55, on 54. Archibald was using data published in the *Vade-Mecum*.
[142] J. Eliot, 'Sir John Eliot, K.C.I.E., F.R.S., 1839–1908', *Nature*, 143 (20 May 1939), 847.
[143] Sikka, 'The Role of the India Meteorological Department, 1875–1947', p. 395.
[144] J. Eliot, *Instructions to Observers of the India Meteorological Department* (Calcutta: Office of the Superintendent of Government Printing, 1902), 2nd ed., p. 1.

understand their actions as governed by a strict moral economy. It was important, for instance, that observations were taken at their appointed time: 'Observers should understand that it is a deception, a fraud, to put on permanent record the temperature at say 3.30 P.M., as the temperature at 4 P.M.' Paying an untrained 'messenger or peon' to take the observations was, similarly, to 'act dishonestly'. Even following Eliot's instructions 'blindly and by rote' was problematic. Rather, the ideal observer was expected to understand why certain rules and methods were imposed on them and to 'take the observations intelligently'.[145]

The mapping out of a strict observational hierarchy was reinforced with the imposition of four observatory classes. First-class observatories kept continuous records using self-registering instruments. Second-class observatories had to make three sets of observations a day. Sites of a third class were those where only one set of observations was made, and then forwarded by telegraph to the first-class observatories at Simla, Calcutta, Bombay or Madras, or where observations were made twice a day. Finally, the fourth class was where only temperature, wind and rain data were recorded.[146] All of Eliot's observatories were expected to contribute to the production of a national picture of India's weather and climate by reproducing as closely as possible their local situation. Although usually placed near a government building, such as a hospital, dispensary or post office, observatories were meant to 'be representative, as far as possible, of the open country in the neighbourhood of the station'. This meant avoiding buildings and keeping bushes, trees and flowerbeds at bay. Simulating local conditions while adhering to universal practices was crucial because the 'the most important use which is made of the observations at any observatory in India is the determination of the normal meteorological or climatic conditions of the district in which the observatory is situated and which it is believed to represent'.[147] Standardised observational practices performed in typical locales were to be fed to Calcutta, where computers would correct and reduce data so that an accurate picture of India's weather and climate could be illuminated.

Conclusion

The Colaba Observatory was one of a small number of sites that were established or repurposed by the British to contribute to a colonial observatory experiment, in turn part of the Magnetic Crusade. Beginning its

[145] Eliot, *Instructions*, p. 11. [146] Eliot, *Instructions*, p. 10.
[147] Eliot, *Instructions*, p. 4 and p. 3, respectively.

life as an astronomical observatory, Colaba tried to adopt the principles of the model observatory as laid out by Lloyd and Herschel, although the abandonment of astronomy at Bombay was probably not quite what Herschel had in mind when he argued that terrestrial physics should gain a greater share of public support from the pattern science. At Colaba Herschel's harmless dream of a model observatory turned into something of a nightmare. Despite the support of the East India Company and various scientific actors in Britain, the observatory's early years illustrate Sen's point that practical contingencies could be quite different at the peripheries of empire. Neither instruments, staff, practices nor data were well managed, such that no one had much faith in the scattered results that did survive into the 1860s.

A review of the physical infrastructure and the instruments in the 1860s helped to put the observatory to rights, but even more important was Chambers's overhaul of its observational and computational regimen. The labour of instruments, observers and computers became more closely regulated. In his attempt to turn Colaba into a counting house, Chambers ensured that constant vigilance was maintained through repetition, most significantly on the pages of the registers and day books. He also sought to establish the site's personal equation even down to the effects of the nearby artillery battery on the observatory's sensitive instruments. The observatory's reorientation towards Kew supported the claim that Colaba was a space of exact practice. Removing the responsibility of computation from the observatories altogether and centralising it in Alipore was an extension of this computational culture and the best means of guaranteeing that knowledge could travel across India's vast distances with minimal friction.

The colonial observatories operated at the end of long and fragile supply chains, along which instruments travelled outwards from metropolitan manufacturers and standard observatories, and routinely broke en route. Broken observers, themselves damaged by unforgiving philosophical and environmental climates, travelled in the other direction in the hopes of being mended. Data moved too – sometimes on the person of convalescent superintendents – but too often degraded or became so dislocated from their contexts that their value was thrown into question.[148] Suspicions of shoddy observational regimes, most notably Buist's mixing of Göttingen and local time, put great pressure on geographies of trust that were already stretched to their limits. As meteorological observatory networks expanded across the Bombay Presidency and the subcontinent, new geographies of trust had to be created and

[148] Raj, *Relocating Modern Science.*

managed. According to Chambers, trustworthiness deteriorated away from Colaba. His successors hoped to manage this sort of deterioration though guidebooks that laid out what exact practice had to involve at all sites. Although it continued to matter whether or not the wind turning an anemometer was disturbed by nearby buildings, or a rain gauge was sheltered by a tree, the expansion of India's observatory network and the development of long series of weather data assuaged fears of data corruption. The mapping of the India's climate relied on a data empire that smoothed out difference statistically, echoing Herschel's views on the value of the manipulation of large data sets. Like India's regional climates, its meteorological observatories were different, but ideally only within one probable error of the mean.

3 Mountain Meteorology on Ben Nevis

This chapter moves us from experiments in colonial meteorology to experiments on mountain tops. The focus is on mountain meteorology in Scotland but begins in the European Alps. Writing in his popular 1909 account *British Mountaineering*, Claude E. Benson stated that the story of mountaineering 'comes back, in the long run, to the story of the Alps'. Benson claimed that the Alps had, until the eighteenth century, been regarded as objects of terror and 'horror', 'unattainable, forbidden wastes, the abodes of demons, dragons, and malevolent spirits', but this 'superstition gave way to common sense, [and] the beauty, the sublimity, the attraction of the mountains began to be recognized, and a chosen few learnt to worship them with distant reverence, and sought to impart to others their new-found emotions'.[1] For many mountaineers and writers about mountains, Horace Bénédict de Saussure was the most important of those chosen few. Born near Geneva in 1740, Saussure devoted his life to the scientific study of the European Alps and was closely associated with Mont Blanc, which he encouraged others to ascend and which he himself climbed in 1787. Edward Whymper, the notable Victorian mountaineer, suggested in his own guide to the Mont Blanc range that Saussure's ascent gave an impetus to mountain exploration and started the fashion for Alpine mountaineering: 'No sooner did he return to Chamonix than a tourist who was there went off and followed de Saussure's track. He was almost the first of the mountaineering race.'[2] Saussure helped to transform the Alps into a theatre for both sublime sensations and scientific investigations. His own wide-ranging inquiries into high-altitude meteorology, botany, glaciology, electricity and magnetism positioned the Alps as a laboratory of nature, while he encouraged an aesthetics of mountain sublimity.[3]

[1] C. E. Benson, *British Mountaineering* (New York: George Routledge, 1909), p. 1.

[2] E. Whymper, *A Guide to Chamonix and the Range of Mont Blanc* (London: John Murray, 1900, 5th ed.), p. 36.

[3] P. Felsch, 'Mountains of Sublimity, Mountains of Fatigue: Towards a History of Speechlessness in the Alps', *Science in Context*, 22 (2009), 341–64, 353; C. Bigg,

In 1851, the English journalist, writer and playwright Albert Richard Smith ascended Mont Blanc and wrote a book about it, *The Story of Mont Blanc*, published in 1853. He also put on a play that described the ascent at the Egyptian Hall in London's Piccadilly, illustrated by dioramic views. The show was tremendously popular, running for six years and 2,000 shows.[4] It was even performed to Queen Victoria and Prince Albert at their home on the Isle of Wight in 1854. Whymper argued that Smith's performance helped to reinvent mountaineering as 'exceedingly popular "entertainment"', although in Smith's hands 'the whole thing was a joke – a piece of sport'.[5] (Whymper attended the show as an eighteen-year-old and apparently enjoyed himself.[6]) Smith was a founding member of the first Alpine Club, established in London in 1857. The Club accepted members who had experience of climbing in the Alps or could provide evidence of other relevant mountain accomplishments. It was predominantly made up of men from the professional middle classes, many of whom were university educated.[7] The longevity of Smith's performance and the growth in membership of the Alpine Club reflected the widespread appeal of Alpine landscapes to the Victorian public. Tourism to the Alps increased markedly in the mid-nineteenth century, particularly to Switzerland. Hotels, railways and other tourist infrastructure developed along with guidebooks, accounts of ascents and scientific studies, including the Alpine Club's 1859 *Peaks, Passes and Glaciers*, John Tyndall's *Hours of Exercise in the Alps*, Leslie Stephen's *Playground of Europe* and Whymper's *Scrambles Amongst the Alps*, all three of which were published in 1871.[8] The 1850s and 1860s were regarded as the golden age of Alpine mountaineering, during which time Europe's mountain regions were turned into an accessible playground for British and other middle-class tourists.[9] The anonymous reviewer of Whymper's *The Ascent of the Matterhorn*, writing in *Chambers's Journal* in 1880, noted that

D. Aubin and P. Felsch, 'Introduction: The Laboratory of Nature – Science in the Mountains', *Science in Context*, 22 (2009), 311–21, 311.

[4] A. McNee, *The New Mountaineer in Late Victorian Britain: Materiality, Modernity, and the Haptic Sublime* (London: Palgrave Macmillan, 2016), p. 8.

[5] Whymper, *Guide to Chamonix*, p. 43.

[6] I. Smith, *Shadow of the Matterhorn: The Life of Edward Whymper* (Herefordshire: Carreg, 2011).

[7] P. H. Hansen, 'Albert Smith, the Alpine Club, and the Invention of Mountaineering in Mid-Victorian Britain', *Journal of British Studies*, 34 (1995), 300–24.

[8] D. Aubin, 'The Hotel that Became an Observatory: Mount Faulhorn as Singularity, Microcosm, and Macro-Tool', *Science in Context*, 22 (2009), 365–86, 371; M. S. Reidy, 'Mountaineering, Masculinity, and the Male Body in Mid-Victorian Britain', *Osiris*, 30 (2015), 158–81.

[9] C. Bigg et al., 'Introduction', 311; F. S. Smyth, *Edward Whymper* (London: Hodder and Stoughton, 1940).

'under the auspices of our modern Alpine Clubs, it has become a favourite and fashionable resort for those adventurous spirits who are fain to fill up the intervals of fox-hunting at home in winter, by the more hazardous sport of Alpine climbing abroad in summer'.[10]

Noting with approval the establishment of the Alpine Club, Benson asserted that

> [i]n the tracks of these leaders came hundreds of tourists of varying sensibility, and scores of vigorous enthusiasts, daring, but alas! imprudent and unskilled. The former vandalised the valleys, the latter attacked the peaks, and on them the mountains took a natural and terrible revenge, insomuch that, for a time, the sport of mountaineering was, in the eyes of the undiscerning, discredited. Then came the tardy recognition that the pastime was indeed 'a reputable pursuit of sensible men,' and that, in the great majority of cases, the blame of the accident must be laid, not on the mountain, but on the man. Lastly, alas and alas! followed the inevitable concomitant of the advance of civilisation, the caterer, the hotel proprietor, the tripper, and, worst of all, the 'lift' and 'funicular' railways, whereby the august mountains, the great world's altar stairs, are defaced and vulgarised, so far as it is possible to vulgarize the sublimest features of nature.[11]

Benson articulated the view that tourism to the Alps had ruined the landscape and spoiled the experience of the lone traveller and mountaineer. Others thought the widespread Victorian enthusiasm for mountains and mountaineering to be a more positive development. Mountaineering cultivated manliness, incorporating elements of physical vigour, health, patriotism, military qualities, chivalry and moral and spiritual codes of conduct. The Victorian middle classes saw climbing as a character-building exercise, which 'actively constructed an assertive masculinity to uphold their imagined sense of Britain's imperial power'. Middle-class mountaineers adopted languages of exploration, conquest and adventure from explorers in Africa and the Arctic. Climbing was even conceived as training for national and imperial service.[12] The Alps, just like the Arctic, was an 'overpowering environment where human capacities could be tested and masculine national character could be dramatized'.[13]

The pursuit of scientific inquiries helped to justify what could be highly risky enterprises or more trivial pursuits. Hevly points out that the culture of Alpine science in the mid-nineteenth century incorporated a 'rhetoric of adventure', where glaciologists like Forbes and John Tyndall gained authority as scientific eyewitnesses because of the risks and hardships they

[10] Anon, 'The Ascent of the Matterhorn', *Chambers's Journal*, 17 (1880), 24–6, 24.
[11] C. E. Benson, *British Mountaineering* (London: George Routledge, 1909), pp. 1–2.
[12] Hansen, 'Albert Smith', 304.
[13] B. Morgan, 'After the Arctic Sublime', *New Literary History*, 47 (2016), 1–26, 3.

had undertaken in challenging environments.[14] In his *Alpine Adventure*, the journalist W. H. Havenport Adams asserted that '[t]he ascents of Mont Blanc have not always been made from a thirst for adventure or to gratify a commonplace curiosity, but in the highest interests of science'.[15] In his preface to *Peaks, Passes and Glaciers*, John Ball, the first president of the Alpine Club, argued that the best sort of mountaineer combined a 'passion for Alpine scenery ... blended with a love of adventure, and some scientific interest in the results of mountain-travel'.[16] Ball's own substantive chapter, 'Suggestions for Alpine Travellers', similarly urged tourists to take various scientific observations pertaining to glaciology and meteorology, reminding them that '[s]crupulous and minute accuracy is the condition that can alone give scientific value to observations'.[17] Also writing in *Peaks*, the Rev G. C. Hodgkinson urged travellers to ensure accurate records by using instruments that were not of 'extreme delicacy or complication, or soon liable to derangement, but such as with a little practice and ordinary care may be used to good purpose'.[18] Alpine mountaineering aspired therefore to fuse elements of scientism, athleticism, bodily hardship and romanticism, whereby sport and scientific study 'dovetailed in emerging configurations of bourgeois life'.[19] McNee suggests that a 'scientific sensibility' informed and influenced mountaineering practice and writing through the second half of the nineteenth century, even if many mountaineers' actual scientific contributions were minimal.[20] Inkpen suggests that this sensibility continued into the twentieth century: there remained a 'continued social relevance of recreation in the field during an era of scientific professionalization – and, specifically, to the importance of an ethics of gentlemanly masculinity, taken from mountaineering, in glaciology'.[21]

[14] B. Hevly, 'The Heroic Science of Glacier Motion', *Osiris*, 11 (1996), 66–86, 67.

[15] W. H. Davenport Adams, *Alpine Adventure; or, Narratives of Travel and Research in the Alps* (London: Thomas Nelson, 1878), pp. 13–14.

[16] J. Ball (ed.), *Peaks, Passes and Glaciers: A Series of Excursions by Members of the Alpine Club* (London: Longman, 1859), p. vi.

[17] J. Ball, 'Suggestions for Alpine Travellers', in J. Ball (ed.) *Peaks, Passes and Glaciers: A Series of Excursions by Members of the Alpine Club* (London: Longman, 1859), pp. 482–508, p. 483.

[18] G. C. Hodgkinson, 'Hypsometry and the Aneroid', in E. S. Kennedy (ed.), *Peaks, Passes and Glaciers; Being Excursions by Members of the Alpine Club*. Second Series (London: Longman, 1862), pp. 461–500, p. 465.

[19] D. Inkpen, 'The Scientific Life in the Alpine: Recreation and Moral Life in the Field', *Isis*, 109 (2018), 515–37, 518.

[20] McNee, *The New Mountaineer*, p. 13. On the prosecution of science in the very highest mountains, see L. Fleetwood, *Science on the Roof of the World: Empire and the Remaking of the Himalayas* (Cambridge: Cambridge University Press, 2022).

[21] Inkpen, 'Scientific Life in the Alpine', 519.

Writing in the second edition of *Peaks, Passes and Glaciers* in 1862, the Rev. Hodgkinson suggested: 'As there is worthy occupation on the mountain side for the artist, the botanist, the geologist, so is there also for the meteorologist that which will combine harmoniously with his enjoyment of the scene, and even enhance it.'[22] From the 1860s meteorologists increasingly argued that the vertical structure of the atmosphere held vital clues to large-scale weather trends and there began a concerted effort to study weather in the mountains.[23] Studies intensified in the 1880s with the first permanent high-altitude weather stations. Bigg suggests that the move to semi-permanent high-altitude stations in the late nineteenth century was one of the most distinctive features of mountain science, whereby the mountain was 'colonised' into the scientific landscape on a permanent basis.[24] What did 'high altitude' mean in these debates? In late-nineteenth-century European science, an observatory's situation on a free peak was more valued than height. About 1,000 metres sufficed for the study of thermal inversions. Coen notes that many nineteenth-century commentators prioritised the 'aesthetic ideal of the panorama', where a truly free peak was valued over a higher mountain that was one part of a mountain range.

In the fifth edition of his *Guide to Chamonix* Whymper documented the establishment of two observatories on Mont Blanc: 'one between the Dome du Gouter and the Bosses du Dromadaire at the height of 14,320 feet, and the other upon the Summit'.[25] The first, a wooden chalet built on solid rock, had been established by Joseph Vallot, after he had camped on the summit for three days and taken meteorological observations. The Commune of Chamonix gave its consent to the observatory's erection on condition that a refuge was also built. Materials were carried up the mountain on the backs of porters and guides, and builders camped on the mountain for a week during its construction. Pierre Janssen, the French astronomer, president of the French Academy of Sciences and director of the astrophysical observatory at Meudon, visited Vallot's observatory and was 'struck with the advantages to science which might be expected from working in a pure atmosphere'.[26] Janssen raised the money needed for an observatory on Mont Blanc's summit, which was built in 1891 on compacted snow and ice. The principal instrument was a self-recording 'meteorographe' that recorded various meteorological observations. A large telescope was added in 1896. The building began to show marked signs of subsidence in 1897, whereby the building itself

[22] Hodgkinson, 'Hypsometry and the Aneroid', 465. [23] Coen, 'The Storm Lab', 463.
[24] Bigg et al., 'Introduction', 316. [25] Whymper, *Guide to Chamonix*, p. 66.
[26] Whymper, *Guide to Chamonix*, p. 69.

was turned into an experiment, 'affording a practical demonstration that the snows at the top of Mont Blanc are constantly descending to feed and maintain the glaciers below'.[27] Other observatories were established across the Alps, including the use of a chalet and a hotel on Mont Faulhorn in Switzerland. The Sonnblick Observatory was built on the Hoher Sonnblick in the Austrian Central Alps in 1886, at a height of 3,106 metres.[28] Europe's Alpine observatories functioned as branches of well-established urban institutions or as 'stations in national and international (usually meteorological or geodetic) networks'.[29] The establishment of observatories in remote regions and challenging environments helped to incorporate mountains into national frameworks and stories of national identity. For instance, Coen argues that the Sonnblick Observatory was an important aspect of the Habsburg empire's symbolic projection of itself: a modernist triumph and part of a 'broader climatological project of a panoramic vision of empire in all its natural and cultural diversity'.[30]

Scotland's Mountain Meteorology

In *British Mountaineering*, Benson argued that attitudes to Britain's mountains followed a similar pattern to the Alps: at first treated with horror and terror, but then slowly embraced. This change in attitude gained pace in the early Victorian era, when people began

> to take a more intelligent view of our mountainous districts and to recognize that, far from being frightful and repellent, they offered a holiday ground of rare beauty. First in tens and twenties, and then, with the march of mind, the steamship and the railway, in hundreds, came tourists, and lastly, with cheap fares and increased facilities, the tripper, the bun-bag, and the ginger beer bottle.[31]

British mountains were valued by more than just bun eaters and ginger beer drinkers: climbers and walkers used British mountains as 'a practice ground, a sort of gymnasium, a kindergarten gymnasium it is true', where they could 'learn and practice . . . the art of mountaineering'.[32] Britain's mountains were seen to be particularly valuable outside of the summer Alpine season, when British climbers could hone their skills and maintain their fitness.

[27] Whymper, *Guide to Chamonix*, p. 74.
[28] D. Aubin, 'The Hotel that Became an Observatory'; Coen, 'The Storm Lab'.
[29] Bigg et al., 'Introduction', 316. [30] Coen, 'The Storm Lab', 482.
[31] Benson, *British Mountaineering*, pp. 5–6.
[32] Benson, *British Mountaineering*, p. 7 and p. 13, respectively.

Writing in *British Hills and Mountains*, James H. B. Bell, the leading Scottish mountaineer, described Scotland as 'essentially a country of mountains and every Scot has some degree of interest or contact with them'.[33] Benson described Scotland as the 'happy hunting-ground of the home mountaineer', claiming that 10–12 peaks (rather than the actual 9) exceeded 4,000 feet and more than 250 were over 3,000 feet.[34] The first British mountaineering club was widely recognised to be Glasgow's Cobbler Club, established in 1866 and named after the Cobbler, or Ben Arthur, a mountain at the head of Loch Long. The Cobbler Club was superseded by the Scottish Mountaineering Club. The latter was also established in Glasgow, after William Wilson Naismith wrote a letter to the *Glasgow Herald* newspaper in 1889 proposing the establishment of a Scottish Alpine Club. The Ladies' Scottish Climbing Club formed separately. Scottish mountaineering developed as a distinct entity from Alpinism, while further south the Yorkshire Ramblers' Club formed in 1892, followed by the English Climbers' Club in 1898.[35]

Benson described the shape of many British mountains as 'the section of plum pudding or the cocked hat', where the mountain swells up from one direction and falls precipitously on the other side. He judged Ben Nevis to be of the plum pudding variety.[36] Like Benson, Graham MacPhee deemed Ben Nevis 'the monarch of British mountains'.[37] Accounts of the mountain were rare until the mid-eighteenth century and most early studies were botanical in nature. The Napoleonic Wars encouraged tourism to the Scottish Highlands and nineteenth-century geologists came to explore the region, while Ordnance surveyors conducted work on Ben Nevis in 1867. Ben Nevis was slow to develop as a climbing ground and centre of tourism due to its relative inaccessibility – the West Highland Railway only reached Fort William in 1894. However, on 1 May 1880, William Naismith climbed the mountain, which was covered in the local newspaper with the title 'First Ascent of Ben Nevis – Without Guides'.[38] From the 1880s, Ben Nevis developed as a climbing ground and a draw for tourists, particularly those coming from England. For those more concerned about the views of the mountain rather than the routes up it, there was the potential for disappointment. Ben Nevis was isolated from the surrounding mountains but its neighbours hid the

[33] J. H. B. Bell, 'The Mountains and Hills of Scotland', in J. H. B. Bell, E. F. Bozman and J. Fairfax Blakeborough (eds.), *British Hills and Mountains* (London: B. T. Batsford, 1940), p.1.

[34] Benson, *British Mountaineering*, p. 12. [35] Benson, *British Mountaineering*, p. 7.

[36] Benson, *British Mountaineering*, p. 8 and p. 9, respectively.

[37] G.G. MacPhee (ed.), *Ben Nevis* (Edinburgh: Scottish Mountaineering Club, 1936), p. 2.

[38] K. Crocket, *Ben Nevis: Britain's Highest Mountain* (Glasgow: Gray & Dawson, 1986), p. 18.

Ben from view from most directions, making it difficult for the tourist to get a good impression of its mass and height. MacPhee acknowledged that '[t]he usual impression of Ben Nevis is disappointing, and one hears it described as a shapeless, uninteresting mound, which, but for its altitude, has few attractive features. Yet, save in the wild recesses of the Coolin [sic], Ben Nevis has no rival in the British Isles for the savage grandeur of its rock scenery.'[39] In contrast, Archibald Geikie described the Ben as rising 'majestically' over the surrounding 'mountains and moor, glen and corry, lake and firth'.[40] Writing in *Leisure Hour* in 1894, Whymper said of the north face of the Ben: 'This great face is one of the finest pieces of crag in our country, and it has never been climbed, though every now and then adventurous ones go and look at it with wistful eyes.'[41] Bell suggested that the

> north-east face of Ben Nevis in spring provides the most truly Alpine conditions to be met with in Britain. Every year at Easter time large parties of climbers, most of them English, disport themselves on the ridges and gullies. . . . The old Ben always gives a great welcome to his votaries ... there is always something fresh and fascinating about the mountain. There are new routes to be made, and old ones to be repeated in entirely novel conditions of rocks, snow or weather.

Fort William, a small town on the shore of Loch Linnhe only four miles from Ben Nevis, became a popular centre for mountaineering. A contemporary newspaper article summarised the popularity of the town's Alexandra Hotel for tourists and mountaineers:

> The umbrella stand is innocent of a single umbrella, but is filled instead with a few dozen ice axes, while coils of Alpine climbing rope depend from the hat pegs. The demand for hob nails at the local cobblers to keep the Alpine boots in order is an entirely new departure in the requisites for visitors to Fort-William, while the village tailor has a busy time in repairing the ravages of the Ben Nevis porphyry. About ten years ago such an instrument as an Alpine ice-axe was almost unknown in Scotland.[42]

While Ben Nevis provided sublime views and a domesticated Alpinism to Scotland's tourists and climbers, it also offered an opportunity for Scotland's meteorologists – as the site for a high-altitude weather observatory. The idea was the brainchild of members of the Scottish Meteorological Society. The Scottish Meteorological Society was the outcome of a discussion amongst Scottish landowners and men

[39] MacPhee, *Ben Nevis*, p. 3.
[40] A. Geikie, *The Scenery of Scotland Viewed in Connection with its Physical Geology* (New York: Macmillan, 1901), p. 145.
[41] Quoted in Crocket, *Ben Nevis*, p. 21.
[42] Anon (initials W. T. K), 'Mountaineering on Ben Nevis', no title, date or page, National Records of Scotland, MET1/5/3/7.

of science immediately after a meeting of the Highland Society of Edinburgh in July 1855.[43] The discussion was prompted by Mr Pitt Dundas, registrar-general of Scotland, who requested help in determining the possible relations between weather, health and mortality rates. David Milne-Home, the Edinburgh lawyer, landowner and earthquake observer, was approached. Aided by Sir John Forbes, the Aberdeenshire landowner and agriculturalist, and James Stark, Scotland's first superintendent of statistics at the General Register Office for Scotland (also established in 1855), Milne-Home worked on the establishment of an association for the purpose of establishing stations for meteorological observations. In their prospectus the men acknowledged the isolated position of Scotland's weather observers and the resultant 'chaos of accumulated facts'.[44] Scotland needed its own meteorological society 'to organise and extend local and individual exertions' and ensure that observations could be usefully put to the study of storms and tides, the development of agriculture, safety at sea and sanitation on land.[45] Others threw their weight behind the endeavour: Sir David Brewster, the optics experimentalist, historian of science and principal of the University of St Andrews, noted the meteorological work being done elsewhere in Britain – notably in England – and the need for Scottish meteorologists to complement these endeavours. Sir Thomas Macdougall Brisbane offered the help of the observer, Mr Hogg, at his magnetic observatory at Makerstoun; the naturalist Sir William Jardine, Forbes, Lloyd, Thomas Anderson, the analytic chemist at the University of Glasgow and Henry James at the Ordnance Survey all wrote to provide their support for the scheme, as did the Highland and Agricultural Society of Scotland, the Royal College of Physicians in Edinburgh and various major Scottish landowners. The proposal for a Scottish Meteorological Society was proposed at the British Association's Glasgow meeting in 1855, where it was approved and the offer was made to have its instruments verified at Kew. A quarterly journal was instituted and a grant was applied for from the British government, which gave £150 yearly to Scotland's Astronomer Royal for the examination and reduction of observations from the Society's fifty-five stations. The Society was given rooms in the General Post

[43] The Society operated under a number of names, including the Meteorological Society of Scotland and the Scottish Meteorological Association, before settling on the Scottish Meteorological Society.

[44] Anon, *Prospectus of an Association for Promoting the Observation and Classification of Meteorological Phenomena in Scotland*, 11 July 1855, bound in Pamphlets, National Library of Scotland, Edinburgh, p. 1.

[45] Anon, *Prospectus of an Association*, p. 2.

Office in Edinburgh for this work.[46] By 1865 it had 570 members and 80 observing stations, making it one of the largest and best-organised meteorological societies in Europe.[47]

Early meetings of the Society reported increasing numbers of meteorological stations in Scotland. Their observations were registered on schedules supplied by the Society, transmitted monthly to the office in Edinburgh and then forwarded to Edinburgh's Royal Observatory on Calton Hill for correction and reduction by the Astronomer Royal's staff. An abstract was then produced for the Registrar General's Office for publication in its Tables of Deaths and Diseases. Once that process was complete, the Scottish Meteorological Society received the station returns, complete with corrections and reductions, for its own purposes.[48] It was also possible for observers to bring their instruments to Edinburgh and have them compared to the Society's standards, which was described as 'an object of primary importance'.[49] Part of the Society's assistance to the registrar general's investigations involved the study of the climate of Scotland, which was assumed to affect human health as well as fix the flora and fauna of the country. Members of the Society also assumed that its network of stations, particularly those on the Scottish coasts and islands, could help with the study and prediction of storms – at a meeting in June 1867 Thomas Stevenson speculated about the use of

> special electric wires connecting an observatory on the west coast of Ireland and one on the north of Scotland with that at Kew, so as to show *constantly*, by an automatic appliance, the *differences* of the barometric readings. Those differences of reading might perhaps be announced to the observer by the sound of a bell or some other similar alarm, so that warnings of approaching tempests might be timeously given.[50]

Connecting Scottish, English and Irish meteorological stations in a single network was an abiding concern of the Society, as was the study and prediction of storms. Alexander Buchan brought the two points together in his popular *Handy Book of Meteorology*. Buchan had been the Scottish Meteorological Society's meteorological secretary since 1860

[46] G. Milne-Home, *Biographical Sketch of David Milne-Home* (Edinburgh: David Douglas, 1891).

[47] Anderson, *Predicting the Weather*, p. 238.

[48] Anon, *Report by the Council of the Meteorological Society of Scotland to the General Meeting held on the 18th January 1859, with the Minutes of General Meeting, Laws and Regulations of the Society and List of Office-Bearers for the Year* (Edinburgh: Murray and Gibbons, 1859), p. 6.

[49] Anon, *Report by the Council*, p. 3.

[50] T. Stevenson, 'On Ascertaining the Intensity of Storms by the Calculation of Barometric Gradients', extract from paper read at the General Meeting of the Scottish Meteorological Society, June 1867, National Library of Scotland, Edinburgh, p. 7.

and was appointed the Meteorological Council's Scottish agent and inspector in 1877.[51] He argued that it was

> a duty incumbent on European meteorologists ... to examine, analyse, and carefully study in detail the storms which have traversed Europe during the last few years ... with the view of ascertaining the course storms follow and the causes by which that course is determined, so as to deduce from meteorological phenomena observed not only the certain approach of a storm, but also the particular course it will take in its passage over Europe.[52]

In relation to the British Isles, Buchan suggested that the shortness of time between a storm hitting the west of Ireland and its arrival over Britain demanded the sharing of pressure and wind observations six to eight times a day via the telegraph so that the storm's movements could be followed and timely warnings sent to various ports.[53]

In the mid-1870s, Milne-Home and Thomas Stevenson turned their collective attention to the issue of high-level meteorology. In 1875, Stevenson proposed to obtain 'vertical meteorologic sections of the atmosphere' by establishing stations at the bottom and top of steep mountains.[54] In her biography of David Milne-Home, Grace Milne-Home claimed that her father became interested in the topic after hearing about the United States Signals Corps' station on Pike's Peak, which stood at 14,110 feet above sea level and was the highest station in the world at the time of its erection in 1873.[55] In his Address to the General Meeting of the Scottish Meteorological Society in 1877, Milne-Home extolled the virtues of the American telegraphic network of weather stations. He noted that three of the Scottish stations sat at over 1,000 feet above the sea and mentioned Ben Nevis on the west coast and Ben Macdui in the Cairngorms as possible additions to these sites.[56] He ascended Ben Nevis in October 1878 to judge for himself the site where the observatory could be erected. At a meeting of the Council of the Society in March 1879 it was reported that the Council had applied to

[51] Anon, 'Minutes of Committee on Land Meteorology', Meeting at Meteorological Office, 24 July 1877, National Archives, BJ 8/10.

[52] A. Buchan, *A Handy Book of Meteorology* (Edinburgh: William Blackwood, 1867), p. 9.

[53] Buchan, *Handy Book of Meteorology*, p. 145.

[54] Anon, *Ben Nevis Meteorological Observatory: An Account of its Foundation and Work* (Edinburgh: Directors of the Ben Nevis Observatory, 1885), p. 10.

[55] G. Milne-Home, *Biographical Sketch of David Milne-Home* (Edinburgh: David Douglas, 1891), p. 139; P. Smith, *Weather Pioneers: The Signals Corps Station at Pikes Peak* (Athens: Ohio University Press, 1993).

[56] Anon, 'Address by D. Milne-Home, Chairman of the Council of the Scottish Meteorological Society, to the General Meeting of the Society, held on 26th July 1877', *Journal of the Scottish Meteorological Society*, 5, new series (1877), 110–16, 114; J. Paton, 'Ben Nevis Observatory 1883–1904', in Anon (ed.), *Ben Nevis Observatory* (London: Royal Meteorological Society, 1983), pp. 1–18, p. 2.

Parliament for public funds to erect an observatory and purchase instruments, noting the

> superiority of high-level stations, for many purposes, over those situated at or near the sea. If it be desirable to discover more of the constitution of the Earth's atmosphere, and more of the laws which determine atmospheric movements and changes; and if it be also desirable to obtain early warning of these movements and changes, the higher the station the better.[57]

The rationales put forward to stage an observatory experiment on Ben Nevis were familiar ones to those made elsewhere across Europe. Mountain meteorological observatories would help to track the passage of ridges of high pressure between centres of low pressure, using telegraphy to link them to observatories at lower levels. Mountains also provided the opportunity to study cyclones and anticyclones vertically. Observatories were better than balloons due to their ability to take continuous measurements of air pressure, to operate in bad weather, and to maintain constant height. Mountain stations also worked in the service of storm warnings, again through telegraphic connections to central meteorological offices.[58] Ben Nevis presented some specific advantages to science. It rose from almost sea level to 4,406 feet. It was only four miles horizontally from the town of Fort William, which sat close to sea level on Loch Linnhe. The value of Ben Nevis as a high-level station that rose from sea level rather than from a mountain plateau was constantly reiterated: 'therein lies the great superiority of Ben Nevis over almost every other mountain of equal or greater height in the world', said one commentator in the *Glasgow Herald* newspaper in 1882.[59] The mountain's summit was 2,000–3,000 feet clear above the mountain ridge that lay between it and the Atlantic. Writing in *Nature* in 1881, Buchan noted that the mountain 'raises its head in the very midst of the west-south-westerly winds from the Atlantic, which exercise so preponderating an influence on the meteorology of Europe'.[60] Important results to come were

> those which relate to the greater movements of the atmosphere, particular the upper currents in the relations to the cyclones and anti-cyclones of Europe, the data for the investigation of some of the laws regulating these movements being obtained by a comparison on the one hand of observations made on Ben Nevis with those made at the other high-level stations of Europe, and on the other with

[57] Anon, 'Report of the Council, held on 7th March 1879', *Journal of the Scottish Meteorological Society*, 5, new series (1879), 276–81, 277.

[58] Coen, 'The Storm Lab', 478.

[59] Letter to *Glasgow Herald* from M. MacKenzie, Tuesday September 5, 1882, no page, newspaper clipping in Report to Charles Greaves, National Records of Scotland, MET1/5/3/14.

[60] A. Buchan, 'Meteorology of Ben Nevis', *Nature*, 25 (3 November 1881), 11–13, 11.

those made on lower levels, and published in the different Daily Weather Reports.[61]

Also writing in *Nature*, Hugo Hildebrandsson, director of the Meteorological Observatory at Uppsala, Sweden, and expert on the upper currents of the atmosphere, said that an observatory on Ben Nevis was 'of *the utmost importance* for the development of modern meteorology. No better situation for a mountain observatory can be imagined.'[62] He also noted the position of the peak in the middle track of depressions or storms of northwest Europe and predicted that it would be of great importance for the development of theories of cyclones.

Clement Wragge and Ben Nevis's Summer Observatory

In 1879, the Council of the Society reported that the Meteorological Council in London had promised an annual grant of £100 towards the maintenance of an observatory on Ben Nevis but that the application to the Royal Society for a grant of £400 for the erection of the building had been turned down. The Council of the Scottish Society noted their deep regret at the decision and deferred their plans.[63] Renewed efforts to realise the scheme were made in 1880, when the young English meteorologist Clement Wragge came forward and offered to assist the Society. Wragge promised to organise and participate in daily ascents of Ben Nevis in time to take observations at the summit at 9 am, while simultaneous observations would be taken at close to sea level in Fort William. The rationale for Wragge's work on the Ben was to use the mountain's unique physical geography to contribute to wider investigations in meteorology – to turn the mountain into a natural, experimental, observatory.

Clement Wragge (born William Lindley Wragge but known by his father's name) was born and grew up in the English Midlands and studied law at Lincoln's Inn in London. He had wide-ranging scientific interests, including geology, natural history and meteorology. His parents died when he was young and he was raised by his grandmother and, after her death, his aunt and uncle. He came into his inheritance at the age of twenty-one as well as a legacy from his aunt and he used them to travel widely in Europe, Egypt, the eastern Mediterranean and North America. He decided law was not for him and trained as a midshipman, sailing to Australia in 1876, where he worked in the surveyor-general's department

[61] Buchan, 'Meteorology of Ben Nevis', 11.
[62] Anon, 'Notes', *Nature*, 27 (2 November 1882), 18–20, 18, original emphasis.
[63] Anon, 'Report of the Council held July 21st 1879', *Journal of the Scottish Meteorological Society*, 5, new series (1879), 368–76.

in South Australia and took part in several field surveys. Wragge married Leonora Thornton in 1877 and they had two sons and a daughter, the last born in Scotland. The Wragge family returned to England in 1878, to Farley, Staffordshire, where Wragge had spent some of his upbringing. He began meteorological observations at various heights, including at Oakamoor railway station and Beacon Stoop, a 120-metre climb.[64] Wragge's passion for fieldwork and his willingness to undergo physical exertions and deprivations in the name of science were captured in a report on a North Staffordshire Naturalists' Field Club excursion in the *Staffordshire Sentinel* in 1880:

> [T]he members probably did more real honest work than is customary on every such pleasant occasion, for the leader [Wragge] being in all things energetic, as he is practical and painstaking, not only took them over a wide area, every foot of which he is familiar with, but was most exact and elaborate in his descriptions of those things to which their attention was directed.[65]

Wragge was 'ever ready and anxious, without fee or reward, to impart to others the outcome of his experiences, gained often under the most untoward circumstances'. The article went on to note that Wragge proved 'he belongs to the workers amongst scientists; his whole life and fortune, in fact, being devoted to study'. He wrote energetically about geology, natural history and meteorology for various Victorian publications, including local newspapers, scientific periodicals and religious magazines.

Wragge arrived in Fort William on Saturday 28 May 1881, in time to set up the summit and sea-level observatories and begin observations on 1 June. Instruments were provided by the Scottish Meteorological Society and £450 of expenses were provided to cover two years of operations. A £200 grant from the Meteorological Council of London, £100 from the Royal Society of Edinburgh, £50 from the British Association and £50 from the British Government Research Fund to help reduce the 1881 observations were all offered. The summit observatory was established on 31 May 1881, when Wragge, several local scientific enthusiasts, carpenters and workmen trekked up the mountain through the night in time to arrive at the summit at 7.30 am.[66] Wragge and his helpers carried the barometer,

[64] After his move to Scotland, Wragge made arrangements for James Hall, of Oakamoor, to continue his meteorological observations at Beacon Stoop until 1 July 1881 to complete a year's observations at the station. He also ensured that the local railway station at Oakamoor became a permanent meteorological station. Letters from Clement Wragge to William Marriott, 4 January 1881 and 4 May 1881, National Meteorological Archive, RMS/1/5/13.

[65] Anon, untitled newspaper clipping, *The Staffordshire Sentinel* (4 September 1880), National Records of Scotland, MET1/5/3/14, p. 62.

[66] C. L. Wragge, 'Watching the Weather on Ben Nevis', *Good Words*, 23 (1882), 343–47 and 377–85, 344.

rain gauges and thermometers. A cairn was built to house the Fortin barometer, made by Negretti and Zambra in London for express use on the mountain. Stones were cleared for the Stevenson screen and several rain gauges were also placed across the summit plateau. A small stone hut was also erected to provide the observer some shelter during their time at the summit, with a tarpaulin for a roof. He and his assistants took unbroken daily observations until the great storm of 14 October 1881, which prevented the ascent of the mountain. The instruments were removed on 27 October for the winter. (The second season began officially on 1 June 1882 and continued until 1 November 1882.)

From 1 June, summit observations were taken at 9 am, 9.30 am and 10 am, as well as observations on the way up and down the mountain (Figure 3.1). Eleven sets of observations were made at Fort William by Wragge's wife, from 5 am to 9 pm, many timing with those taken while Wragge ascended and descended the Ben and took readings at the summit. During the first few ascents Wragge had a guide – a local man, Colin Cameron – but soon dispensed with the help except during particularly

Figure 3.1 Photograph of Clement Wragge seated on the summit of Ben Nevis, 1881, taken by Peter MacFarlane. (Source: National Records of Scotland, MET1/5/3/14. Reproduced with permission of National Records of Scotland.)

treacherous weather and occasional winter ascents. He initially left home at 5 am, using a Highland pony and accompanied by his dog, Renzo. He later found the ascent could be done in three hours, so left at 6 am. He used the pony up to 1,900 feet, where it was hobbled. Upon return to Fort William in the afternoon, Wragge despatched a telegram to the Meteorological Office in London and the *Glasgow Herald* newspaper with the 9 am observations, took the low-level observations at 3 pm, had dinner and a pipe and then slept until 9 pm, when he took further observations at the sea-level station. He claimed that his afternoon rest was 'absolutely necessary' because he then worked until midnight 'writing notices for the newspapers'. He took a second sleep until 4 or 5 am, when his day began again.[67] Wragge did four or five ascents a week and Mr William Whyte, his assistant, did the remaining days. Wragge noted that: 'Occasionally, as the posting-up of the observations for the Society was of itself heavy work, I was obliged to send my assistant to the Ben four days in one week; but I made a point of making up for it, and in consequence have sometimes climbed the Ben on eight or nine days in direct succession.'[68] Wragge was bound by the Scottish Meteorological Society not to share any of his observations apart from the 9 am readings. Due to his demanding schedule these were the only observations that Wragge reduced. The 1882 season also ran from 1 June until the end of October. During this season, Wragge improved the stations along his route to the summit, with Stevenson screens added at a number of intermediate sites, arranged on the scheme proposed by Thomas Stevenson in 1875.[69] Wragge noted that the work in 1882 was much heavier than 1881, with larger sets of observations along the route: at Achintore, Peat Moss, The Boulder, The Lake, Brown's Well, Red Burn Crossing, Buchan's Well, and the summit. 'The great value of the intermediate observations is,' Wragge argued, 'that they enable disturbances in the varied stratum of atmosphere between Ben Nevis and Fort William to be localised and examined in discussion. We hope largely to increase the value of forecasts' (Figure 3.2).

Wragge provided accounts of his work in a daily 'observation-book', one of which covered the Ben Nevis observations, and another the low-level readings taken at Achintore, Fort William.[70] These logs recorded

[67] Wragge, 'Watching the Weather on Ben Nevis', 380.

[68] C. L. Wragge, 'Ascending Ben Nevis in Winter', *Chambers's Journal of Popular Literature, Science and Art*, 19 (1882), 265–8, 265.

[69] A. Buchan, 'The Meteorology of Ben Nevis', *Transactions of the Royal Society of Edinburgh*, 34 (1890), xvii–xxi, xix.

[70] C. L. Wragge, Ben Nevis Meteorological Observatory 'Rough' Observation-book No. 11, 12 June 1881, National Records of Scotland, MET1/5/2/2/1.

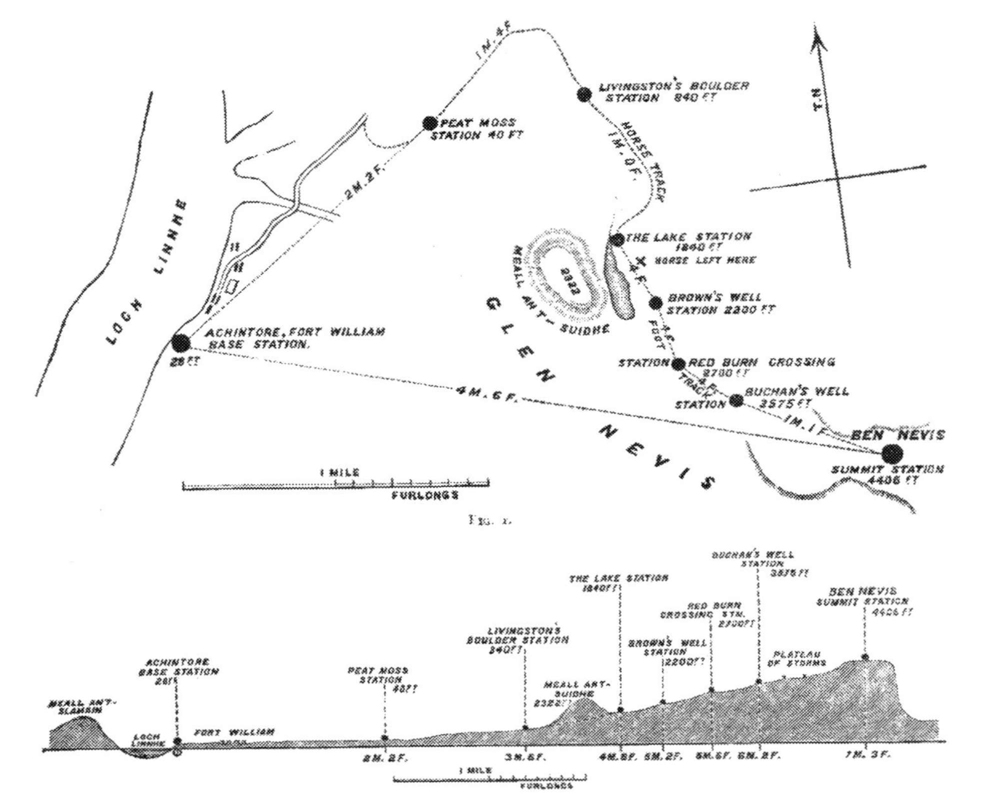

Figure 3.2 Map and cross section of Clement Wragge's route to the summit of Ben Nevis. (Source: C. L. Wragge, 'Ben Nevis Observatory', *Nature*, 27 (1883), 487–491, 488.)

numerical observations taken by instruments, eye observations, documentations of incidents during ascents and descents, as well as brief descriptions of landscapes and views. Although Wragge referred to the books as 'rough' accounts, they were clearly compiled at home rather than on the mountain itself. Wragge would note, for instance, where he couldn't decipher original observations or where the daily record had been written up by his assistant from Wragge's notes. Wragge provided more lengthy, discursive reports for various Scottish newspapers, scientific periodicals – including *Nature* and *Symons's Monthly Meteorological Magazine* – and popular magazines like *Good Words* and *Chambers' Journal of Popular Literature, Science and Art*. Wragge was keen to promote his work in the press. In July 1881, he brought up a reporter from the *Inverness Advertiser*, although the latter failed to reach the summit.[71] On 26 August 1881, he took a *Times* newspaper journalist up so he could report on Wragge's work, noting in the observation book that they 'had to crawl along in parts owing to the terrific force of the wind . . . Sleet pelting heavily . . . Bitterly raw + cold'.[72] On another occasion Wragge invited a local photographer, Peter MacFarlane, to accompany him. MacFarlane took a number of photographs of Wragge working on the summit[73] (Figures 3.1 and 3.3). The photographs were converted into wood-engraved illustrations and incorporated into many of Wragge's publications.[74] When a British Meteorological Society inspector, Charles Woolnough, visited Fort William in September 1882 to review Wragge's 'meteorological system', Wragge was said to be 'highly delighted at the interest' in his work and presented Woolnough with copies of his articles in *Good Words*, *Chambers' Journal* and *Nature*. Wragge also claimed that Buchan's report on the Ben Nevis observations in *Nature* in November 1881 was in large part written and the sketches provided by him, even though it was signed by Buchan and 'purports to be that gentleman's inspection of the Ben Nevis system'.[75]

Wragge's accounts of his work on the Ben tended to mimic wider conventions employed by Victorian Alpinists and explorers, mechanically and discursively. Similar to contemporary travel writers, Wragge's published accounts were the products of several iterations of inscription

[71] Wragge, Ben Nevis 'Rough' Observation-book No. 5, 3 July 1881, National Records of Scotland, MET1/5/2/2/1.
[72] Wragge, Ben Nevis 'Rough' Observation-book No. 12, 26 August 1881, National Records of Scotland, MET1/5/2/2/1.
[73] Wragge, Ben Nevis 'Rough' Observation-book No. 17, 1 October 1881, National Records of Scotland, MET1/5/2/2/1.
[74] MacFarlane in turn sent a set of his photographs to Queen Victoria, claiming them to be the first and only photographs that had been taken on the top of the mountain.
[75] Woolnough, National Records of Scotland, MET1/5/3/14, p. 4 and p. 9, respectively.

Figure 3.3 Photography of Clement Wragge standing on the summit of Ben Nevis, 1881, taken by Peter MacFarlane. (Source: Wragge, 'Ben Nevis Observatory', 489.)

practices, from rough notes kept on the mountain to the daily observation book compiled in the evenings and the tabular accounts dispatched to the newspapers and the Meteorological Society.[76] Wragge was keen to draw attention to his expeditionary experience. In his interviews with the inspector from the Meteorological Society, Wragge emphasised his time overseas and linked that to his physical persona. Woolnough noted:

> I was astonished to hear of Mr. Wragge's determination and nerve and his almost unbounded mental and physical energy. He is a young thin and somewhat delicate looking man (about 30) and was when I first saw him dressed in a waterproof with an undercoat of a decidedly nautical pattern with the accompanying buttons – I understand he has acquired this hardihood through being so much at Sea and various travels in Australia, Africa and the Holy Land.[77]

Like the Arctic explorer, Wragge was keen on naming particular features of his terrain. A spring was renamed Buchan's Well in August 1881 in

[76] I. Keighren and C. W. J. Withers, 'The Spectacular and the Sacred: Narrating Landscape in Works of Travel', *Cultural Geographies*, 19 (2012), 11–30, 12.
[77] Woolnough, National Records of Scotland, MET1/5/3/14, p. 8.

memory of Alexander Buchan's visit and inspection on 28 July 1881.[78] In early September 1881, he gave the plateau between Buchan's Well and the summit the dramatic title 'The Plateau of Storms' due to the north-easterly gales that had made it almost impossible to cross when the *Times* reporter had accompanied him.[79] Wragge's popular accounts often focused on individual ascents of the Ben and narrated them as thrilling adventures which emphasised displays of athleticism, physicality and bodily suffering, the picturesque and sublime nature of the landscape and the precision and reliability of his observations. Wragge's reports were similar to those of the so-called new mountaineer of the late-Victorian period, whose accounts promoted what McNee calls the 'haptic sublime', that is an 'emphasis on direct physical experience and embodied understanding of mountain landscapes'.[80] He frequently drew attention to his own bodily experiences and suffering while ascending and descending the mountain and observing the weather on the summit. He noted:

> [M]y under-clothing was chiefly of thick lamb's-wool, with a sailor's jersey and my oldest suit over all. Very seldom did I carry an overcoat, and never a waterproof or gloves. Hence by my own negligence I was often wet through for many hours, with my hands swollen by the cold and biting wind, so that I could only just legibly scrawl down the observations holding the pencil in my clenched fist; and yet I preferred all this to encumbering myself with an overcoat or waterproof.[81]

Cold hands also made the handling of keys and instruments 'difficult matters'.[82] When Wragge was forced to make the journey to the summit in quick time, he would often 'reach the summit streaming with perspiration; and, with a mean temperature of three or four degrees above freezing, and enveloped in a thick rain-cloud, I have remained there nearly two hours; and yet I experienced no ill effects beyond a passing sense of fatigue in the muscles, and my general health remained excellent throughout'.[83]

The rough terrain and the weather were narrated as significant challenges to be overcome during journeys up and down the mountain. The ruts and swamps were 'very treacherous and deep'.[84] When a depression

[78] Wragge, Ben Nevis 'Rough' Observation-book No. 9, 1 August 1881, National Records of Scotland, MET1/5/2/2/1.
[79] Wragge, Ben Nevis 'Rough' Observation-book No. 13, 1 September 1881, National Records of Scotland, MET1/5/2/2/1, original emphasis.
[80] McNee, *The New Mountaineer*, p. 4.
[81] Wragge, 'Watching the Weather on Ben Nevis', 379.
[82] C. L. Wragge, 'Resumption of the Ben Nevis Meteorological Observatory', *Meteorological Magazine*, 17 (1882), 81–4, 84.
[83] Wragge, 'Watching the Weather on Ben Nevis', 381.
[84] Wragge, 'Resumption of the Ben Nevis Meteorological Observatory', 84.

dominated, the gales on the upper slopes forced Wragge to 'fight every foot of my way across this plateau, crawling along, and pushing on from boulder to boulder, to obtain breathing-time, and some little shelter under their friendly lee'.[85] Snow was a year-round problem for travel and navigation but allowed Wragge to draw connections between his own experiences and those of the Arctic explorer. He noted for instance that the '[s]now lay deeply during the better part of June in great shelving masses in this ravine; and I had to plod my way across it as by a series of steps, giving a peculiarly Arctic zest to my experiences'.[86] On 5 June 1881, a depression centre crossed Ben Nevis and Wragge reached the summit late. He reported:

> [A] blinding snow-squall came on, cloud-fog enveloped the mountain, the wretched track (hardly discernible in clear weather, and with which during the first few days I was not very well acquainted,) was now covered with snow, and I took two hours in groping my way through the fog to the summit, building cairns in true Arctic fashion as I pushed along.[87]

During a winter ascent in late March 1882, accompanied by Colin Cameron, Colin Livingston, the headmaster at Fort William's Public School, and a visiting friend, Wragge reported 'vast shelving masses of snow many feet deep. It was evident that great difficulties were before us; and before pushing on, we determined to rest and refresh ourselves with a snack of luncheon; for the pure mountain air – though bitterly cold and raw – was yet most exhilarating and our appetites had become keen.'[88] Despite having left home at 5.40 am, the men only reached the summit at 10.40 am and returned to Fort William at 3 pm 'in a pitiable plight, but nothing the worse for our adventures'.[89] The one occasion when the terrain and the weather got the better of Wragge was during the storm of 14 October 1881. Despite forewarning that a dangerous storm centre was approaching from the south-west, Wragge set off for the summit with Colin Cameron. The two men managed to reach 2,100 feet above sea level before admitting that it was impossible to proceed further and that 'any further attempt would have been courting destruction'. By then their clothes 'were hard frozen and coated with ice and ice lumps like large eggs had formed in our beards'. They headed back down the mountain arm in arm, falling every few paces and sheltering behind boulders 'for breath and shelter from the tearing drifts'. Upon reaching Fort William Wragge

[85] Wragge, 'Ascending Ben Nevis in Winter', 267.
[86] Wragge, 'Watching the Weather on Ben Nevis', 378.
[87] Wragge, Watching the Weather on Ben Nevis', 379.
[88] Wragge, 'Ascending Ben Nevis in Winter', 267.
[89] Wragge, 'Ascending Ben Nevis in Winter', 268.

recorded that Cameron 'declared it was the very worst weather he had ever experienced and in all my experience at home and abroad, by land & sea, "up north" and "down south" in the "roaring forties" I have never seen weather equalling this in fury'.[90]

The mountain's boulder fields, cliff faces, bogs and ravines were not always narrated only as a stage for the performance of 'manly values of adventure and sport'.[91] Like Forbes in the Alps, Wragge exploited the 'picturesque and sublime character of his subject' in his popular accounts.[92] In one of his articles in *Good Words*, Wragge claimed:

No amount of writing can adequately describe the grandeur of the view from the summit of Ben Nevis on a clear day. The rough old platform of sombre grey rock on which we stand contrasting with the bright plots of snow gleaming in dazzling whiteness in the blazing sun, the awful abysses of the precipices thrown out in greater majesty by the huge black shadow of some opposite crag, the adjacent dykes, sombre moors, ancient valleys, deep glens, blustering torrents in the distance below – surrounded by the noble display of mountains, range behind range, peak behind peak, bounded on the west by the lochs and sea, with the whole capped by the clear blue vault above – all impress the mind profoundly with a sense of the majesty of Nature.[93]

This style of landscape writing was not confined to the pages of popular magazines. Wragge occasionally included notes on the mountain's sublime and picturesque views in his observation books. For instance, a few minutes before his 9 am readings on 8 June 1881 he noted:

Above the dark uniform loose cumulus cloud is a canopy as of dark dirty 'gloom'; below, adjacent, appear the lesser mountains – in shape reminding me of the lunar wall craters now with their shelving mossy slopes, streaked here and there with snow, catching a gleam of sunlight (but soon to come out in shadowed contrast) while, in the far distance the pale blue mountain peaks and cones tower upwards towards the hanging dirty loose cumulus 'stuff' before mentioned.[94]

On another occasion near the end of his year's observations, he wrote:

[C]onditions of the atmosphere majestically grand ... and cloud having a dark inky blue colour, its appearance being almost awful. ... About 9.33 [am] 'cloud-fog' and 10 [denoting 'whole sky entirely overcast'[95]] noted; but the fog lifted

[90] Wragge, Ben Nevis 'Rough' Observation-book No. 18, 14 October 1881, National Records of Scotland, MET1/5/2/2/1.
[91] Hevly, 'Heroic Science of Glacier Motion', 86.
[92] Hevly, 'Heroic Science of Glacier Motion', 70.
[93] Wragge, 'Watching the weather on Ben Nevis', 346.
[94] Wragge, Ben Nevis 'Rough' Observation-book No. 2, 8 June 1881, National Records of Scotland, MET1/5/2/2/1, original emphases.
[95] Wragge, Ben Nevis 'Rough' Observation-book No. 1, 1–7 June 1881, Appendix, National Records of Scotland, MET1/5/2/2/1, original emphases.

occasionally about this hour showing snow-capped mountains, sun-lit slopes and Loch below, with dirty loose 'cloud-stuff' in squalls and 'streamers' overhanging, forming a magnificent picture.[96]

Writing about the mid-Victorian Alpine glaciologists, Hevly argues that their scientific measurements gained force from their 'links to the aesthetics of nature, the sublime or the picturesque elements of natural settings, appreciated via the exertions associated with the manly values of adventure and sport'.[97] Wragge was doing something similar when he combined descriptions of his own bodily trials and moments of landscape appreciation with accounts of his meteorological observations. Wragge's willingness to put up with significant discomfort and to place himself in risky situations on the mountainside were meant to enhance his authority as an adventurous eyewitness. He claimed, in other words, to provide 'reliable perception on the basis of authentic, rigorous, manly experience'.[98] Wragge further legitimised his work on the Ben by repeatedly emphasising observation as laborious but regimented work. In the most general terms he noted: 'The work is very heavy, but well under control, and punctuality and method will carry it through'; that today was 'A heavy day's work'; and 'Great labour was expended today.'[99] He described the establishment of instruments on the summit in May 1881 as 'indeed toilsome work, owing to our being encumbered with the delicate instruments'.[100] Turning the mountainside into a site of observation demanded that the high standards of an observatory routine had to be met, regardless of the challenges presented by terrain or weather. Reaching the summit for the 9 am observations was given the upmost priority in this regard, while 'the outward and homeward observations are but of secondary importance ... altho' of course taken with all possible accuracy'.[101] The establishment of Stevenson screens at intermediate stations up the mountainside in 1882 improved this situation, removing the need to carry instruments or retrieve them from their hiding places: 'punctuality ensured, and accuracy also. The entire observing system goes like clockwork.'[102] In 1882, Wragge asked Woolnough, the Meteorological Society inspector, what he thought of the Ben Nevis

[96] Wragge, Ben Nevis 'Rough' Observation-book No. 18, 12 October 1881, National Records of Scotland, MET1/5/2/2/1, original emphases.
[97] Hevly, 'Heroic Science of Glacier Motion', 86.
[98] Hevly, 'Heroic Science of Glacier Motion', 68.
[99] Wragge, 'Resumption of the Ben Nevis Meteorological Observatory', 83, 82 and 82, respectively.
[100] Wragge, 'Watching the Weather', 345.
[101] Wragge, Ben Nevis 'Rough' Observation-book No. 3, 16 June 1881, National Records of Scotland, MET1/5/2/2/1.
[102] Wragge, 'Resumption of the Ben Nevis Meteorological Observatory', 84.

system. Woolnough 'remarked that I thought the system worked like clockwork and that I was very pleased with the great pains taken to render the observations correct in every way. Any time lost was always recorded.'[103]

Wragge concluded his two seasons' work in terms that weighed his personal suffering against that of the scientific gains:

Whatever hardships I have endured – and I delight in an active, open-air life – were self-imposed; and I have been well repaid by the stimulating knowledge that I was working under the auspices of a Society that appreciated my labours and so cordially seconded my efforts, and that I have been of some service in the cause of physical research. Regular winter observations on Ben Nevis would, I am convinced, prove of immense value to the country in the matter of weather forecasting; but these cannot be insured until an observatory-house has been erected there.[104]

At a public meeting in Glasgow's Council Chambers on Wednesday 14 February 1883 a group of Scottish industrialists, academics and civic leaders gathered to discuss the renewed proposal to build just such an observatory house on the top of Scotland's highest mountain. The history of European high-level observatories was summarised, with 'some feeling of shame' noted with regard to the lack of work done in Britain on high-level meteorology. It was assumed that unlike in other countries – notably France – the British government was not going to assist in this endeavour, so appeals were made to the people of Glasgow, and especially to its mercantile and shipping interests and its agriculturalists. Glasgow was being approached as the largest city and the richest community in Scotland and because the Scottish Meteorological Society originated there. William Thomson, Professor of Natural Philosophy at Glasgow University, drew attention to Wragge's work over the last two years, which he said had been conducted 'with great skill, endurance, and enthusiasm. (Applause.) That seemed to him (Sir Thomson) a stronger testimony than any other consideration that could be offered as to the importance of such a work.' Thomson noted: 'It appeared to him that Mr Wragge's perseverance during two seasons on the top of Ben Nevis – the great amount of labour and personal endurance and self-sacrifice he had given – on the one hand, and the hearty way in which the Scottish Meteorological Society on the other, cognisant of his labours, had taken up the subject, showed how it ought to be appreciated. (Applause.)'[105]

[103] Woolnough, National Records of Scotland, MET1/5/3/14, 33.

[104] Wragge, 'Watching the Weather on Ben Nevis', 385.

[105] Anon, 'The Proposed Observatory on Ben Nevis', *Glasgow Herald* (15 February 1883), 75–6, all quotes from 75.

Building the Ben Nevis Observatory

The Scottish Meteorological Society launched an appeal for funding for a permanent observatory in early 1883 and had raised £5,000 by October that year. Subscriptions varied from almost £200 to a halfpenny, the subscribers 'representing all ranks, from Her Majesty downwards'.[106] The majority of subscribers came from Glasgow and Edinburgh but others were from elsewhere in Scotland, England and Ireland.[107] A Building Committee was established, consisting of John Murray, director of the *Challenger* Expedition Commission, David Milne-Home, Thomas Stevenson, James Sanderson, Alexander Buchan, the Deputy Inspector-General of Hospitals, and Arthur Mitchell, Commissioner in Lunacy.[108] Building of the observatory and the road to it began during the summer months, before construction was prevented by poor weather in the autumn. In the meantime, the daily summer observations continued, beginning on 1 June and stopping on 31 October 1883. The summit observations were taken by Wragge's former assistants, Messrs William Whyte and Angus Rankin, due to Wragge's planned visit to Australia in the autumn of 1883. Colin Livingston took five sets of eye observations at the Public School in Fort William from 8 am to 10 pm.[109]

A Board of Directors was established to manage operations of the new observatory, made up of officer-bearers of the Scottish Meteorological Society; Professors Peter G. Tait and George Chrystal – representatives of the Royal Society of Edinburgh; Sir William Thomson, representative of the Royal Society of London; and Dr Muirhead, representative of the Philosophical Society of Glasgow. In late July Wragge wrote to the Meteorological Society to offer himself as Observatory superintendent. Another was received from Robert T. Omond, who was employed by Tait at the University of Edinburgh to work on the corrections to the *Challenger*'s deep-sea thermometer readings. After advertisements for the post were placed in the *Scotsman* and *Glasgow Herald* newspapers and in *Nature*, seventeen further applications were received and the post was

[106] A. Buchan, 'The Meteorology of Ben Nevis', Special Issue of *Transactions of the Royal Society of Edinburgh*, 34 (1890), i–lxiv and 1–406, xix.

[107] Anon, 'Subscriptions in aid of Ben Nevis Observatory', *Transactions of the Royal Society of Edinburgh*, 34 (1890), ix–xvi.

[108] A. Mitchell, J. Sanderson and J. Murray, 'Introductory Note by the Council of the Scottish Meteorological Society', *Transactions of the Royal Society of Edinburgh*, 34 (1890), iii–v, iii.

[109] R. T. Omond, 'Abstract of Paper on a Comparison of Observations at the Observatory and at the Public School, Fort-William', *Transactions of the Royal Society of Edinburgh*, 42 (1892), 537–40, 537.

offered to Omond.[110] The salary was a very modest £100 a year, which Omond supplemented with his own money.[111] He held the post until 1895, retiring due to ill health, although he continued as honorary superintendent.[112] He was to be assisted by Angus Rankin and James Miller. Wragge was disappointed not to be offered the post and left for Australia soon afterwards, where he continued to pursue a career in meteorology, appointed government meteorologist for Queensland in 1887. He went on to establish high-level observatories in Australia – on Mount Wellington in Tasmania[113] and on Mount Kosciuszko in the Australian Alps, New South Wales, where he also established a corresponding low-level station at Merimbula.[114]

The western side of Ben Nevis was part of the Fassifern and Callart estates, which were the property of Mrs Cameron Campbell of Monzie. Although Campbell was keen to help the Scottish Meteorological Society, difficulties arose in her powers to grant a conveyance. However, Lord Rosebery and the Lord Advocate inserted a clause in the 1882 Entail (Scotland) Bill, then passing through Parliament, which authorised entailed proprietors to grant a feu – a form of land tenure in Scotland – not exceeding three acres, for public or scientific utility. A feudal title of one acre on the summit was granted to the Council of the Royal Society of Edinburgh, which, being incorporated by Royal Charter, was entitled to hold heritable property for scientific purposes.[115] The feu of a path from the Achintee farm, Glen Nevis, to the summit was also granted.[116] The building of the bridle path began before the observatory, following a route suggested by Colin Livingston. Six feet wide and with a gradient of no more than one in five, the path was said to make the trip from Fort William to the summit a three-hour journey, 'without much effort'.[117] It cost almost £800 to build. Once the path was completed, construction began on the observatory. The walls of the observatory – up to ten feet

[110] Anon, 'Minutes of Meetings', 1 August 1883, National Records of Scotland, MET1/5/3/11, pp. 2–3. Wragge's young family was apparently one of the reasons he was not offered the post, although presumably Tait's patronage of Omond and Omond's expertise in data management were also crucial factors in the latter's employment.

[111] By way of comparison, Alexander Buchan was paid an annual salary of £150 for his role as Scottish inspector and agent of the Meteorological Council, which was additional to his salary from the Scottish Meteorological Society. Anon, Minutes of Committee on Land Meteorology, 24 July 1877, National Archives, BJ 8/10.

[112] Paton, 'Ben Nevis Observatory', 15.

[113] Letter from Wragge to Royal Meteorological Society, 15 April 1897, National Meteorological Archive, RMS/1/5/13.

[114] Letter from Wragge to Royal Meteorological Society, 6 December 1901, National Meteorological Archive, RMS/1/5/13.

[115] J. Paton, 'Ben Nevis Observatory', 5.

[116] Anon, *Ben Nevis Meteorological Observatory*, 13.

[117] Anon, *Ben Nevis Meteorological Observatory*, p. 13.

thick at the base – were constructed of unmortared blocks of stone taken from the summit plateau. All other materials were carried up from Fort William on the backs of Highland ponies, while the workmen camped in tents provided by the War Office.[118] The stone walls were double-lined with wood and felting. The single-storey building had a small living area and office, three small bedrooms, store rooms and a coal cellar. The flat roof was covered in lead and snow boarding. The cost of building the observatory came to £2,620.93. An old, heavily reinforced marine telegraph cable was laid on a ridge of stones to connect the observatory with Fort William's Post Office.[119] An annual rent of £357 was paid to the Post Office for the cable rental.[120] Various events took place in October 1883 to mark the opening of the observatory, including a procession to the summit, a public dinner and the presentation of a silver key to the observatory to the landowner.[121] The Meteorological Council's earlier offer of an annual maintenance grant of £100 was taken up by the observatory directors.

The stormy and cold winter of 1883–84 exposed some of the building's inadequacies (Figure 3.4). The living area had proved to be far too small to accommodate the tasks of reducing observations, testing instruments, as well as cooking, eating and washing. Deep snow drifts blocked the door and extreme winds prevented the observers from leaving the building, which left gaps in the meteorological record. For instance, on 26 January 1884 it was noted in the logbook that

> at 1pm Mr Omond and Mr Rankin went out tied together, but found it impossible to go further than the end of the snow porch with safety: at 4 pm they got as far as the [thermometer] box but could not see instruments as the drift blew up into their faces. No temperature observations were taken till 10pm when wind moderated & observations were resumed.[122]

A tower was added to the building in the summer of 1884, with an emergency thermometer screen attached to it. The stems of the thermometers projected through holes in the walls, allowing temperature readings to be taken without leaving the building. A separate laboratory and two additional bedrooms were also added. These additions placed

[118] Anon, 'Minutes of Meetings', 6 September 1883, National Records of Scotland, MET1/5/3/11, p. 9.
[119] Anon, 'Minutes of Meetings', 15 November 1883, National Records of Scotland, MET1/5/3/11, p. 14.
[120] Paton, 'Ben Nevis Observatory'.
[121] Anon, 'Minutes of Meetings', 10 October 1883 and 15 November 1883, National Records of Scotland, MET1/5/3/11, p. 12 and pp. 13–14, respectively.
[122] Anon, Ben Nevis Observatory Log Book, No. 1, 1883–1886, National Records of Scotland, MET1/5/2/3/1, p. 32.

Figure 3.4 Clearing ice off the instruments on the observatory tower. (Source: Anon, *Ben Nevis Meteorological Observatory*, facing p. 22.)

unwanted pressures on observatory staff: 'The spectacle of twenty or thirty shaggy Highland ponies making their daily trip up the hill loaded with lead for the roof, or trailing heavy beams and bundles of planks may have been picturesque in the eyes of the casual tourist, but was somewhat trying to the tempers of those in charge'[123] (Figure 3.5).

The directors laid out a formal scheme for the observations to be taken at the summit observatory. Eye observations were to be taken of the barometer, dry and wet bulb thermometers and of the state of the wind and cloud cover, on an hourly basis. Rain and snow was to be measured every twelve hours. Ozone measurements and spectroscopic observations were also to be taken regularly. Automatic barometric, dry and wet bulb, black bulb and sunshine observations were also taken. In an appendix to the observatory's first logbook, a full catalogue of instruments was provided.[124] There were two barometers by Negretti and Zambra;

[123] Anon, *Ben Nevis Meteorological Observatory*, p. 18.
[124] Anon, 'Appendix I: Catalogue of Instruments, 8 April 1886', MET1/5/2/3/1, pp. 245–8. On the relations between instruments and observatory buildings see Higgitt, 'A British National Observatory'.

Figure 3.5 The observatory during the summer, with visitors. (Source: Anon, *Guide to Ben Nevis: With an Account of the Foundation and Work of the Meteorological Observatory* (Edinburgh: Directors of the Ben Nevis Observatory, 1893), p. 6.)

a large number of ordinary and maximum and minimum thermometers by Negretti and Zambra and Adie and Wedderburn, some of which had Kew certificates. The observatory also had an aneroid barograph and thermograph, various hygrometers, rain gauges of several sizes, a number of anemometers of different designs, louvred instrument screens of different sizes and patterns and a sunshine recorder. The catalogue also recorded surveying equipment, including measuring rods, snow posts, a sextant, compasses, a spirit level and sights, as well as more unusual instruments, including a pocket rainband spectroscope, a stephanome and fittings for measuring fogbows and halos, ozone test papers and scales and a Thomson galvanometer, on loan from Tait. The latter was presumably for experimental work with the marine telegraph cable. The peculiar climatic conditions on the summit of the mountain tested the instruments to their limits and prevented the use of self-registering instruments during much of the year. The high winds, often above 100 miles an hour, affected the registering thermometers. Drifting snow clogged up and buried the thermometer screens. Fog crystals encrusted instruments, including the anemometers and the screen louvres. Solutions included mounting screens on ladder-like stands so that they could be adjusted to keep a uniform height above the snow. A frozen screen would also be brought into the observatory to thaw and be replaced by a clear one. The Robinson hemispherical cup anemometer was declared useless in the winter months and human observation was relied on instead. The demanding atmospheric conditions and height did make the observatory an ideal site to test the efficacy of meteorological instruments. For instance, Edward Whymper visited the summit in 1892 to test various pocket aneroid barometers.[125]

The observational regimen was divided into a number of watches, shared between the two observers. The first watch started at 3 am and ran until the 11 am observations. The second watch covered 12 until 2 pm, the third from 3 pm to 5 pm and the fourth from 6 pm until 2 am.[126] The third member of the observatory, the cook, kept a normal clock and divided his time between the two observers to reduce their solitude to a minimum. He also assisted with observations when required. It was estimated that only five to ten minutes were needed to take all the hourly observations, although various accounts emphasised the hardships and challenges endured by the observers during that short time. In carrying out their rounds, observers were compared to Arctic explorers or seafarers. One newspaper account provided an inventory of the observer's 'semi-Arctic attire', which included an old suit, long sea boots, a thick reefer jacket, a complete suit of oilskins, a muffler, woollen gloves and sou'wester hat, suggesting that the figure resembled 'a voyager in Polar seas'.[127] Although observatory staff often pointed out the significant differences between the weather on Ben Nevis and conditions experienced in the polar regions, the observatory was used as proxy for that environment. Robert Falcon Scott was granted permission to overwinter at the observatory in 1900–01 so that he could test the equipment for his National Antarctic Expedition, but was eventually unable to go and Lieutenant Charles Royds went in his stead.[128] William Speirs Bruce served as an observer on Ben Nevis before he led the Scottish National Antarctic Expedition. He viewed his time there as good training in running a polar meteorological station. Two of the four scientists on the Antarctic expedition – Robert Mossman and David Wilton –

[125] Smith, *Shadow of the Matterhorn*, p. 248.

[126] A. Drysdale, 'Some Experiences of High Life IV', *The Ilkley College Times* (no date), National Records of Scotland, MET1/5/3/7, p. 6. It appears that this schedule was an amendment on an earlier practice of an 8, 4, 4, 8 hour watch scheme described in the official account of the Observatory: Anon, *Account of the Foundation and Work at the Ben Nevis Observatory*, p. 16.

[127] Anon, 'Life and Work in Ben Nevis Observatory', newspaper clipping (no title, date, or pages), National Records of Scotland, MET1/5/3/7.

[128] Paton, 'Ben Nevis Observatory'.

had been observers on Ben Nevis. When the expedition established a meteorological observatory on Laurie Island, South Orkney, they based its design on Ben Nevis and named it Omond House.[129]

Alexander Drysdale, one of the relief observers, said that after taking readings in the winter, the observer would come back into the building, heeding

> not the howling of the wind and the rattle of the drift on the chimney, as with heavy garments doffed, knee-boots kicked off, and fine snow-drift shaken from the innermost recesses of his clothing, he takes his comfortable chair before the fire, the fine glow of combat and sense of moral rectitude making him the happiest as well as the loftiest inhabitant of the British Islands.[130]

Work was not over for the observer at that point. His other tasks included filling up the daily record sheet and drawing up daily and monthly averages of the readings of the instruments and checking the results. A summary of the observations was telegraphed to Fort William in the evening, for transmission to the newspapers and the Scottish Meteorological Society's Edinburgh office. Other activities included general maintenance and repairs, clearing away snow from windows and doors, and disposing of refuse, which was done by hurling it over the precipitous cliff that fringed the plateaux, nicknamed 'Tin-can Gully'.[131] Personnel would get exercise from digging snow holes, tobogganing, snowshoeing and Norwegian skiing, and fill any remaining free time playing cards and musical instruments or reading.[132] As well as a range of well-known meteorological texts and periodicals, the observatory library included a few texts about Scottish topography and local history, as well as the Complete Works of Shakespeare and poetry by John Keats.[133] Breakfast was taken at 9 am, dinner at 2 pm – shared by all observatory personnel – tea at 6 pm and supper was at 10 pm. In similar manner to polar expeditionary camps and Arctic voyages of exploration, the Ben Nevis Observatory held in tension a set of activities conducted in a treacherous external environment and in a highly domesticated interior (Figure 3.6).

[129] G. N. Swinney, 'William Speirs Bruce, the Ben Nevis Observatory, and Antarctic Meteorology', *Scottish Geographical Journal*, 118 (2002), 263–82, 274–5.

[130] A. Drysdale, 'Some Experiences of "High" Life; or, "Observations" on Ben Nevis Part VII', National Records of Scotland, MET1/5/3/7, p. 51.

[131] A. Drysdale, 'Life at Ben Nevis Observatory: Notes for an illustrated lecture given by Alexander Drysdale in Dollar, National Records of Scotland, MET1/5/3/7, p. 5.

[132] Anon, 'Life and Work in Ben Nevis Observatory', *The Scotsman*, no date or page, Ben Nevis Observatory *The Scotsman* Clippings, National Records of Scotland, MET1/5/3/3.

[133] Anon, Appendix II Catalogue of Books and Pamphlets, National Records of Scotland, MET1/5/2/3/1, pp. 249–50.

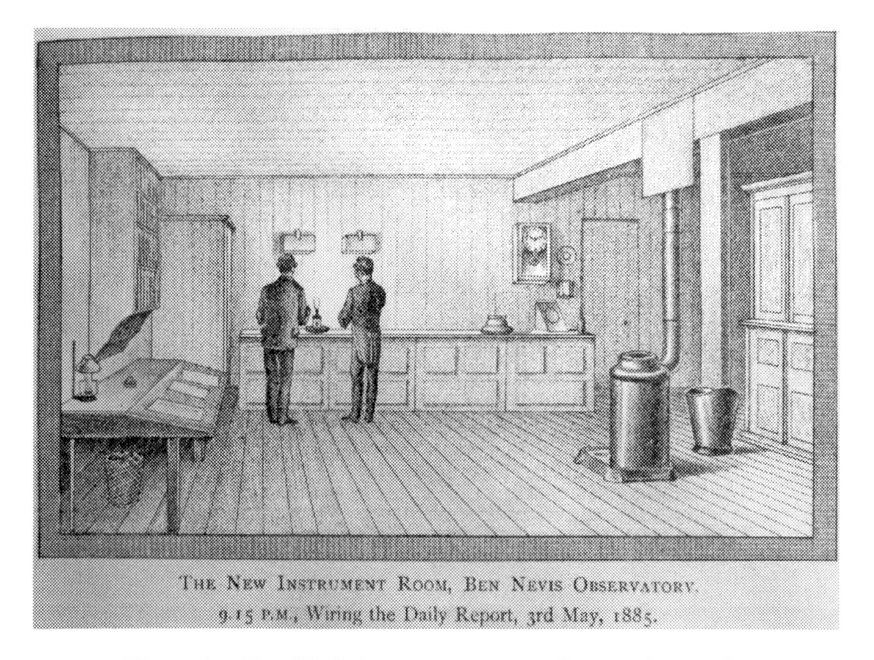

The New Instrument Room, Ben Nevis Observatory.
9.15 P.M., Wiring the Daily Report, 3rd May, 1885.

Figure 3.6 Ben Nevis instrument room. (Source: Anon, *Guide to Ben Nevis*, facing p. 10.)

The principal threat to the routine and proper running of the observatory, according to its staff, was the tourist. The observatory was meant to be a space unaffected by unwanted external stimuli. This ideal was both especially important and easily met on a mountain top, where observers set about recording a rarefied environment while assuming they would not be bothered by the inconveniences of everyday life. Despite the promise of a pristine, isolated, summit observation site, many actors comingled on mountains, while the remoteness, isolation and extreme conditions demanded a reliance on guides, porters, volunteers, telegraph operators and provisioners.[134] The supposed threat posed by unenlightened tourists had been a concern even before the erection of the permanent observatory in 1883. Wragge posted notices in various hotels in Fort William:

Those who visit the mountain are most earnestly requested for the good of the country to co-operate with the Society in the cause of science and to keep at a distance of 25 feet from the various exposed instruments, thermometer box, and

[134] Bigg et al., 'Introduction'.

other Observatory erections in order to prevent the vitiation of results by artificial means. The rain gauges are kept perfectly level and the Observer begs that the various instruments may not be touched. Persons causing damage will be prosecuted with the utmost vigour of the Law.[135]

The road to the summit made the journey up the mountainside easier and the new building attracted curious visitors. The directors reported that the observatory was 'besieged by a well-nigh innumerable multitude of strangers' during the building work in 1884. About 4,000 were estimated to have climbed Ben Nevis that summer, half of those in the month of August. Drysdale reckoned that in a normal year 2,000–3,000 visitors visited the summit in the summer months and complained they 'were often a plague':

Their unvarying string of questions, the determination of many to see the 'Telescopes and Things', regardless of the convenience of the observers, their peering in at windows, their fingering the thermometers and displacing the rain gauge, their lighting their pipes with the sunshine recorder, their dancing on the roof over a slumbering observer, and last enormity of all, their tipping the observers on departing![136]

Another relief observer, William Kilgour, noted that tourists expected observers to provide information on local topography as well as weather forecasts.[137] Although some commented favourably about the views from the summit, far more complained that the weather concealed the surrounding landscape. One tourist from Edinburgh wrote, '1st time + the last I hope horrible day no view'.[138]

The official observatory guide complained that many visitors 'waxed wrathful when they discovered that food and a night's lodging could not be procured on any terms. The more sensible of the visitors were satisfied, though sometimes a little astonished, by the explanation that the building on Ben Nevis was primarily a Scientific Observatory and not a hotel.'[139] One tourist from Leeds noted in the Observatory Visitors' Book in August 1884 that '[i]t would add to people's interest in the observatory, + assist in procuring subscriptions, if luncheon were provided'.[140] This

[135] Woolnough, National Records of Scotland, MET1/5/3/14, p. 7 and pp. 8–9, respectively.
[136] Drysdale, 'Life at Ben Nevis Observatory', National Records of Scotland, MET1/5/3/7, p. 9.
[137] W.T. Kilgour, 'An Original Holiday', *Pearson's Magazine*, no date, National Records of Scotland, MET1/5/3/9, pp. 244–8, p. 245.
[138] Anon, 8 Sept 1885, Ben Nevis Observatory Visitors' Book 1, 1884–1888, Lochaber Archive Centre, L/D155.
[139] Anon, *Ben Nevis Meteorological Observatory*, p. 19.
[140] Anon, Entry on 25 August 1884, Lochaber Archive Centre, 1884–1888, L/D155. The visitor nonetheless donated £1 to the Observatory fund.

Figure 3.7 Collection of photographs of visitors to the summit observatory. (Source: National Library of Scotland, Phot.med.89. Reproduced with permission of National Library of Scotland.)

pressure was relieved in the summer of 1885, when a Fort William hotelier, Robert Whyte, opened the Ben Nevis Observatory Hotel on the summit, described by Drysdale as 'a strongly-built wooden shanty, containing a public room and two or three sleeping bunks where strangers can be accommodated for the night at "Highland" charges'.[141] The Observatory Visitors' Book started to fill with comments on the restorative power of the Hotel's coffee although some complained that it did not serve alcohol (Figure 3.7).

Some visitors were more welcome than others at the observatory. Drysdale said that members of the 'Scots Mountaineering Club' were 'always welcome guests when they reached the Observatory after conquering some new and apparently impossible route to the summit'.[142] The difficulty

[141] Drysdale, 'Some Experiences of "High" Life', National Records of Scotland, MET1/5/3/7, p. 434.

[142] Drysdale, 'Life at Ben Nevis Observatory', National Records of Scotland, MET1/5/3/7, p. 10.

of the ascent highlighted the mountaineers' attributes of 'good fellowship, self-reliance, endurance, and courage', while the more casual tourist followed the path, 'up which, the mountaineers say, any man with vigour enough may push a wheel-barrow'.[143] The implication was clear – the mountaineer and the weather observer shared a set of common values, including traits of masculinity, heroism and fortitude, which were otherwise missing in the day tripper. Kilgour deliberately blurred the two vocations into one figure in his account of the ascent undertaken by new relief observers: 'At length the plateau near the summit is reached, and when the remaining short rise beyond this is scaled, the eyes of the mountaineer can dimly discern, as the scud-cloud lifts, the straggling outline of his abode during the period of his sojourn as a man of science.'[144] Both the observer and the mountaineer were held in contrast to the tourist who ventured up the new path from Fort William, improperly dressed, without provisions, possibly even riding a pony, who then complained about their lot in the Visitors' Book. It went unobserved that the same path that opened the mountain to the casual tourist also enabled the running of the observatory. The road even provided an income stream in that its users had to pay for a permit – one shilling for walkers and three shillings for those on horseback. The permit allowed tourists to use the visitors' room at the observatory and to send a telegram, where the observatory received a percentage of the charge to do so. The observatory also raised money through the sale of its official guidebook, first published in 1885 and reissued in 1893. Various of the observatory staff and directors also wrote newspaper articles and other popular accounts of observatory life which promoted the observatory as an accessible site of public science rather than an isolated and inaccessible spot beyond the reach and imagination of the majority of the population.

After early interruptions in its observations and its expansion in 1884, the Ben Nevis Observatory operated successfully through the 1880s, while Colin Livingston continued to take observations at his school house in Fort William. A legacy of £500 and a donation of almost £1,000 from the Association of the Edinburgh International Exhibition in 1886, along with a mortgage, allowed the directors of the observatory to fulfil their long-term plan for a purpose-built observatory close to sea level where continuous observations could be made. Building work for the so-called low-level observatory in Fort William began in 1889 and was completed in July 1890. The new observatory was furnished with self-recording instruments supplied by the Meteorological Council, which

[143] Anon, undated newspaper clipping, National Records of Scotland, MET1/5/3/7, no page.
[144] Kilgour, 'An Original Holiday', 244.

Figure 3.8 The low-level observatory in Fort William. (Source: Anon, *Guide to Ben Nevis*, p. 34.)

provided continuous records of air pressure, dry and wet bulb temperature, rainfall and sunshine. It was also outfitted with the usual set of instruments found at an ordinary station: barometer, maximum, minimum, dry and wet bulb thermometers in a Stevenson screen, solar and terrestrial radiation thermometers and a rain gauge.[145] Observations began on 14 July 1890 (Figure 3.8). As one of the Meteorological Office's 'first-class' stations, the Fort William Observatory received an annual grant of £250 from the Meteorological Council. The Fort William Observatory became one of five meteorological observatories across the British Isles that received funding from the Meteorological Council, the others being Valencia, Falmouth, Aberdeen and Kew, while observatories at the University of Glasgow and Stonyhurst College in Lancashire also operated to the same standard and supplied observations.[146] The Scottish Meteorological Society established one of their own stations in the grounds of the observatory, for which they provided the instruments. Livingston continued with his observations until the end of 1891 to allow a comparison between his observations and those of the new site

[145] Anon, 'Preface', *Transactions of the Royal Society of Edinburgh*, 42 (1902), viii–ix.

[146] S. Naylor 'Nationalising Provincial Weather: Meteorology in Nineteenth-Century Cornwall', *British Journal for the History of Science*, 39 (2006), 407–33.

(the schoolhouse and the new observatory were only 150 yards apart). Livingstone's observations had been used for the original determinations of the differences of pressure and temperature between Fort William and Ben Nevis and any discrepancy between the observations made at the schoolhouse and at the observatory would necessitate a readjustment of these constants.[147] The new low-level observatory was believed to be close enough to the summit station in terms of horizontal distance 'as to be virtually one Observatory', capable of offering unique insights into the relations of height to temperature, humidity and pressure in the free atmosphere.[148] Fort William was used to train new observers before they headed to the summit station. Observers were also regularly moved between the two observatories and the Edinburgh offices due to concerns about their physical and mental health.[149] In 1896, the Directors decided to establish a temporary intermediate station at 2,322 feet above sea level. A small wooden hut was secured and improved, and instruments installed. Volunteer observers kept hourly observations there during the summer months with the aim of further improving understanding of the vertical gradients of temperature, pressure and humidity[150] (Figures 3.9 and 3.10).

Staff at the two observatories and in Edinburgh were encouraged to pursue original research alongside the more routine work of collecting and reducing the observations, even though it was initially acknowledged it was hard to achieve much more than the determination of hourly constants of the various meteorological elements.[151] Nonetheless, the directors hoped that in time it would be possible to investigate the relations of their own observations to larger-scale weather patterns across north-western Europe. Significant computing time was invested in reworking the two observatories' hourly observations to this end.

[147] Omond, 'Abstract of Paper', 537.

[148] Anon, 'Meteorology of Ben Nevis', *Nature*, 43 (9 April 1891), 538.

[149] Observers complained that the lack of exercise, the isolation and the demands of observing through the night led to 'langour and depression of spirits'. *The Scotsman* newspaper clippings, undated, National Records of Scotland, MET1/5/3/3. They also complained about the so-called Ben Nevis cold, which afflicted observers when they came down from the summit: 'In the opinion of medical men, it is caused by the change from the purer atmosphere above to the comparatively grimy state of the air below, and not, as might be supposed, the difference of temperature.' Anon, 'Life and Work in Ben Nevis Observatory', untitled newspaper clipping, no date or page, National Records of Scotland, MET1/5/3/7.

[150] Anon, 'Report of the Committee Appointed for the Purpose of Co-operating with the Scottish Meteorological Society in Making Meteorological Observations on Ben Nevis', 1898, National Records of Scotland, MET1/5/3/8, p. 5.

[151] Anon, 'Report of the Committee Appointed for the Purpose of Co-operating with the Scottish Meteorological Society in Making Meteorological Observations on Ben Nevis', 1888, National Records of Scotland, MET1/5/3/8, p. 4.

(a)

(b)

Figure 3.9 (a): Photograph of thermometer screens and observers at the Half-Way Station; (b): Photograph of bridle path to summit of Ben Nevis. (Source: National Library of Scotland, Phot.med.89. Reproduced with permission of National Library of Scotland.)

Figure 3.10 Photograph of Half-Way Station, meteorological station on Ben Nevis. (Source: National Library of Scotland, Phot.med.89. Reproduced with permission of National Library of Scotland.)

A £100 grant was provided by the Royal Society of London in 1895 to cover the costs of routine computational work in Edinburgh and Fort William so that Buchan and Omond could be freed up to lead these inquiries.[152] Omond later devoted much of his time to a comparison of the Scottish observations with those taken at twenty-seven other high-level observatories across Europe.[153] Omond also carried out investigations into the diurnal variations in the direction of summer winds at sea level and at the summit, and Rankin assisted by developing a thermic wind rose. From the early years of the observatory's establishment its staff had paid special attention to the occurrence of glories, haloes, coronae and other atmospheric effects at the summit. The logbooks often contained sketches of various phenomena. These observations were improved in the late 1880s when a grant from the Government Research Fund was used to purchase apparatus to photograph clouds

[152] Anon, 'Report of the Committee Appointed for the Purpose of Co-operating with the Scottish Meteorological Society in Making Meteorological Observations on Ben Nevis', 1895, National Records of Scotland, MET1/5/3/8, p. 6.

[153] Anon, 'Report of the Committee Appointed for the Purpose of Co-operating with the Scottish Meteorological Society in Making Meteorological Observations on Ben Nevis', 1901, National Records of Scotland, MET1/5/3/8, pp. 5–6.

and other optical features.[154] A further grant of £50 from the Fund in 1890 enabled the observatory to purchase instruments to measure the quantity of dust particles in the atmosphere, one of which was fixed to the summit observatory tower, the other being portable. John Aitken, the Falkirk-born engineer and physicist, constructed the dust counter and supervised the research. His investigations into the 'meteorological and industrial conditions governing the production of dust particles in the air, the influence of locality and altitude, the effect of prevalent winds and of cyclonic and anticyclonic distributions' were published in a series of papers in the *Transactions of the Royal Society of Edinburgh*.[155] Work was done in 1892 to establish the relationships between mists and fogs and the quantity of dust. Also in that year the University of Edinburgh awarded a two-year scholarship to Andrew J. Herbertson to conduct research on Ben Nevis and Fort William, and then in the south of France, on the hygrometry of the atmosphere at different heights. Herbertson published his findings in the *Transactions of the Royal Society of Edinburgh* and the *Journal of the Scottish Meteorological Society*, results the Observatory Committee regarded 'as of great importance, not only as furnishing data towards a better knowledge of the hygrometry of the atmosphere, but also as leading to much needed improvements in the methods of reducing the readings of the dry and wet bulb thermometers'.[156]

Writing in their 1887 British Association report, the observatory committee suggested that they had 'from the very outset insisted with some earnestness and strength of language on the absolute necessity of combining the observation at the top of Ben Nevis with that made at the same instant at Fort William'.[157] Sir Arthur Mitchell, the notable Scottish medic and historian (later president of the Scottish Meteorological Society), argued that meteorologists tended to operate horizontally

> at the base of the atmosphere, – in other words, at the bottom of what has been called the *ocean of air*. But it is now accepted that *upward soundings* of the *atmospheric ocean* must also be made in order to increase the hope of obtaining additions to our

[154] Anon, 'Report of the Committee', 1888, National Records of Scotland, MET1/5/3/8, p. 1.

[155] C. G. Knott, *Collected Scientific Papers of John Aitken. LL.D. F.R.S.* (Cambridge: Cambridge University Press, 1923), p. x.

[156] Anon, 'Report of the Committee Appointed for the Purpose of Co-operating with the Scottish Meteorological Society in Making Meteorological Observations on Ben Nevis', 1893, National Records of Scotland, MET1/5/3/8, p. 4. Herbertson pursued a university career after his scholarship, ultimately becoming Professor of Geography and head of the Department of Geography at the University of Oxford.

[157] Anon, 'Report of the Committee Appointed for the Purpose of Co-operating with the Scottish Meteorological Society in Making Meteorological Observations on Ben Nevis', 1887, National Records of Scotland, MET1/5/3/8, p. 4.

knowledge of meteorology, since the vertical condition of the atmosphere may perhaps govern weather changes even more than the horizontal.[158]

This study of the so-called atmospheric ocean was increasingly referred to as the 'great experiment'. Speaking to a meeting of the Scottish Meteorological Society in 1902, Mitchell referred to the Society's 'big and costly experiment in atmospheric physics'.[159]

With the establishment of the low-level observatory and then the intermediate weather station, increasing emphasis was placed on the investigation of the vertical gradients of temperature, pressure and humidity of the atmosphere. It was assumed that there was a relationship between different vertical distributions of temperature and pressure on the one hand and the presence of cyclones and anticyclones on the other.[160] Work on barometric pressure intensified in 1900. It was recognised that the distribution of pressure was not arbitrary and that physical laws governed its movement, increase and diminution. It was also assumed that areas of high and low pressure had a vertical quality, extending upwards through the atmosphere, and were 'largely determined by the conditions surrounding them in the upper regions of the atmosphere'.[161] Observatory personnel committed to the study of the vertical aspects of atmospheric pressure on the basis that it would lead to a 'science of forecasting', 'it being by vertical gradients, and not by horizontal gradients, that the observations at high-level observatories can be turned to their proper and fullest account in forecasting weather'.[162] Funding from the Royal Societies of London and Edinburgh for the publication of the Ben Nevis and Fort William observations was interpreted as confirmation that the 'fundamental data' the observatories provided was a 'necessary preliminary to the scientific study of forecasting'.[163] In 1901, the Observatory Committee argued that '[i]t is just these different distributions of pressure in the higher layers of the atmosphere from what prevails at sea-level at the same time which is most likely to aid the forecaster of weather in seeing the most probable

[158] A. Mitchell, 'Remarks appended to Memorandum by Directors of Ben Nevis and Fort William Observatories in connection with their closure', June 1902, National Records of Scotland, MET1/5/3/5, p. 1.

[159] Arthur Mitchell, 'Remarks appended to Memorandum', MET1/5/3/5 p. 3.

[160] Anon, 'Report of the Committee', 1898, MET1/5/3/8, p. 5.

[161] Anon, 'Report of the Committee Appointed for the Purpose of Co-operating with the Scottish Meteorological Society in Making Meteorological Observations on Ben Nevis', 1900, National Records of Scotland, MET1/5/3/8, p. 7.

[162] Anon, 'Report of the Committee Appointed for the Purpose of Co-operating with the Scottish Meteorological Society in Making Meteorological Observations on Ben Nevis', 1887, National Records of Scotland, MET1/5/3/8, pp. 4–5.

[163] Anon, 'Report of the Committee', 1900, National Records of Scotland, MET1/5/3/8, p. 7.

distribution of the sea level pressure one day, two days, or even three days in advance', and they expressed the hope that 'the examination and discussion of the work of the two observatories will be thorough, and will have scientific utility in the general study of the phenomena of weather, and a practical utility in its bearing on weather forecasting'.[164] The issue of weather forecasting did turn out to be of crucial importance for the fate of the observatories, but ultimately not in their favour.

Funding the Ben Nevis Observatories

Debates about the state support for science was a fraught one in the nineteenth century. Macdonald argues that the principle of laissez-faire economics – that is, the doctrine that the government should not interfere in an economy that was presumed to be self-regulating – defined government attitudes to science funding.[165] The state did fund science but tended to do so by fostering individual enterprise. Challenges to that system were made from the 1860s as the costs of the physical sciences escalated, with more elaborate equipment and the need for larger numbers of staff. Increasing calls for fuller state support for science led to the Devonshire Commission on Scientific Instruction and the Advancement of Science, which sat from 1870 to 1875. The Devonshire Commission has been seen by some as the beginning of a shift to state-funded organised science.[166] For instance, the Meteorological Office had received an annual budget of £10,000 since 1867, which had been managed through the Royal Society of London. A review of the work of the Office led to the establishment of the Meteorological Council and an increase in the annual budget for meteorology to £14,500, which later increased to £15,300.[167] That said, the general impacts of the Devonshire Commission remained quite small initially. Macdonald challenges the idea that 'laissez-faire – and the physical sciences' consequent reliance on private sources of patronage – went out of fashion before the end of the nineteenth century.'[168] Even the Kew Observatory – arguably the most important physical science observatory in Britain in the nineteenth century – remained an exemplar of the laissez-faire system up to the end of the nineteenth century and into the early years

[164] Anon, 'Report of the Committee', 1901, National Records of Scotland, MET1/5/3/8, p. 6.

[165] Macdonald, *Kew Observatory*.

[166] R. MacLeod, *Public Science and Public Policy in Victorian England* (Ashgate: Aldershot, 1996).

[167] R. MacLeod, 'Science and the Treasury: Principles, Personalities and Policies, 1870-85', in G. L'E Turner (ed.), *The Patronage of Science in the Nineteenth Century* (Leiden: Noordhoff International Publishing, 1976), pp. 115–172, p. 164.

[168] Macdonald, *Kew Observatory*, p. 12.

of the twentieth, continuing to rely heavily on fees charged for testing instruments for manufacturers and government bodies. The Ben Nevis and Fort William observatories existed and struggled under the same laissez-faire system throughout their entire existence. From the very outset the Meteorological Council had been sceptical about providing financial support to the Scottish meteorologists, especially when it came to risky and demanding ventures like the erection of observatories. As early as 1879, Henry J. S. Smith, the chair of the Meteorological Council, suggested any observatory on Ben Nevis

> must inevitably prove a costly experiment, and may possibly prove a costly failure. If the experiment is to be made under conditions favourable to economy of working, it must be allowed to depend for its success on the amount of interest which the proposal inspires in Scotland, and on the efficiency of the local management which can be provided there. It would seem to follow that, in the case of such an undertaking, assistance from the State should be employed to supplement and stimulate local efforts, not to supersede them.[169]

The questions of what the observatories were for, who they were for and how they should be funded brought the directors and employees of the observatories into periodic – and often quite heightened – conflict with the Meteorological Council, most notably in 1887 when the Council accused the Ben Nevis Observatory of failing to deliver observations useful to forecasting; and in 1903–04, when the British Treasury conducted an inquiry into the subject of the government's meteorological grant. From the perspective of the directors of the Ben Nevis observatories these moments revealed a failure to appreciate the value of high-level meteorology and a systematic bias against Scottish science.

In January 1884, the directors of the Ben Nevis Observatory confirmed an agreement with the Press Association to supply it with daily weather reports, at a cost to the Association of £100 a year. This agreement gave the Association exclusive rights to publish the information it received as a daily weather report. The directors also committed to supply the Meteorological Office with hourly observations from Ben Nevis, but these would only be sent down to London every two to six weeks, although the observatory superintendent, Omond, was free to use his discretion in telegraphing to London any 'striking change of conditions or special phenomenon of great interest. It is understood however, that such information as may thus be occasionally given will be for use in

[169] Letter from H. J. S. Smith to President of the Royal Society of London, 27 March 1879, Archives of the Royal Society, MS/775.

scientific work and not for immediate publication.'[170] The Meteorological Council, which was also supplying an annual grant of £100 to the observatory, took issue with this arrangement and the priority it gave to the Press Association, calling the arrangement 'utterly ludicrous' and a breach of the contract between the Council and the directors.[171] In reply Buchan complained that the Council had never expressed a wish to publish the Ben Nevis observations immediately and in their bare form. Buchan stated that the principal aim of the directors was to make the Ben Nevis Observatory 'useful in scientific work. For this end they are prepared to make considerable sacrifices.'[172] These sacrifices included supplying observations to serious meteorologists at little or no cost where they were not in a position to pay for them. As far as the directors were concerned, this did not include a body like the Meteorological Council, given its receipt of nearly £15,000 a year from the state. Buchan ended his letter with a reminder that the Meteorological Office could certainly be supplied with a daily weather report for use in its office, just not for immediate publication, 'though inferences to which they may have contributed in forecasting may at once be published'.[173] He also called for more money from the Council to support the new observatory's work.

Correspondence between London and Edinburgh continued to be antagonistic. The directors claimed that the Council was imposing increasing demands 'of great magnitude' on the observatory, which would turn it into 'a mere observing station for the Meteorological Office'.[174] The directors reported that the Council also refused to increase the grant it gave to the observatory, on the grounds that the 'special value' of the observations for forecasting were not yet proven. The directors replied that the government grant was for the 'advancement of Meteorology generally' and so there was no need for the Council to be assured of the Ben Nevis observations' value to forecasting specifically

[170] Anon, 'Minutes of Meetings', 14 January 1884, National Records of Scotland, MET1/5/3/11, p. 27.

[171] Letter from A. Buchan to R. Scott, including excerpts from Scott's letters on 31 March, 14 April, 2 May, in 'Minutes of Meetings', 14 May 1884, National Records of Scotland, MET1/5/3/11, no page. The quote came from a letter from the Meteorological Office to Omond on 15 February 1884.

[172] Letter from A. Buchan to R. Scott, 14 May 1884, reproduced in 'Minutes of the Proceedings of the Meteorological Council 1884–5', 21 May 1884, National Archives, BJ 8/11, p. 14.

[173] A. Buchan, letter to R. Scott, dated 14 May 1884, reproduced in 'Minutes of the Proceedings of the Meteorological Council 1884–5', 21 May 1884, National Archives, BJ 8/11, p. 15.

[174] Anon, 'Minutes of Meetings', 4 July 1884, National Records of Scotland, MET1/5/3/11, no page, including extract from Minutes to Council, 25 June 1884, and letter from R. Scott to Directors, 1 July 1884.

before increasing its annual grant. Noting the Council's previous support for the summit observatory and its awareness of the great costs of such a venture, while refusing to provide adequate financial support, meant that the Council 'practically ask the Directors to make a costly experiment for them'. The directors warned that if the Council did not increase the grant or at the very least proved unwilling to continue it, they would regard that 'as injurious to the country and to science'.[175]

Debates about government support for the Ben Nevis Observatory surfaced again in 1887. In August that year Parliament discussed the government vote for Britain's learned societies. Discussion in the newspapers the next morning reported that Mr Jackson of the Treasury, Sir John Lubbock and others had argued against any grant to the Ben Nevis Observatory on the grounds that the Meteorological Council, 'composed of men of the very highest scientific standing, had given it as their opinion that the practical results to be obtained from the Ben Nevis Observatory did not warrant the grant asked for from the Treasury'.[176] A Meteorological Council memo in 1887 on the issue of the telegrams received from Ben Nevis in London had concluded that '[i]n their existing form the telegrams (from Ben Nevis) are absolutely useless' for the purposes of forecasting.[177] The observatory directors took issue with this conclusion, which they interpreted to be a comment on the value of the observatory more generally. Writing to the Scottish press, the directors publicly spelled out the recent history of relations between the observatory and the Meteorological Council, including the £100 annual grant from the Council's own government grant, the directors' offer to send daily weather telegrams to London and the Council's refusal due to the cost of daily transmission and what could be done with the observations in terms of computation. The Council asked instead for monthly sheets as well as telegrams from the observatory superintendent on 'any striking changes of meteorological conditions, or interesting special phenomena'.[178] Writing to Alexander Buchan in late July 1887, Robert Scott, the secretary of the Meteorological Council, stated: 'During this period [1886–7] we issued from the Meteorological

[175] All quotes from Anon, 'Minutes of Meetings', 4 July 1884, National Records of Scotland, MET1/5/3/11, including extract from Minutes to Council, 25 June 1884, and letter from R. Scott to Directors, 1 July 1884.

[176] Anon, Extract from 'Report to the British Association for the Advancement of Science, 1903, by committee appointed to co-operate with the Scottish Meteorological Society in making meteorological observations at Ben Nevis and Fort William', National Records of Scotland, MET1/5/3/5, p. 2.

[177] Anon, Extract from 'Report to the British Association ... 1903', National Records of Scotland, MET1/5/3/5, p. 3.

[178] Anon, newspaper cutting from *The Scotsman*, 26 August 1887, National Records of Scotland, MET1/5/3/3, no page.

Office eighty-six warnings to the Scottish coasts, but we received from Ben Nevis only nineteen telegrams. Of these, two arrived before we issued the warnings, and on these occasions the receipt of the telegrams in no way influenced us in determining to warn.'[179] This private letter was then published in the press, with Buchan's private reply published immediately below, where he claimed that Scott's criticism had not 'kept in view' the original instructions issued to the observatory from London.[180] There was nothing in the original instructions, he claimed, pertaining to the issuance of storm warnings or the study of storms and no sense that the Meteorological Office wanted Ben Nevis to send telegrams reporting the threat or onset of a storm. Rather, Buchan contended, the request had been for more general changes of conditions or special phenomena. In fact, Omond reported on the change of conditions 'which occurred that prognosticated settled weather'.[181] These memos from Omond were classed by the Meteorological Office as if 'to be prognostic of storms' and hence deemed useless, a statement that was both 'unquestionably correct' and 'void of all meaning'. Omond claimed: 'What has been done [by the Meteorological Council in London] is not an investigation, and it is not science.'[182] Speaking in the Scottish press, the directors complained:

> It sounds incredible, yet it is the case, that the Council now make a point of the uselessness of the Ben Nevis telegrams for purposes of storm warnings. Because nineteen telegrams in the course of sixteen months, only two of which arrived before the warnings were issued have not proved helpful to the Meteorological Council, these sapient gentlemen gravely deliver themselves of the opinion that the Ben Nevis telegrams are 'absolutely useless.' How could they be useful unless transmitted regularly, and regularly compared with other observations, so that their significance might be understood? ... The fact that these gentlemen made such a useless request, and now base on its futility a general charge of uselessness, almost suggests a suspicion of sinister motive. Such a suspicion would doubtless be unjust. The truth simply is, that the Council is composed of English men of science who, like Sir John Lubbock, are very good men and admirable scientists, but have a tendency, like a natural instinct, to seek the centralisation of everything in London.[183]

Calls followed for more Scots on the Committees of the Royal Society of London and a 'fair proportion' of scientific grants distributed through the

[179] Anon, newspaper cutting from *The Scotsman*, no day or month, 1887, no page, National Records of Scotland, MET1/5/3/3.
[180] Anon, newspaper cutting from *The Scotsman*, 8 August 1887, no page, National Records of Scotland, MET1/5/3/3.
[181] Anon, newspaper cutting from *The Scotsman*, 8 August 1887, no page, National Records of Scotland, MET1/5/3/3, original emphasis.
[182] Anon, Extract from 'Report to the British Association ... 1903', National Records of Scotland, MET1/5/3/5, all quotes from p. 4.
[183] Anon, newspaper cutting from *The Scotsman*, 26 August 1887, no page, National Records of Scotland, MET1/5/3/3.

Scottish Office. These calls were by no means new. Government support for science had long been seen as geographically selective, with a disproportionate number of grants awarded to individuals based in London and the south-east of England, while those based elsewhere in England and in Scotland felt overlooked. Peter G. Tait, Professor of Natural Philosophy at the University of Edinburgh, was never made Fellow of the Royal Society of London and had several applications to the Government Grant fund rejected. As a result he was hostile towards London's scientific circle.[184] Charles Piazzi Smyth, the Astronomer Royal for Scotland – until he resigned in 1888 in protest at chronic underfunding – also held the view of a 'distant government, resisting the legitimate concerns of its citizens and badly advised by scientific officials'.[185] Some years later, in 1902, Arthur Mitchell pointed out that no request from the Meteorological Office for information on summit observations had been received since 1890. Even though the directors had done all they could to render the observations useful in forecasting, they 'could not themselves issue forecasts. This, indeed, can only be done from a Central Office receiving information by wire, at short intervals, from a great many stations, near and remote.'[186]

The financial pressures on the Ben Nevis and Fort William observatories never went away. It was estimated that the annual cost of running the two observatories was £1,000. The directors spent much of their time trying to raise the £650 shortfall after the receipt of the £100 for the Ben Nevis and £250 for the Fort William observatories. Writing to the Meteorological Council in 1898, the directors asked for further financial support, saying that there was a strong claim for assistance from the nation due to the large amount of private money put into the project, which was 'a striking example of courageous private enterprise in the search after additions to knowledge'. The observatories also made a unique claim to a significant portion of the annual £15,300 grant awarded to the Council and tried to reassure the Council that this did not constitute special treatment, on the grounds that

> some one division of the kingdom may contain better localities and conditions for meteorological research than others, and may thus, in the interest of the study of the meteorology of the United Kingdom broadly, be entitled to an expenditure

[184] R. MacLeod, 'The Royal Society and the Government Grant: Notes on the Administration of Scientific Research, 1849-1914', *Historical Journal*, 14 (1971), 323–58, 335.

[185] Anderson, *Predicting the Weather*, p. 242.

[186] A. Mitchell, Remarks appended to Memo by Directors of Ben Nevis and Fort William Observatories in connection with their closure, June 1902, National Records of Scotland, MET1/5/3/5 p. 3.

within its borders of more of the grant than its proportion as calculated on population. For instance, as regards the observations at Ben Nevis, Scotland is in a position which is altogether exceptional.[187]

The directors asked the Council to forward their request on to Robert Hanbury MP, the Financial Secretary to the Treasury. The Council took this as an opportunity to put its own case to the Treasury, noting the need for a superannuation scheme at the Meteorological Office and its necessary drain on the Parliamentary grant unless the grant was increased. The Council reported that the Royal Society of London's advice was to reduce spending to fund the pension scheme, which made it 'impossible' to contemplate an increase in the grant to the Ben Nevis Observatories and might in fact necessitate a reduction.[188] In reply to Scott, Hanbury left it to the Council to decide what course of action to pursue: 'My Lords have no doubt that you will weigh impartially the relative claims of the several branches of Meteorological enquiry to a share in the funds at your disposal.'[189] At a meeting of the observatory directors on 11 March 1899, a letter by Buchan, Scotland's Meteorological Council representative, was read which contained extracts of this correspondence between the Council and the Treasury. Nonetheless, further requests for financial support were made – for £1,000 from the Council and, later in 1899, £500 from the Treasury. In late March, Scott rejected the former request, drawing attention to the fact that Fort William received the same amount of financial support as Valencia and Falmouth, and that 'in Scotland the grants to Aberdeen, Ben Nevis and Fort William aggregated £625'.[190] In July, the Treasury turned down the latter request, saying that it would be 'inexpedient to depart from the present well-established practice under which state aid to the study of Meteorology is given solely through the medium of the Meteorological Council, a representative of experts selected and designated for this purpose'.[191]

The directors of the observatories therefore had to rely heavily on private donations to keep the stations open. In July 1898, Mr J. Mackay Bernard of Kippenross, the Edinburgh brewing magnate and enthusiastic amateur meteorologist, offered £500, 'so that the directors may at least be

[187] Letter from Directors of the Ben Nevis observatories to the Meteorological Council, 2 December 1898, National Records of Scotland, MET1/5/3/4.

[188] Letter from R. Scott to R. W. Hanbury, 2 January 1899, reproduced in 'Minutes of Meetings', National Records of Scotland, MET1/5/3/11, no page.

[189] Letter from R. W. Banbury to Scott, 30 January 1899, reproduced in 'Minutes of Meetings', National Records of Scotland, MET1/5/3/11.

[190] Letter from R. Scott to Scottish Meteorological Society, 29 March 1899, no pages, National Records of Scotland, MET1/5/3/4.

[191] Letter from R. W. Hanbury to A. Mitchell, 25 July 1899, reproduced in 'Minutes of Meetings', 31 July 1899, National Records of Scotland, MET1/5/3/11.

able to obtain continuous and complete observations for the eleven years of a sunspot period. This would mean the making of an important addition to knowledge by Scotland, and in that aspect Mr Bernard is patriotic as well as liberal.'[192] Reporting on this news, the Scottish press asked: surely 'the only high-level Observatory in our islands, should be established on foundations that will prove permanent. It serves a national purpose. Why should it not receive adequate national support?' The article went on to draw attention to the £15,000 given annually from Parliament to the Meteorological Council, saying:

> But apart from the fact that official bodies stationed in London – even when they are also scientific bodies – cannot be entirely trusted to do justice to the claims of Scotland and Scottish research, the Meteorological Council are understood to withhold more aid to Ben Nevis, not because it is not doing good work or because they are not interested in it, but because the claims upon them are great and pressing, and that the Government grant is insufficient to meet them all.[193]

The observatories continued to rely on donations from private benefactors and grants from scientific bodies, including £2,000 from Bernard over several years, £1,875 from the trustees of the late Earl of Moray and significant sums from James Key Caird, the Dundee jute baron, and James Coats, the Paisley mill owner. By June 1902, the directors estimated they had spent £24,000 since 1883 and that £17,000 of that had come from subscriptions.[194]

In the first years of the twentieth century various members of the Scottish Meteorological Society made increasing reference to the correspondence between the observations collected at Fort William and Ben Nevis and the eleven-year sunspot cycle, even if Buchan had been interested in the possible relations between weather and sunspot activity since the 1870s.[195] Writing to Arthur Mitchell in 1901, Lord John McLaren, the astronomer and Scottish Liberal MP, asserted that the reason for collecting Ben Nevis and Fort William observations for eleven years 'was to test the correspondence of variations of temperature with the prevalence of sunspots, a problem of great interest to astronomers and physicists'.[196] Mitchell also wrote to the observatories' major donors to

[192] Letter from Arthur Mitchell to *The Scotsman*, 27 July 1898, no pages, National Records of Scotland, MET1/5/3/3.

[193] Anon, cutting from *The Scotsman*, no date or pages, National Records of Scotland, MET1/5/3/3.

[194] Anon, Memo by Directors of Ben Nevis and Fort William Observatories in connection with their closure, June 1902, National Records of Scotland, MET1/5/3/5.

[195] A. Buchan, 'Sun-Spots and Rainfall', *Nature*, 17 (25 April 1878), 505–6.

[196] Letter from Lord J McLaren to A. Mitchell, 15 June 1901, National Records of Scotland, MET1/5/3/5.

make the same point – noting to James Coats that '[t]his was the scientific experiment which we set before ourselves, and when it was finished we hoped that the Government would take the Observatories into their own hands'.[197] Mitchell acknowledged that without government support there was every reason for Coats and other philanthropists to withdraw their support. If Coats and Bernard continued to provide annual support, the observatories could get two more years' observations 'but there is no special outstanding advantage or desideratum in adding two years' to the eleven.[198] Mitchell noted the cost of the South African and Chinese wars and saw no reason why the chances of the government taking the observatories into their own hands were better in two years than now. In these circumstances, is it right to take money from donors, he asked? Mitchell was personally in doubt and thought it probably best to end the work with the sunspot period.

Walker notes that questions were raised in the House of Commons in 1902 over government support for the threatened observatories. John Dewar, the Liberal MP for Inverness-shire and manufacturer of Dewar's whisky, asked the Treasury and the Conservative East Lothian-born Prime Minister Arthur Balfour what could be done to support the observatories. Balfour promised an investigation into funding for meteorology.[199] In November 1902, after no signs of action, Sir John Stirling-Maxwell, the Conservative MP for Glasgow College, asked the Treasury for a report on progress. A committee was appointed soon afterwards, chaired by Sir Herbert Maxwell, the Conservative MP for Wigtownshire, president of the Society of Antiquaries of Scotland and, later, chairman of the National Library of Scotland. John Dewar was invited onto the committee, along with representatives from the Board of Trade, Board of Agriculture, the Treasury, National Physical Laboratory and several scientists. The directors of the observatories continued to raise money to fund their operation in anticipation of the committee recommending that they receive full financial support from the state. The committee eventually reported in May 1904.

Given the amount of time the committee took to issue its report, debate continued over the continuance of the Ben Nevis Observatories. Omond wrote to Herbert Maxwell in July 1903 complaining about press reports that suggested the committee had already resolved to report against their continuance. He also suggested that the Meteorological Council had

[197] Letter from A. Mitchell to J. Coats, 1 June 1901, National Records of Scotland, MET1/5/3/5.

[198] Letter from A. Mitchell to J. Coats, 1 June 1901, National Records of Scotland, MET1/5/3/5.

[199] Walker, *History of the Meteorological Office*.

done 'little or nothing' to investigate the value of high-level observatories and that the committee had as yet consulted 'no scientific experts', by which he meant 'men of eminence in Physical Science who have been practically engaged in weather forecasting, and who have been in a position to utilise observations made at a high level'.[200] Maxwell replied saying that he found the remark about scientific experts strange because, amongst others, Omond himself had given evidence, along with Alexander Buchan and Lord Kelvin.[201] Omond wrote back, arguing that

> neither myself, nor Dr Buchan, nor Sir John Murray, nor Lord Kelvin can be classed among those 'who have been practically engaged in weather forecasting, and who have been in a position to utilise observations made at a high level', noting that others on the list are 'of great eminence as physicists but not known to have been practically engaged in weather forecasting with the power to utilise high level observations'.[202]

Omond was also in correspondence with Stirling-Maxwell, again appealing against the long-standing judgement that the Ben Nevis observations were of little practical use to weather forecasting, noting on 24 November 1903 that 'there have been no practical results at all, not even small, because there has been absolutely no effort to turn the observations to practical account in forecasting by the only body in the kingdom [the Meteorological Council] which possessed the necessary machinery and funds'.[203]

In March 1904, Lord McLaren, then the Royal Society of Edinburgh's vice president, gave a speech – subsequently circulated in newspapers and by the Scottish Meteorological Society – in which he criticised the way Maxwell had conducted his inquiry and highlighted the Meteorological Council's neglect of telegraphic reports from Ben Nevis in preparing their forecasts. He went so far as to argue that Maxwell's Committee had allowed time 'to run on' until money for the observatories ran out.[204] Maxwell responded publicly to McLaren in late April by way of a letter in *The Scotsman*, where he drew attention to the history of the supply of observations from Ben Nevis to London. Maxwell pointed to the observatory directors' decision to grant exclusive rights to the Press

[200] Letter from R. T. Omond to H. Maxwell, 23 July 1903, including clipping from *Evening Despatch*, 9 June 1903, National Records of Scotland, MET1/5/3/5.
[201] Letter from H. Maxwell to R. T. Omond, 24 July 1903, National Records of Scotland, MET1/5/3/5.
[202] Letter from R. T. Omond to H. Maxwell, 29 July 1903, National Records of Scotland, MET1/5/3/5.
[203] Letter from R. T. Omond to J. Stirling-Maxwell, 24 November 1903, National Records of Scotland, MET1/5/3/5, original emphasis.
[204] H. Maxwell, 'Sir Herbert Maxwell and Lord McLaren', *The Scotsman*, 23 April 1904, National Records of Scotland, MET1/5/3/5, no page.

Association, noting that '[i]t has always been a rule of the Council to publish with their daily forecasts the data on which such forecasts have been founded'. The deal with the Press Association therefore delayed the Council from conforming to this practice with the observatories' data and so debarred the Council from using it in framing forecasts. Maxwell repeated the Council's claim that this departed from the conditions of the observatory's annual £100 grant. Maxwell concluded his letter noting that copies had been sent to the Scottish press: 'As so much publicity has been given to your criticism of the proceedings of my Committee and upon the discretion of the Meteorological Council in conducting their business, I think it is right that the public should be made aware that there is another side to both questions.'[205] McLaren responded, also in the press, asserting that the Meteorological Council 'were unrestricted as to the use they might make of the Ben Nevis observations for weather forecasting'.[206]

Maxwell's report, laid before Parliament in June 1904, seemed to include amongst its more general recommendations the proposal to continue to provide £350 to the directors of the Ben Nevis Observatories. However, many commentators claimed that the report was unclear – Cargill Gilston Knott, the Scottish seismologist, pointed to the report's 'unintelligibility' regarding the £350 grant. He was castigating of the claims that the observatories' data were of no practical value, arguing that the report's 'utilitarian argument is essentially unsound. It would be interesting to know how much money is spent by our Government through the Admiralty and otherwise in "accumulating" magnetic observations. Many of them are made with an accuracy far surpassing the practical needs of the navigator.'[207] John Aitken was particularly critical. Writing to Omond on 28 June he declared:

> They don't – English like – want it to succeed or be useful. They seem to propose – but don't count your chickens till they are hatched – to give you £350, and dictate all conditions + take all they want?!! I know [William Napier] Shaw [Secretary of the Meteorological Council from 1900] fears the responsibility of having men on the Ben lest anything should happen. So they won't take it over even if offered. They are a hopeless, helpless lot. . . . That beastly report has found me very much out of temper.[208] [Aitken signed his letter, 'Yours disgustedly'.]

[205] H. Maxwell, 'Sir Herbert Maxwell and Lord McLaren', *The Scotsman*, 23 April 1904, National Records of Scotland, MET1/5/3/5, no pages.
[206] J. McLaren, 'Lord McLaren and Sir Herbert Maxwell', *The Scotsman*, 27 April 1904, National Records of Scotland, MET1/5/3/5, no pages.
[207] Letter from C. Gilston Knott to R. T. Omond, 22 June 1904, National Records of Scotland, MET1/5/3/5, no pages.
[208] Letter from J. Aitken to R. T. Omond, 28 June 1904, MET1/5/3/5, no pages.

The directors of the observatories wrote to various MPs, including Herbert Maxwell, pointing out that the £350 grant nowhere near met the full annual cost of £1,000. Writing to the prime minister, Omond argued that the report was 'inexplicably erroneous' in its assumption that £350 was enough to keep the two observatories running. Asking for at least one more year's money, he begged 'that you will not look on them as pleading for a Scottish Institution. The Observatories are only Scottish because Ben Nevis is situated in Scotland. The work of the Observatories is national. It belongs indeed to all nations, though it has clearly a special importance to our Island Nation, with its distant colonies and dependencies all over the world.'[209]

Despite statements of support from various British and European scientists, the directors failed to extract further financial support from the Meteorological Council and Omond wrote to the Council stating that they would no longer be taking up the offer of the £350 grant. In September 1904, the process began of closing up the summit observatory and removing the instruments, fittings and supplies. Observatory staff were instructed to stop hourly observations at noon on 1 October 1904 and self-recording instruments were to be stopped at 10 am on the same day. Weather reports also ended then. Instruments, fittings, books and records, clothes and supplies were to be brought down to Fort William by horse and stored at the low-level observatory. The cook's contract was terminated but the superintendent, Angus Rankin, stayed on as a salaried employee at Fort William while he supervised the dismantling of the summit observatory and continued to run the low-level observatory.[210] The Ben Nevis Observatory became the property of the landowner in October 1904. The Fort William Observatory continued to operate until the end of the year, after which it was sold and the instruments returned to the Meteorological Council.[211] Work continued on the principal legacy of the two observatories: the processing and publication of the hourly meteorological observations. The first two volumes of data had been published in volume 34 (1890) and volume 42 (1902) of the *Transactions of the Royal Society of Edinburgh*, which covered the period 1883–92. The third volume (volume 43) was published in 1905 and the final volume (volume 44) in 1910, which came in two

[209] Letter from R. T. Omond to J. Balfour, 14 July 1904, National Records of Scotland, MET1/5/3/5, no pages.
[210] Letter from Directors of the Scottish Meteorological Society to Angus Rankin, 14 September 1904, National Records of Scotland, MET1/5/3/5.
[211] Letter from R. T. Omond to Mr. D. McLeish, 28 October 1904, National Records of Scotland, MET1/5/3/2.

parts, the first covering the period 1893–1902 and the second covering 1903–04.[212]

The *Meteorological Magazine* regarded the closure of the observatories as a 'national calamity'. The *Magazine* argued that its editors could

> only avoid the old and unsatisfactory explanation that someone has blundered, by adopting the older and less satisfactory explanation that several people have blundered. The blunder has been made, and it has been intensified by some well-meaning, but ill-advised, writers and newspapers trying to make the cessation of the observatories a Scottish grievance. The Directors of the Ben Nevis Observatories very properly repudiate any such suggestion. It was a splendid and public-spirited act on the part of the Scottish people to render so substantial and considerable a service to science as the founding and support of a high level observatory at the highest possible level; but the fact that they did so does not make the observations of more value to Scotland than to England. ... We look upon the loss as a loss to science.[213]

The report went on to blame Parliament and government for 'looking at scientific research in a local or a merely utilitarian light'.[214] The German Weather Bureau, who had been receiving regular weather reports from the observatories for several years, also expressed 'great regret' at the closure: 'The stoppage of the observatories on the mountain top, which occupies such a unique position, and at the associated low-level station, will indeed be found to be a heavy loss to meteorological research.'[215] Meanwhile, Archibald Geikie, the joint secretary of the Royal Society of London, wrote to the Treasury to report that the £350 saving from the withdrawal of the grants to the observatories at Fort William and Ben Nevis would help to cover the increased expenditure arising from the Meteorological Office's new pension scheme. Given this new demand Geikie expressed his pessimism about the ability of the government

[212] A. Buchan (ed.), 'The Meteorology of Ben Nevis', Special issue of *Transactions of the Royal Society of Edinburgh*, 34 (1890), i–lxiv and 1–406; A. Buchan and R. T. Omond (eds.), 'The Ben Nevis Observations 1888–1892', Special issue of *Transactions of the Royal Society of Edinburgh*, 42 (1902), i–xiv and 1–552; A. Buchan and R. T. Omond (eds.), 'The Ben Nevis Observations 1893–1897', Special issue of *Transactions of the Royal Society of Edinburgh*, 43 (1905), 1–564; A. Buchan and R. T. Omond (eds.), 'The Ben Nevis Observations 1898-1902', Special issue of *Transactions of the Royal Society of Edinburgh*, 44, part I (1910), i–v and 1–463; A. Buchan and R. T. Omond (eds.) 'The Ben Nevis Observations 1903, 1904 and Appendix', Special issue of *Transactions of the Royal Society of Edinburgh*, 44, part II (1910), 464–714.

[213] Anon, 'Meteorological Notes and Letters', *Meteorological Magazine*, 39 (November 1904), 206–10, 206.

[214] Anon, 'Meteorological Notes and Letters', 206.

[215] German Meteorological Office to Scottish Meteorological Society, 30 September 1904, National Records of Scotland, MET1/5/3/5. The Office conceded that the Sonnblick Observatory in Austria was only made possible through the work of the Sonnblick Society, which placed private funds at the disposal of the station.

meteorological grant to support much scientific research and wondered if it would be better 'to circumscribe the operations of the Council to routine rather than to expect them to undertake investigation for which they have not adequate means'.[216]

There remained hope that the Ben Nevis Observatory would be reopened. The building continued to be used as an annex to the hotel into the 1910s. There was some discussion of using it during the International Polar Year of 1932–33, when an inspection revealed the building to be in good condition.[217] The Meteorological Office was reasonably enthusiastic, the director noting: 'In my opinion the failure of Ben Nevis observatory before was due to the fact that nobody knew what to look for in the observations. Now that we have fronts and air masses I believe that we could make use of the observations.'[218] However, the costs were deemed impossible to meet and gradually the effects of the weather and of climbers and tourists meant that the observatory decayed beyond use. The building was set on fire on several occasions and in 1950 the flat lead roof was stripped by a party of climbers, who rolled the lead down the mountain to a waiting lorry. The Scottish Meteorological Society eventually paid off the debts it incurred from running the observatories. After suffering a drop in membership, the Scottish Meteorological Society amalgamated with the Royal Meteorological Society and the Ben Nevis Observatories dropped off its agenda.

Conclusion

In her work on Austrian meteorological observatories Coen argues that mountains were variously conceived as laboratories and as fieldsites – either 'a space for producing generalizable atmospheric effects under extreme conditions' or 'a place where distinctive atmospheric conditions are collected like specimens, as clues to broader patterns of geographic variation'.[219] As laboratories, mountains were treated as generic sites of scientific experiment. As fieldsites their unique attributes were studied and revealed. In the history of the Ben Nevis Observatories, the mountain

[216] Letter from A. Geikie to H. M. Treasury, 8 November 1904, Archives of the Royal Society, NLB/23/2/557. Geikie, well known for his geological and geographical work, had actually spent the night at the Ben Nevis observatory in 1895 and wrote about it in his autobiography: A. Geikie, *A Long Life's Work: An Autobiography* (London: Macmillan, 1924), p. 264.

[217] A. Goldie, 'Ben Nevis Observatory', Report to Director of Meteorological Office, 22 July 1931, National Archives, BJ 5/25.

[218] Director of the Meteorological Office to A. Goldie, 25 July 1931, National Archives, BJ 5/25.

[219] Coen, 'The Storm Lab', 465.

began as a fieldsite and ended up as an experiment, although ultimately a failed one.

In his work for the Scottish Meteorological Society Clement Wragge treated Ben Nevis as a fieldsite; as a challenging environment and a demanding journey to be confronted and conquered on a daily basis. Wragge conducted himself as if he were an Alpine mountaineer or an Arctic adventurer. Mountain meteorologists, just like the explorer or climber, came to know their place of inquiry through forms of embodiment – what McNee calls the haptic sublime. Specifically, the conduct of mountain meteorology involved adventurous eyewitnessing and rigorous manly experiences. The labour of observation on the slopes and summit of the Ben mimicked the temporal regimen of the observatory and was lent further credence through the labours it took to reach those spots in good time. The emphasis that Wragge gave to place as a factor in his quest for credibility in claim making was also emphasised by his articulation of a mountain sublime in his reports and popular writing. Wragge's self-denial, bodily suffering and observational labour in turn helped Milne-Home, William Thomson and others to sell the concept of a summit observatory to funders and donors.

The prospect of a purpose-built observatory on the summit of Ben Nevis promised to alter the epistemological ground on which the building was erected – from field to laboratory – but in part the new station internalised many of the virtues that Wragge emblemised. The inability of self-recording instruments to function adequately in the trying climate found on the summit demanded hourly eye observations, routine instrument checks and constant repair work. These demands perpetuated a culture of physical hardship. Newspaper reports of the hourly trials of taking observations in gale-force winds, the challenges of keeping the building from getting buried in snow in the winter, the 'langour and depression of spirits' the observers suffered during their tenures on the summit, all referenced an experience practically interchangeable with those of the likes of Bruce, Scott, Peary or Nordenskjöld. Observers continued to suffer and continued to make a virtue of it. The presence of day trippers and overnight tourists on the summit disrupted observatory routine but also undermined the station's culture of suffering, denial and isolation, which goes some way to explain the extent of the frustration expressed by the observatory personnel at their presence. Staff were happy to compare themselves to the members of the mountaineering clubs, who eschewed the bridle path and suffered too.

The Ben Nevis summit observatory was treated in a number of specific ways as a weather laboratory – as a site for the testing of new instruments for instance – but was more often conceived of as an experiment itself. The low- and high-level observatories and intermediate station were directed to investigate the vertical components of the so-called atmospheric ocean, specifically the vertical interrelations of temperature, pressure and humidity. In the final years of the observatories' life, the directors claimed retrospectively that the rationale for the experiment was to investigate the effects of an eleven-year sunspot cycle on weather patterns. Ultimately though, the observatories *were* the experiment. Under scrutiny was the value of mountain meteorology itself. Observatory personnel and directors claimed to have found various answers – scientific, educational, even patriotic – but whether they liked it or not the observatory experiment came to be judged on its contribution to forecasting, and by extension its value to the nation. According to the Meteorological Council, the Ben Nevis forecasting experiment failed on the grounds of flawed data flow. The Council complained that the communication of weather intelligence was useless for weather warnings, while its deal with the Press Association prevented the Meteorological Office's ability to receive and use the data in the first place. Meanwhile, the directors insisted that the fault lay with the Meteorological Office. Only it had the resources and data networks to turn weather information into weather predictions, while its attempts to assess the observatories' contributions to forecasting as part of the Maxwell inquiry were themselves flawed.

What the observatory was for got mixed up with who it was for and who should pick up the bill. This aspect of the experiment came down to a geographical question – did the Ben Nevis Observatories serve a Scottish, British or European constituency? The directors insisted that the answer was all three, for different reasons: they were culturally Scottish, economically British and scientifically European. The observatories demonstrated Scottish self-sufficiency, public spiritedness and national pride; they were an integral and exceptional node in a network of British weather stations and deserved to be funded accordingly; and they contributed to continental understandings of Europe's large-scale weather patterns.[220] The Meteorological Council agreed with the first two of these but for different reasons and with different consequences: the observatories were clearly a source of Scottish pride and should ultimately be supported financially by those who found an interest in them; they were part of a British network of stations and were due to no more support than any other first-order station, while their value to understandings of high- and low-pressure systems over north-western Europe remained unproven. The Council's repeated refusal to increase the annual grant to the observatories

[220] C. W. J. Withers, *Geography, Science and National Identity: Scotland Since 1520* (Cambridge: Cambridge University Press, 2001), p. 210.

was a reflection of these decisions, its wider commitment to the laissez-faire funding model for science and a consequence of the other pressures on its government grant, most notably the Meteorological Office's pension scheme. Ultimately, the Council defended its refusal to increase funding for Ben Nevis on the grounds that it ended up being somewhere between a fieldsite and a laboratory, both too local culturally and not local enough economically. Its attempt to be both while not managing to be emphatically one or the other, turned, in Henry Smith's prophetic phrase, a 'big and costly experiment' into 'a costly failure'.

4 Geographies of the Rain

> The observations of rainfall in the British Isles not having hitherto been properly classified, the Author has undertaken their collection and arrangement, but finds the number of observations so large, and the care requisite to verify them so great, that no time can be fixed for *their* publication. ... Accuracy ... being of the first importance, it may be desirable to state that the rainfall has in almost every instance been forwarded in monthly, as well as annual depths, by the persons whose names are given as 'Authorities' ... The collection of old observations being still in progress, any information respecting rain registers, past or present, will be esteemed a favour.[1]

In 1862, a young Londoner, George Symons (who we first met in the Introduction chapter), published at his own expense an eighteen-page report on the rainfall of the British Isles. Symons's aim was to generate a detailed picture of British rain, historically and geographically. It was recognised that rainfall could vary quite markedly over short distances, even if the reasons for this variation were the subject of conjecture and debate. The solution to the mapping of Britain's rain and an explanation for local rainfall variation could not be provided by analysis of rain gauge data from a small handful of observatories widely spaced across the territory. What was needed was a rainfall network comprised of hundreds of observing stations. To that end Symons searched out old rainfall records and encouraged the establishment of new ones. This chapter traces the work of Symons and his successors as they worked to construct what was in effect a distributed observatory dedicated to recording Britain's rain. Located necessarily in the relatively modest confines of the domestic garden and staffed by volunteer labour, these rainfall stations nonetheless had to conform to the same principles and strictures as the other observatories discussed in this book. Just like the ships discussed in Chapter 1, Symons's rain station network was part of an experiment in

[1] G. J. Symons, 'British Rainfall', *British Rainfall. On the Distribution of Rain over the British Isles, during the Years 1860 and 1861* [hereafter, *British Rainfall*] (1862), 3.

the study of the weather and climate *beyond* the confines and control of central observatories and scientific institutions. Some of the sites in the network became the focus of intense experimental work into the causes of rainfall variations and as a result became embroiled in meteorological controversy.

The British Rainfall Organisation

Born in August 1838, Symons had an interest in the weather from an early age, making observations with home-made instruments. He joined the British Meteorological Society in 1856, aged seventeen. Although he did not study for a university degree, he did take a course in natural philosophy at the School of Mines on Jermyn Street in central London, taught by John Tyndall. Symons passed the course with distinction. In 1860, Tyndall helped him secure a position as a clerk in the Meteorological Department of the Board of Trade, working under Admiral FitzRoy. His early meteorological interest was in thunderstorms, publishing his first paper on the subject in 1857, which was extended and developed in a paper in *The Builder* magazine in 1859 and in a paper read at the British Association meeting in Oxford in 1860. The study of thunderstorms inevitably involved the study of rainfall. As part of his research in this field, Symons established a small network of rainfall observers. During his work at the Meteorological Department, Symons became increasingly dissatisfied with the extent of rainfall observation. In his leisure hours he sent a circular to all the observers he knew across the British Isles, expressing his intention to collect ongoing and historical, published and unpublished, rainfall observations. His modest campaign was relatively successful and he committed to publish a report of British rainfall data on an annual basis, the first of which appeared in 1862, titled *British Rainfall. The Distribution of Rain over the British Isles, during the Years 1860 and 1861, as Observed at about 500 Stations in Great Britain and Ireland.*

The information Symons received from observers across the British Isles revealed significant divergence in terms of instruments used, observational practices and recording protocols – what constituted a 'rainy day' was not even clear. Although he originally assumed the establishment of a rainfall observation network would be a relatively simple task, he quickly discovered how demanding a task he had set himself, especially with regard to the imposition of a uniform observation system upon a network of volunteer labour. In 1863, Symons was elected to the Council of the Meteorological Society. In the same year he resigned his position at the Meteorological Department to devote himself full-time to the development of a volunteer rainfall observation network.[2] Symons dedicated himself completely to this project – a project that became known as the British Rainfall Organisation. Alexander Buchan, the notable Scottish meteorologist, described Symons's establishment of rain stations, the annual publication of rainfall returns and the study of the distribution of the British Rainfall Organisation as the 'great work of his life'.[3]

From the outset Symons fostered an inclusive culture amongst his observers. In a letter to *The Times* in November 1863, where he petitioned for new observers, Symons pointed out to potential recruits that the role was open to all: 'my correspondents are of both sexes, all ages, and all classes'.[4] In his popular account of rainfall investigations, Symons reported that he had received reports from 'Peers and Peasants, Dignitaries of Law, State, Church and Medicine, Lighthouse-keepers, Gardeners, aye, and a Policeman'.[5] Reflecting back on the history of the British Rainfall Organisation in 1910, Symons's assistant and, later, successor Hugh Robert Mill also celebrated its social inclusivity. Inspections of stations allowed him to develop friendships 'in all strata of society, through which observing provided a truly representative cross-section'.[6] He met shepherds, railway porters, domestic servants, gardeners, farmers, as well as medical officers, sanitary inspectors, local authority officials, engineers, country gentlemen, members of the clergy and leading figures in the British establishment.[7] Symons reinforced this ethos by referring to all his observers as 'authorities' (see the epigraph that opened the chapter) and was wary of imposing rules upon them without their consensus. The collection of meteorological data, he argued, 'must be done by authorities of unquestioned position, men who are not only beyond all bribery, but who would and could give personal care to see that absolute impartiality and absolute justice ruled throughout'.[8] Writing to Mill in December 1900, several months after Symons's death, Herbert Sowerby Wallis – the British Rainfall Organisation co-director with Mill – observed

[2] Anderson, *Predicting the Weather*; D. E. Pedgley, 'A Short History of the British Rainfall Organization', *Occasional Papers on Meteorological History*, 5 (2002), 1–19.

[3] A. Buchan, 'The Monthly and Annual Rainfall of Scotland, 1866 to 1890', *Journal of the Scottish Meteorological Society*, 10 (1892), 3–24, 3.

[4] G. J. Symons, 'Report', *British Rainfall. 1863* (1864), 4–12, 6.

[5] G. J. Symons, *Rain: How, When, Where, Why Is It Measured. Being a Popular Account of Rainfall Investigations* (London: Edward Stanford, 1867), p. 43.

[6] H. R. Mill, *Life Interests of a Geographer 1861–1944: An Experiment in Autobiography* (East Grinstead: privately issued, 1945), p. 97.

[7] Anon, 'British Rainfall Organisation. Transfer to Trustees. Dr. H. R. Mill interviewed', *Morning Post* (9 June 1910), National Meteorological Archive, RMS/2/5/1/1/6.

[8] G. J. Symons, 'English Climatological Stations', *Quarterly Journal of the Royal Meteorological Society*, 7 (1881), 281–3, 282–3.

that Symons never referred to himself as director to his observers, presumably on the grounds that they would have resented it. Many were his superior in social standing and his equal scientifically, Wallis argued, 'and as volunteers he had no right to direct them'.[9] The fact that there was no subscription facilitated participation in the network, as did the promise of free or at-cost copies of *British Rainfall*. The prospect of seeing one's returns featured in the journal functioned as another token of appreciation and was even more of an inducement to observers.[10]

Although Symons's appeal to rainfall observers had been met with significant interest – Symons had received returns from almost 700 stations by late 1863 – he worried that maybe hundreds more remained unreported due to ignorance of his scheme or diffidence on the part of the recorder.[11] His desire for a national network of rain stations remained unfulfilled. In particular, he judged his own network to lack a 'good geographical distribution'.[12] Significant gaps existed in the north, east and south-west of England, in much of rural Wales, Highland Scotland and across most of Ireland. Symons's letter in *The Times* newspaper in November 1863, mentioned earlier in this section, hoped to improve this situation by recruiting observers in more remote districts. At the Newcastle upon Tyne meeting of the British Association in 1863, Symons had been awarded a grant of £20 to cover the costs of making and sending gauges to potential observers in out-of-the-way locations.[13] Symons wrote again to *The Times* in mid-December 1863, reporting an overwhelming response to his initial letter. Correspondence poured in daily from eager observers and applications for 'Association' gauges from isolated districts had 'been so much more numerous than was expected', to the extent that some requests had to be refused.[14]

Symons had to correct the impression that he was not interested in areas of the country not in his list of desirable locations: 'in an amateur system, removal and death are constantly thinning the ranks, and hence it becomes necessary to have a reserve corps ready to fill the vacancies;

[9] H. S. Wallis, letter to H. R. Mill, 4 December 1900, National Meteorological Archive, RMS/2/5/1/1/3.

[10] Vetter, *Field Life*, p. 108.

[11] G. J. Symons, 'Report', *British Rainfall. 1865* (1866), 4–19, 8.

[12] Symons, 'Report', *British Rainfall. 1863*, 6.

[13] Anon, 'Recommendations Adopted by the General Committee at the Newcastle-Upon-Tyne Meeting in August and September 1863', *Report of the Thirty-Third Meeting of the British Association for the Advancement of Science* (1864), xxxix–xliii.

[14] G. J. Symons, 'On the Fall of Rain in the British Isles during the Years 1862 and 1863', *Report of the Thirty-Fourth Meeting of the British Association for the Advancement of Science* (1865), 367–407, 369.

besides which, they act as checks on one another'.[15] Hundreds of letters were also written to the regional and local press across the British Isles with appeals for rainfall records and new observers. At the Birmingham meeting of the British Association in 1865 it was resolved to send a circular to every newspaper in the United Kingdom, containing a list of those stations in the relevant region that Symons was aware of. Symons begged the local reader

> to think for a moment if he or she knows of any one who keeps, or has kept, a rain-gauge; or who has any tables of rainfall (or old weather journals) in their possession. And if they do know of such persons, I ask them on behalf of science, or my fellow-observers, and on my own behalf, to use every effort to secure their assistance. ... We want old records, we want records for the present year, and from many parts of the country we want returns for the future.[16]

Symons received hundreds more replies to his appeal in the local press, although only a small number provided rain records that were unknown to him. Several hundred new stations were established, however, contributing to the 1,200 that were reporting to the British Rainfall Organisation in 1866. Pressure to recruit observers was especially acute in the run up to new decades. Symons placed great weight on the generation of complete sets of observations for decennial periods.[17] Long runs of observations provided valuable data for studies of the secular variation of rainfall and could in turn be used to estimate variations in supply for industry and domestic water use.[18]

Given the private and voluntary nature of the organisation, support by the British Association was crucial to the network's early expansion. An annual grant of around £50 covered some of the costs of purchasing and sending gauges to new observers. In return a British Association Rainfall Committee was established, to which Symons acted as secretary. Its members in 1866 included the meteorological superintendent at Greenwich James Glaisher, president of the Royal Society Lord Wrottesley, the glaciologist John Tyndall, presidents of the Meteorological Society Charles Brooke and John Lee, the civil engineers Robert Mylne and John Bateman, and the geologist John

[15] Symons, 'On the Fall of Rain', 369.

[16] G. J. Symons, letter to the Editor of the *Bedford Times*, reproduced in 'Report', *British Rainfall. 1865* (1866), 8.

[17] G. J. Symons, 'Report of the Rainfall Committee for the Year 1868–69', *Report of the Thirty-Ninth Meeting of the British Association for the Advancement of Science* (1870), 383–404, 391.

[18] G. J. Symons, 'On the Secular Variation of Rainfall in England since 1725', *British Rainfall. 1870* (1872), 53–7; Letters from T. S. Crallan to G. J. Symons on the subject of the rainfall year, 30 December 1870 and 23 March 1871, National Meteorological Archive, RMS/1/5/13.

Phillips.[19] An annual report was produced by Symons for the Committee and published in the British Association's annual proceedings. Other funding came from voluntary contributions from network observers, the occasional bequest or Royal Society grant (including £100 payments to cover an assistant's salary), as well as income from Symons's consultancy work for sanitary and hydraulic engineers. Symons put great store in his network's voluntary and independent nature but argued that funding from leading British scientific organisations demonstrated their approval for his venture. He also conceded that paying observers was necessary in places where no volunteers had come forward and where rain data were badly needed. This arrangement threatened to create a tension and a contradiction in the network. Something similar had undermined the Scottish Meteorological Society's own volunteer weather observation network when the Meteorological Office in London began paying an observer in Stornoway on the Outer Hebrides. The observer in question then refused to supply data gratis to Edinburgh, leaving the Scottish Society feeling undermined.[20]

The arrangement between the British Rainfall Organisation and the British Association continued until 1876, when Symons delivered his final rainfall report to the British Association meeting in Glasgow. Writing in his own journal *British Rainfall*, Symons acknowledged that his organisation had benefited financially from British Association grants, but that the relationship had ended because the British Association wished to force the British government to take over the provision of funds for the maintenance of the rainfall system, while Symons refused 'to allow a system created and developed during fifteen of the best years of my life, to be buried in an obscure corner of some Government office'.[21] Symons warned that 'though by taking the present system and applying more red tape it might be possible to make the machinery even more nearly perfect than it is, it would be at the cost of that intelligent independence of thought which so greatly rules the progress of science'.[22] Symons championed Britain's amateur science tradition, claiming that the experimental work conducted by his members was superior to anything equivalent produced in government work. He thought his volunteer observers would not approve of being placed under the supervision of government officials (even if he had complained in 1870 that England was

[19] G. J. Symons, 'Report of the Rainfall Committee for the Year 1865–66', *Report of the Thirty-Sixth Meeting of the British Association for the Advancement of Science* (1867), 281–321, 281.

[20] Anderson, *Predicting the Weather*, p. 104.

[21] G. J. Symons, 'Report', *British Rainfall. 1875* (1876), 6–9, 7.

[22] Symons, 'Report', *British Rainfall. 1875*, 8.

the only country that left the funding of rainfall observation to private enterprise).[23] Doing so would extinguish the esprit du corps that he claimed existed amongst his observers.[24] Nothing, he declared, 'would induce me to forget the feelings or interests of the observers; they have been uniformly kind to me'.[25]

Although Symons warned the British Association that its decision lost the éclat of being a financial supporter of the rainfall system and of having its observers keeping the name of the British Association 'before the world', he expressed his relief at being freed from having to attend its annual meetings and producing an annual report.[26] Meanwhile, the British Association rainfall committee celebrated its successes in its final report. Its funding had corrected the deficient geographical distribution of Britain's rainfall stations, claiming that the Association's grants had funded the placement of nearly 250 gauges and covered some of the costs of site inspections. A definite unit for the term 'rainy day' had been adopted; a complete code of rules had been drawn up for observers to follow; investigations had taken place into a variety of topics, including the secular variation of rainfall across the British Isles and the determination of the average proportion of the total yearly rainfall in each month; and a variety of experiments had been effected.[27] The next section reviews these achievements as they were put in place over the 1860s and 1870s. It begins with Symons's strategies for managing his observers' instruments, their rainfall stations and their instrumental and inscription practices.

The Rain Gauge

As Symons began to build his network of rain observers, he worried about the patterns and quality of the instruments that were being used. Of all the meteorological instruments in use in the mid-nineteenth century, the rain gauge was often treated as one of the simplest, whether in terms of its purpose, its construction or use.[28] Salter summarised the gauge's purpose as 'the interception of the precipitation over a fixed area, and the accurate measurement of the collected water, which is expressed in terms of depth, i.e. the thickness of the layer of water which would have accumulated on

[23] G. J. Symons, 'Report', *British Rainfall. 1870* (1871), 6–8, 8.
[24] Symons, 'Report', *British Rainfall. 1875*, 8.
[25] Symons, 'Report', *British Rainfall. 1875*, 8.
[26] Symons, 'Report', *British Rainfall. 1875*, 8.
[27] G. J. Symons, 'Report on the Rainfall of the British Isles for the Years 1875–76', *Report of the Forty-Sixth Meeting of the British Association for the Advancement of Science* (1876), 172–203, 173.
[28] A. J. Herbertson, *The Distribution of Rainfall over the Land*, p. 1.

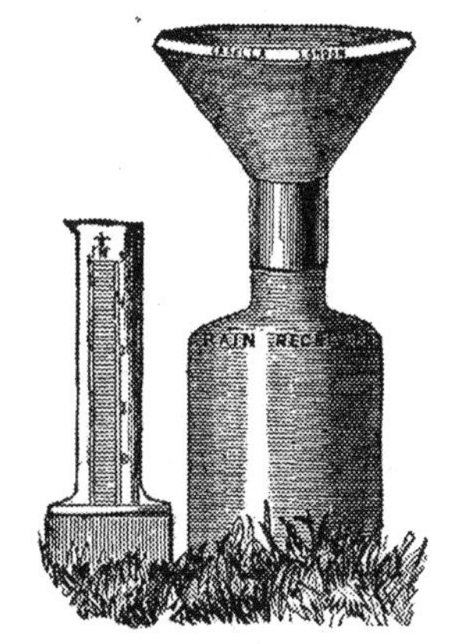

No. 1.—Howard's Bottle Gauge, 5 in. diameter.

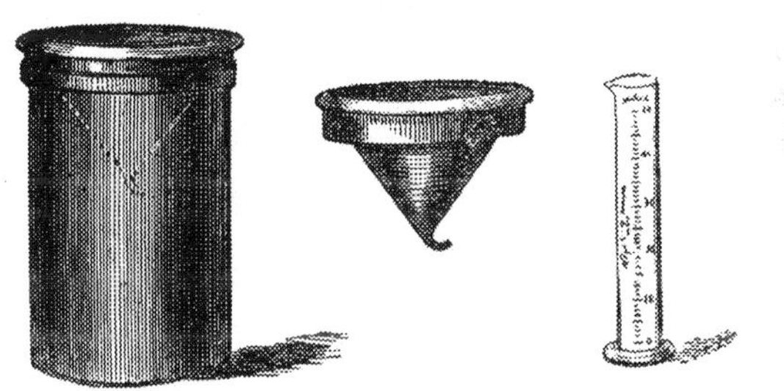

No. 2.—Glaisher's Gauge, 8 in. diameter.

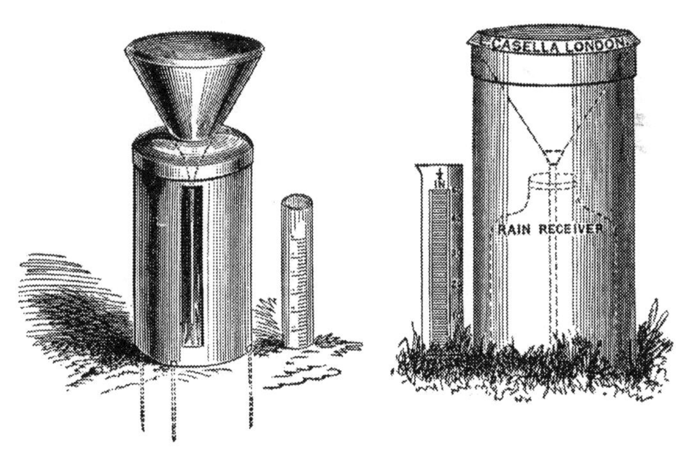

No. 3.—Negretti's 5 in. Gauge.

No. 4.—Casella's 5 in. Gauge.

Figure 4.1 Illustrations of popular patterns of rain gauges in Britain. (Source: G. J. Symons, 'Rain Gauges, & Hints on Observing Them', *British Rainfall. 1864* (1865), facing 11.)

a flat surface if none were re-evaporated nor drained away'.[29] Given that directive, any watertight receptacle could be turned into a gauge simply by leaving it out in the open before a rain shower. The water collected in the receptacle could then either be measured or weighed. The apparent simplicity of the gauge presented both an opportunity and a challenge for meteorologists. The rain gauge was in theory a democratic instrument that asked little of its observer, financially and technically. The collection of rainfall records in remote areas necessitated the use of 'uneducated and sometimes illiterate observers', which required methods and instruments to be 'fool-proof'.[30] However, Symons worried about the proliferation of what he described as 'a very motley group of instruments'.[31] Patching up of instruments by 'country tinmen' was also frowned upon.[32] He called for the production of good-quality, low-cost gauges by reputable instrument makers so as to discourage the use of home-made gauges, 'which nineteen times out of twenty are entirely useless'.[33] In his book *Rain: How, When, Where, Why is it Measured*, Symons summarised the so-called motley group into four loose types: gauges with side tubes, float gauges, mechanical gauges and gauges with measuring glasses.

Gauges with side tubes, or pedestal gauges, were 'luxurious', 'almost handsome', instruments, which Symons suggested had been introduced by the instrument makers Troughton and Simms in 1835.[34] As the gauges were tall they had to be secured with sand or gravel in the base. Measurement was made easy due to the glass observation tube, which meant that observers could technically carry out their daily task of checking the previous day's rain without leaving their house, simply by using a pair of opera glasses.[35] Once read, the water was emptied using a tap. According to Negretti and Zambra, the pattern was intended to have an 'ornamental appearance'.[36] The luxuriousness of this pattern of gauge would 'long secure for them the patronage of those who not only wish to know the amount of rain, but scientifically to ornament their lawns', although the glass tube was vulnerable to frost damage, and the tap could corrode and drip.[37]

[29] M. de Carle and S. Salter, *The Rainfall of the British Isles* (London: University of London Press, 1921), p. 31.
[30] Salter, *The Rainfall of the British Isles*, p. 29. [31] Symons, *British Rainfall. 1864*, 8.
[32] Symons, *British Rainfall. 1866*, 5. [33] Symons, *British Rainfall. 1864*, 9.
[34] G. J. Symons, 'A Contribution to the History of Rain Gauges', *Quarterly Journal of the Royal Meteorological Society*, 17 (July 1891), 127–42, 138.
[35] Symons, *British Rainfall. 1864*, 9.
[36] H. Negretti and J. W. Zambra, *A Treatise on Meteorological Instruments: Explanatory of the Scientific Principles, Methods of Construction, and Practical Utility* (London: Negretti & Zambra, 1864), p. 111.
[37] Symons, *Rain*, p. 5.

Float gauges contained a float and measuring rod, where any rain in the receiver would lift the float and the depth could be read from the rod. Symons conjectured that the pattern was a British one – he remarked on several models in circulation in England in the late eighteenth century, including one described by George Adams, the mathematical instrument maker to King George III, in 1790.[38] The principal benefit of the pattern was that it did not require a glass measure. It also worked well in locations that experienced heavy rain. There were various problems with it, however: the gauge did not record very small quantities of rain clearly; observers needed to remember to add water to set the float to the zero mark; the rod would gradually rise up out of the gauge and would intercept rain that would otherwise have missed the gauge; and like the pedestal gauge it was liable to freezing and evaporation.

Mechanical, or self-registering, gauges were those that employed the collected rain to register its amount, the first of which was claimed to have been designed by Sir Christopher Wren. Wren's design featured a wedge-shaped bucket that tipped and emptied itself when filled. Later designs featured a sort of water wheel with heliacal buckets, where the weight of the water in the buckets caused the wheel to turn and a hand to advance on a clock-like dial.[39] In the late 1820s, a gas-meter maker called Crossley produced a self-registering gauge that was based on a chain of wheels similar to those found in gas meters. The invention of the Crossley gauge had 'been most injurious to the progress of rainfall observation', claimed Symons, due to their appearance of autonomy and self-reliance. Complacent observers failed to notice problems with wheels sticking and water evaporating.[40] Symons wished that other users of the pattern imitated his friend the Rev. W. Steggall, of Thurston, Bury St Edmunds, who arranged for the town's watchmaker to visit the vicarage on New Year's Eve every year, to 'thoroughly clean and oil the gauge, and start it at zero for the new year, finishing up with a substantial supper'.[41] Self-registering gauges were also expensive and so their use tended to be restricted to sizable observatories.

In contrast to the overly complex Crossley gauge, gauges with measuring glasses were considered by Symons cheap, accurate and durable, and so ideal for a volunteer network with a wide social range (see Figure 4.1). This pattern of gauge featured a funnel leading into a rain receiver – often a glass or earthenware bottle – and sheathed in a metal cylinder or box for protection. The water in the receiver was then measured in a separate

[38] Symons, 'A Contribution to the History of Rain Gauges', 137.
[39] Symons, 'A Contribution to the History of Rain Gauges', 140.
[40] Symons, *British Rainfall. 1864*, 9.
[41] Symons, 'A Contribution to the History of Rain Gauges', 140.

vessel with a narrow diameter that would display even small quantities of water. Evaporation was minimised by narrowing the funnel and the small opening at the top of the receiver. Symons noted references to this pattern in the eighteenth century, including by John Dalton in his *Meteorological Observations* in 1788. Luke Howard gave his name to a five-inch gauge, and James Glaisher did the same for an eight-inch gauge. One of the problems of this pattern was its instability in high winds and the fact that water could escape from the shallow funnel due to out-splashing. The Glaisher pattern gauge was distinguished by the inclusion of a projecting flange near the top of the funnel, which was designed to prevent out-splashing. The later Snowdon gauge extended the vertical walls of the funnel to provide a hollow cylinder that could hold a fall of snow and reduce the risk of out-splashing and the formation of wind eddies.[42]

Reflecting back on the history of rain gauges, Salter suggested that the attention Symons drew to the subject of rain observation during his press campaigns had actually led to a diversity of instrument types rather than a movement towards uniformity, as enthusiastic new observers rushed to set up a station without due regard to the instruments they put to use.[43] Prioritising the development of an extensive, voluntary network of rain stations placed pressure on the concerns for uniformity of pattern, especially in the context of the varied social landscape of Symons's network. One way Symons went about ensuring trustworthiness was to recommend the simplest of the gauge designs already outlined – the Glaisher or Snowdon gauge. The simple, inexpensive pattern demanded a level of complexity and promised measurements of a precision that fitted with the busy daily lives of Symons's volunteers. It was a gauge that could be read trustworthily at a glance without much preparatory work and relieved observers of great care in the act of measurement.[44] The pattern's simplicity, portability and robustness distinguished it from the more extravagant mechanical and pedestal gauges, which encouraged complacency and lack of care.

Another way of ensuring trust in instruments was to ascertain who made them, how the makers had calibrated the instruments and whether from personal experience one could trust the testimony of their owners.[45] In 1863, Symons instituted a scheme to examine rain gauges before transmission to their stations. About 100 gauges were examined in this

[42] Salter, *The Rainfall of the British Isles*, p. 34.

[43] Salter notes that the situation changed 'after several series of exhaustive and well-devised experiments', which were a 'definite step towards standardization' and are considered in detail later in the chapter. Salter, *The Rainfall of the British Isles*, p. 32.

[44] Gooday, *The Morals of Measurement*, p. 79.

[45] Gooday, *The Morals of Measurement*, p. 267.

way in that year and another 43 were tested in situ.[46] By 1864 this scheme was already proving a 'somewhat onerous duty' financially and so Symons proposed to imitate Kew's practice of charging recipients for the testing of their instruments. Recipients of the tested gauges were charged half a crown, for which their instrument would be initialled by Symons and accompanied by a printed certificate. Testing of gauges involved filling the measuring glass up to various scale-points and the quantity measured in grains – a single grain weight of water corresponded to 0.008 inch. The difference between the observed quantity and that expected at the particular scale-point for a certain diameter provided that gauge's error.[47] Very small errors, for example 0.001 inch, were deemed less than the ordinary error of observing and so were ignored.[48] Symons went so far as to claim that 'gross errors are now impossible, owing to the number of new tested gauges interspersed among the old ones and their availability as checks upon them'.[49] As with his prioritisation of simple gauge patterns, Symons's tolerance of small degrees of observational error around the scale-point emphasised trustworthiness as a necessarily compromised scientific attribute that had to cohabit with domestic and working lives.

Building a Uniform Network

In order to arrive at any just and true deductions from the observations of rainfall, it is of course first and absolutely necessary that the observations themselves should approach as nearly as possible to perfect correctness. To arrive at this desirable state of things, it seems imperative to eliminate all sources of error, and arrive at tables for correction, for altitude, magnitude, and such other differences as may occur, so as to bring all observations to the same uniform standard.[50]

Choosing an appropriate rain gauge was of vital importance in the development of an observation network. There were similarly important decisions to be made about how to conduct and represent a measurement with that gauge, in a way that lent attributes of trustworthiness to that measurement and its inscription in a table of figures. Care, accuracy and uniformity were crucial terms in this regard. So what did meteorologists mean by uniformity, accuracy and care? Were they or could they become identifiable and even quantifiable attributes

[46] Symons, *British Rainfall. 1863*, 4.
[47] Symons, 'Report of the Rainfall Committee for the Year 1865–66', 317.
[48] Symons, 'Report of the Rainfall Committee for the Year 1865–66', 316–7.
[49] G. J. Symons, 'On the Rainfall of the British Isles', *Report of the Thirty-Fifth Meeting of the British Association for the Advancement of Science* (1866), 192–242, 199.
[50] T. E. Crallan, 'On Rain Gauge Experiments at Framfield Lodge, Hurst Green', *British Rainfall. 1866* (1867), 21–4, 21.

of a rainfall measurement?[51] Symons went about defining these terms and applying them to his network of authorities in several ways: first, through the development of a set of rainfall rules – communicated through written instruction – and, second, through site inspections.

Symons's journal *British Rainfall*, which went out to observers annually, provided him with a means of conveying preferred observational and inscription practices, and descriptions of how he wanted observers to arrange their rainfall station. In the 1864 edition of *British Rainfall*, Symons published the first iteration of his 'Rain gauges, and hints on observing them'. Symons's 'hints' were a pragmatic solution to his inability to pay frequent visits to all stations and inspect them personally, and an alternative way of imposing rules, guidelines and formal structures on his authorities.[52] In 1868, Symons published a formal set of rules for rainfall observers, even if he prefaced it by saying he did not like using the word 'rules' given its 'dictatorial appearance'.[53] His intention in setting out the list of best practice was to ensure accuracy, and particularly uniformity, by regulating practice and precluding error before measurement took place. Symons promoted qualitative meanings of accuracy and precision, which relied on the exercise of care. Decisions had to be taken about the degree of care needed to achieve a certain level of accuracy, in that sufficient care to pre-empt error in experimental work led to 'accuracy qua truthfulness'.[54]

Alongside the use of an appropriate gauge, one of Symons's most fundamental concerns was that observers secured an appropriate site for its placement. Gauges had to be placed on level ground at a good distance from plants, trees, buildings or walls, at least as far away from the base of an object as that object was tall. Securing adequate exposure for the gauge was both an issue of paramount importance and a source of difficulty in a voluntary weather network. The places in which rain stations were established ranged widely, from the lawns of stately homes, public parks, seaside promenades and allotments to private back gardens. Writing in his autobiography, Mill remembered his own time as an inspector, when he saw 'many of the Royal Gardens, from Buckingham Palace to Balmoral . . .; many great houses where members of the peerage took a personal interest, or encouraged their head-gardeners to do so; innumerable rectories, vicarages and manses, Jesuit seminaries and Quaker households; and I found cheerful helpers in many humbler

[51] Gooday, *The Morals of Measurement*, p. 263. [52] J. Vetter, *Field Life, p.* 79.

[53] Symons, 'Rules for Rainfall Observers', *British Rainfall. 1868* (1869), 102–3, 102. Symons did encourage his members to suggest observational improvements, which would be published in *Symons's Monthly Meteorological Magazine* (hereafter *Meteorological Magazine*) and open to discussion.

[54] Gooday, *The Morals of Measurement*, p. 63.

homes'.[55] The open lawns of the rural country house were meant to respond in an intimate way to their site and to nature, principally through their exposures, from which the changing scene could be appreciated. This made them an ideal space of inquiry and underwrote British meteorology's wider conceptualisation of uniformity.[56] However, palace gardens and landed estates were relatively rare sites in volunteer meteorological networks. Although understood as a compromised site of scientific survey, the urban or suburban domestic garden was a much more common location for a rain gauge or other meteorological paraphernalia. Its use was not necessarily problematic, however, in that the private garden was viewed as a space of useful work and familial respectability, while its use as a space to study an appropriately domesticated leisure pursuit like meteorology helped observers to embody virtues of self-denying passivity, self-restraint and automatism, attentiveness and dexterity (Figure 4.2).[57]

Another important means by which Symons monitored and managed his network of observers was through regular site inspections. Symons began to travel around the country to conduct a rigorous examination of rain gauges and their exposure, so that their accuracy, suitable position, height above ground and above sea level could be determined.[58] For instance, in 1862, while still employed at the Meteorological Department, Symons managed to visit around forty sites during his vacations, while a friend visited another thirteen. From his round of visits he found the majority of gauges to be in a good position but a few were judged objectionable, including several where observations had been made for a long series of years. As the network grew, inspections also increased and demanded more of Symons's time, energy and resources. Symons would typically focus on one particular area of the country for his travels. For instance, in 1866, he travelled around Kent and Sussex, visiting most of the gauges there, and then in the autumn he visited Yorkshire, Lancashire, Cumberland and Westmoreland.[59] A visit every five years was only an ideal scenario for most regions, even that becoming difficult as more and more observers were recruited. Nonetheless, even an occasional visit was beneficial in 'removing errors, and stirring up observers to increased care'.[60]

[55] Mill, *Life Interests of a Geographer*, p. 97.

[56] S. Naylor, 'Thermometer Screens and the Geographies of Uniformity in Nineteenth-Century Meteorology', *Notes and Records of the Royal Society*, 73 (2019), 203–21, 212.

[57] L. Daston and P. Galison, *Objectivity* (New York, Zone Books, 2007), p. 39; J. Tucker, 'Objectivity, Collective Sight, and Scientific Personae', *Victorian Studies*, 50 (2008), 648–57, 649.

[58] Symons, *British Rainfall. 1863*, 5. [59] Symons, *British Rainfall. 1866*, 5.

[60] Symons, 'Report', *British Rainfall. 1869* (1870), 6–10, 6.

Figure 4.2 Photograph of weather station at Berkhamsted, 29 July 1896. (Source: National Meteorological Archive, MET/4/2/1–3. Reproduced with permission of the Royal Meteorological Society.)

Symons kept a formal account of his site visits, recording the name of the observer, hour of observation, first year of observation, position of gauge, the azimuth and angular elevation of surrounding objects, the inclination of the ground, the height of the gauge above the ground and the site's height above sea level, the pattern of gauge as well as other general remarks. Those who received a visit 'are frankly told of any departures from ordinary custom or established rules; and the consciousness that "there's a chiel amang ye takin' notes, and faith he'll print 'em," acts beneficially on very many whom I can hardly hope ever to visit personally'.[61] In terms of ordinary custom and established rules, Symons checked that gauges were well exposed and not overshadowed by plants or trees, this usually being the main source of error in rainfall readings. He also checked that gauges were strictly at level – even a tilt of two or three degrees, if inclined towards the wind, would give an error of 5 per cent – and that the gauge or gauges were of an

[61] Symons, 'Report', *British Rainfall. 1867* (1868), 6–32, 7.

appropriate pattern and in good condition.[62] His threat that 'faith he'll print 'em' was not an idle one. In his annual *Reports* to the British Association, Symons published the results of his inspection reports. He praised and admonished in equal measure, employing an unwritten scale, with complete and poor exposure at opposite ends. Comments in 1867 included: 'In garden S.E. of the house and sufficiently exposed. A few trees, but not high ones' and 'No use at all. The gauge was right underneath a sycamore tree, in most ridiculous proximity to the stem. No further observations will be made.'[63] Comments were also made on the quality of the instruments themselves, particularly whether or not they were made by a reputable optician. For instance: 'Clear position, gauge (as usual with privately made ones) very incorrect. Returns have never been published, except under a pseudonym in a local paper; hope they never will be.'[64] Symons urged observers not to move poorly placed gauges. Rather he asked them to place a new gauge and observe the old and new together for a year or two and compare one to another so that corrections could be deduced and the old observations could be made good.

Symons claimed his visits were 'uniformly beneficial'. Writing in his Report to the British Association, Symons said that the suggestions he made during station visits were most cheerfully adopted; that observers 'often warmly express their approval of this inspection, both for its own sake and as tending to secure uniformity of practice, and to increased *esprit de corps* among the observers'.[65] When observers sent in their returns, they were obliged to provide a report on the site itself and it allowed them to provide their own assessment of its suitability. The form demanded similar information to Symons's site visit pro forma, including details on the height of the receiving surface above the ground, height above sea level, diameter of the gauge, a statement on the general position of the gauge and whether it was sheltered by trees or walls. Some observers spent considerable time describing their station and the gauges in use, providing a commentary on the returns, as well as maps and sketches of their site.

As Symons developed his network several questions emerged that posed issues for its uniformity: when to take measurements, and when

[62] G. J. Symons, 'Report of the Rainfall Committee for the year 1867–68', *Report of the Thirty-Eighth Meeting of the British Association for the Advancement of Science* (1869), 432–74, 433.

[63] G. J. Symons, 'Second Report of the Rainfall Committee', *Report of the Thirty-Seventh Meeting of the British Association for the Advancement of Science* (1868), 448–67, 455 and 459, respectively.

[64] Symons, 'Second Report of the Rainfall Committee', 459.

[65] G. J. Symons, 'Report of the Rainfall Committee for the Year 1869–70', *Report of the Fortieth Meeting of the British Association for the Advancement of Science* (1871), 170–227, 172.

to enter them. In terms of taking observations, Symons recommended that gauges be read and emptied daily at either 8 am or 9 am and insisted that in doing so 'uniformity is of the highest importance'.[66] It was crucial that observers continued with their regime unchanged. Symons also warned observers to hold the measuring glass perfectly levelled and not to throw away the water until the reading was entered. He also suggested a second gauge be maintained that could hold a month's worth of rain, which could be used as a check on the daily gauge. To further refine practice, Symons carried out a survey of his observers, which he published in the 1865 issue of *British Rainfall.* It was found that more than half of his observers observed the rain gauge at 9 am and that four-fifths did so between 8 am and 10 am. Only around 20 per cent did so at 9 pm or midnight. The time 9 am was therefore selected as the hour at which all observers were to check their gauges.

When should observers enter the date of the fall of rain? Should they record the rain on the morning the gauge was checked or when it was known to have fallen? This Symons considered to be a 'very knotty point'.[67] Glaisher at Greenwich Observatory argued for the former, as did the British Meteorological Society, while the Scottish Meteorological Society argued for the latter. Symons assumed that most of his observers entered the rain on the day the gauge was read rather than when it fell. Given that this was 'a matter on which uniformity is of more consequence than strict accuracy', Symons sided with his own network, as well as with Glaisher and the British Meteorological Society, but he was also willing to allow a vote on the subject. Symons's position and justifications were later criticised in the pages of the *Proceedings* of the British Meteorological Society, where it was stated that neither Greenwich nor the Meteorological Society in fact followed the practice Symons had described. Rather, it was pointed out that the returns at Greenwich were attributed to the day they fell.[68] Symons wrote in response, expressing his own astonishment at his apparent error. He nonetheless queried the headlines of the Society's observation forms, which seemed to request all rain over the twenty-four-hour period to be entered the following morning. He also cited the recommendations of Admiral FitzRoy in his daily *Meteorological Reports* and of Sir Henry James in his *Instructions for Taking Meteorological Observations* as two notable recommendations that promoted the practice of entering a fall of rain on the date of measurement. This, to his mind, placed the Scottish Meteorological Society in

[66] Symons, 'Rain Gauges, and Hints on Observing Them', *British Rainfall. 1864* (1865), 8–15, 12.
[67] Symons, 'Rain Gauges, and Hints on Observing Them', 12.
[68] Symons, 'Report', *British Rainfall. 1865*, 4.

direct conflict with FitzRoy and James. Resolving this issue had implications not only for daily entries but also for month- and year-end records. Should they stop at midnight on the last day of a particular month or at the morning reading of the first day of the new month? Symons concluded his own defence by striking a pragmatic tone and reiterating the importance of respecting the wishes of his network: 'We must take things as we find them, and be guided by the general practice.'[69]

In terms of when to end a month's worth of readings, Symons pointed out that 9 am on the final day of the month was fifteen hours too soon, while 9 am on the first day of the following month was nine hours too late, and therefore 9 am on the first day was judged more accurate.[70] Observers were therefore instructed to record the rain at 9 am but to enter the reading into the previous day's column. This was communicated to Glaisher at Greenwich, who agreed to recommend that 9 am be adopted, while the fall would be placed against the day preceding that on which it was measured. However, Glaisher requested that a note be placed in the register if rain was known to have fallen between midnight and 9 am on the first of any month. Symons hoped that this practice would be generally adopted and there would be 'few departures from uniformity of practice'.[71]

Another issue where Symons worried about uniformity and accuracy related to the definition of a 'rainy day'. What was it and when should it be recorded as such? He had again sought the opinions of his authorities on the matter. A sample of suggested definitions revealed significantly different approaches. One observer said that a rainy day was one with at least six hours rain; another said when there was at least one hour's rain; a third said when the rain fell 'for the greater part of the day'.[72] In return, Symons proposed to judge a rainy day either on the basis of the number of hours during which rain fell, or on the basis of the fall of a measurable amount of rain (he suggested 0.01 inch of rain). However, the results from his survey showed that 'the observations of hardly any persons are comparable, even (and this is the most provoking part of the matter) when they adopt the same definition'.[73] Symons worried in particular about days of very light rain, which posed a 'test of the carefulness of our observers'. The ideal observer would continue to perform their observations as carefully as normal, while others might decide that dry ground and no rain the previous day meant there was no need to inspect the gauge, so missing

[69] Symons, 'Report', *British Rainfall. 1865*, 5.
[70] Symons, 'Report', *British Rainfall. 1865*, 6.
[71] Symons, 'Report', *British Rainfall. 1865*, 7.
[72] Symons, 'Report', *British Rainfall. 1863*, 11.
[73] Symons, 'Report', *British Rainfall. 1863*, 9.

a small amount of rain in the night. Discordance between proximate stations showed that more care should be given to the measurement and recording of small quantities of rain.[74] The fall of at least 0.01 inch of precipitation was later adopted as constituting a day with rain.[75]

Further instruction was provided for the entry of observations. Two figures were always to be entered after the decimal point, even where the second figure was a cypher, or zero. Neglect of this rule 'causes much inconvenience'.[76] Small quantities of rain between 0.01 and 0.005 inch were to be entered as 0.01. Where measurements were of less than 0.005 inch, a line was to be drawn rather than cyphers inserted. In cases of very heavy rain, Symons suggested that observers take a measurement upon its termination, but to return the water to the gauge so as not to interfere with the morning reading. In cases of snow, observers could either melt what was caught in the gauge's funnel; select a place where snow had not drifted and invert the funnel to scoop up the snow, melt and measure it; or measure the snow depth and take one-twelfth as the equivalent of water.

By 1870 Symons was claiming that his efforts, combined with other leading meteorologists, had led to a rapid awakening of interest in rainfall investigation and that the so-called British rules were being shared and reprinted widely.[77] As well as making claims to widespread uniformity, Symons's evocation of precision measurement to two decimal places mapped out a language of accuracy and its limits. Just as with the late-Victorian electrical practitioners that Gooday discusses, meteorologists like Symons implied the limits of uncertainty or error by the placing of the last digit of a number.[78] Symons used his reports to project a precise, statistical language of accuracy, when his detailed advice about precautions to eliminate error and interference, and demands for precise measurement, actually relied on qualitative understandings of care, credibility and trust. As we will see in the section titled 'Average Rain', the way in which British meteorologists went about managing error was very far from the application of least-squares analysis to the distribution of measurements that was being developed in other national contexts and scientific traditions.

[74] Symons, 'Report', *British Rainfall. 1864*, 6.
[75] Symons, 'Report', *British Rainfall. 1866* (1867), 4–8, 4.
[76] Symons, 'Rules for Rainfall Observers', 103. Stewart and Gee offered similar advice on the dangers of 'over refinement' in the use of decimal places in their contemporary physics textbook. B. Stewart and W. W. Haldane Gee, *Lessons in Elementary Practical Physics* (London: Macmillan, 1885), p. 263.
[77] Symons, 'Report', *British Rainfall. 1870*, 8.
[78] Gooday, *The Morals of Measurement*, p. 12.

Managing the Paper Network

The submission of observer site reports and annual returns were prompted by a circular sent out from Symons's headquarters in Camden in December each year, along with blank forms and spares in case of mistakes or the desire to share data with others. As Vetter points out in his analysis of meteorological networks in the Midwestern United States, the paper reporting form was a crucial instrument of standardisation.[79] Returns started to arrive in Camden in January – in 1868 Symons reckoned he received 500 in the first week of January alone. Annual returns provided data on monthly rain totals, the maximum rain collected in a twenty-four-hour period for each month and the numbers of rainy days. Symons spent a great deal of time going over the returns in the first months of the year. Just like his station visits, inspection of returns involved the identification and management of error as an attribute of care, or indeed deficiencies in its application. There were the errors relating to the instruments themselves, caused by the use of a substandard pattern or home-made instrument or due to a leak. There were also errors engendered by poor position or exposure, or due to shoddy measurement techniques. Transcription errors were a different sort of problem and required different solutions. When managing and compiling observers' annual returns, Symons found the date of the year was often mislabelled; that station names were not always sufficiently precise; the height of the top of the gauge was sometimes entered as on the ground, when the observer meant that the gauge sat on the ground; the height above sea level was often vague given the distance of some stations from an Ordnance benchmark; and observers occasionally used the wrong modes of notation when filling the rain columns'.[80] While these errors could be dealt with relatively easily, the more significant error was to send in one's annual return too early. Symons complained that annually he received three or four returns on the morning of New Year's Day, meaning that, at the very least, part of New Year's Eve was not covered in the return:

> The promptitude of these observers, therefore, involves the necessity of reminding them of the rule. Possibly some of the recipients may think this very tyrannical, but every departure from accuracy and uniformity is a step on the road to uselessness, and I can assure them that it is quite as unpleasant for me to point out errors as it can be for the observers to have them detected.[81]

[79] Vetter, *Field Life*, p. 98.

[80] In 1867 Symons asked 800 observers to carry out barometric measurements three times daily for ten days and then compared calculated heights above sea level to nearby Ordnance benchmarks. Symons, 'Second Report of the Rainfall Committee'.

[81] G. J. Symons, 'Report', *British Rainfall. 1868* (1869), 6–11, 7.

Symons later hardened his approach to the minority of observers who 'seem unable to case a single column of figures' and whose returns were consequently travelling backwards and forwards between the observer and Symons and holding up progress as a result. If observers sent in poor returns in three consecutive years, they were excluded from the network.[82]

Once individual returns had been verified and grouped, they were fastened together and archived away for later reference. As well as chasing observers who had failed to submit their return, Symons was employed in February undertaking the calculations necessary for the compilation of annual tables, writing articles and checking proofs for publication in *British Rainfall*. Letters were only then answered in March and April. By the early 1870s Symons estimated he received several thousand letters a year on top of around 9,000 returns. Although a significant demand on his time, he saw the increase in correspondence as 'a most favourable token, not only of the development of the organization, but of increased uniformity in instruments and observation'.[83] Topics included proposed alteration of a gauge's position, replacement of broken instruments, comparison of old and new gauges, the organisation of new stations and the existence of historical observations.

Editing *British Rainfall* took up an increasing amount of Symons's time across the year. Although the basic format of the journal stayed roughly the same over the course of Symons's life, its scope expanded significantly. The 1862 edition of *British Rainfall* ran to twenty-five pages and included Symons's Report that front-ended the journal; a commentary on the fall of rain, organised by home nation and region; observers' remarks for particular stations, including maximum falls of rain, average falls and other notable events; tabulated data showing the monthly fall of rain for a selection of stations, grouped by home nation; and tabulated information – organised by division (groups of contiguous counties) – on every station, including the name of its authority, gauge details and the station's total annual rainfall. From the late 1860s a detailed account of the meteorology of the year was provided by day and month. The journal also began to feature articles by invited authors as well as by Symons. For instance, the 1865 volume featured an article on rainfall across the Cumberland Mountains by Isaac Fletcher; the 1868 edition had a paper analysing the monthly percentage of mean annual rainfall across the year by Frederic Gaster; the 1870 volume featured a paper by James Glaisher on rainfall observations at Greenwich.[84]

[82] Symons, *British Rainfall. 1870*, 6. [83] Symons, *British Rainfall. 1872*, 6.

[84] I. Fletcher, 'Remarks on the Rainfall among the Cumberland Mountains in 1865', *British Rainfall. 1865* (1866), 9–10; F. Gaster, 'On the Monthly Percentage of Mean Annual

Tucker points out that meteorology from the mid-nineteenth century became an increasingly visual science, which embraced the landscape art tradition, the nascent practice of scientific photography and graphical innovations from the social and natural sciences. From its very earliest volumes, *British Rainfall* reflected wider enthusiasm for the use of visual evidence to support written argument and as a way of translating numbers into easily digestible images; what Maas and Morgan in their analysis of nineteenth-century economics refer to as the 'graphic method', or the 'method of curves'.[85] Illustrations increased in number over the course of the journal's life. They included numerous drawings of gauges and other meteorological instruments; engravings of landscapes containing experimental trials, often taken from photographs; maps of regions showing the positions of its rain gauges, the distribution of isohyets or the progress of a rain band; graphs or curves representing rain data; and abstract diagrams illustrating geometrical proofs. There were also regular tabulated comparisons of mean rainfall for the year in question as against recent decadal averages for selected stations. By 1872 *British Rainfall* was running to over 200 pages and it stayed that way until the close of the century – the 1900 volume ran to 254 pages.

The remainder of Symons's year was spent conducting site visits and testing gauges, producing the British Association report, maintaining the observers' address book and the county book – a list of stations in geographical order – and identifying new observers to replace those either deceased or retired. He also spent a portion of his year searching out, collecting, cataloguing and analysing historical rain records. He petitioned members of his network to supply their family's records of past rain alongside their current observations, worrying that old observation books would otherwise be lost: 'too often they are for a few years jealously treasured as family relics, then they get dusty and are put aside, and finally are either burned or sold for waste paper'.[86] Symons surmised that most rainfall observations remained unpublished and in manuscript form, and reported spending months in the Advocates' Library at Edinburgh and at the British Museum in London copying them out.[87] Both Symons and Glaisher disabused observers of the notion that historical observations were useless and could not be trusted. The opposite was often the case,

Rainfall', *British Rainfall. 1868* (1869), 50–4; J. Glaisher, 'Rainfall at the Royal Observatory, Greenwich', *British Rainfall. 1870* (1871), 42.

[85] H. Maas and M. S. Morgan, 'Timing History: The Introduction of Graphical Analysis in 19th-Century British Economics', *Revue d'Histoire des Sciences Humaines*, 7 (2002), 97–127, 120.

[86] G. J. Symons, 'Report', *British Rainfall. 1871* (1872), 6–15, 7.

[87] Symons, *Rain*, p. 43.

they argued: in the seventeenth and eighteenth centuries, 'to measure the fall of rain was esteemed a serious undertaking, only to be accomplished by first-class men'.[88] James Jurin, the secretary of the Royal Society, promoted the use of rain gauges amongst members of the Society in a paper in the *Philosophical Transactions* in 1723.[89] Pioneering work had been carried out by Mr Townley in Lancashire and by Dr Heberden at Westminster Abbey from the 1760s; Dianes Barrington observed the rain in North Wales in the eighteenth century; and John Dalton did the same in Manchester. Other luminaries with an interest in rainfall included Benjamin Franklin and James Hutton.[90] The period from the 1760s to the beginning of the nineteenth century was positioned as a golden age of rain studies, while the first forty years of the nineteenth century were said to witness a decline in interest from meteorologists, although Henry Mill claimed that the 1818 publication of Luke Howard's *The Climate of London* made keeping a rain gauge 'a common hobby of scientific men and garden lovers'.[91] Other meteorologists evoked the same cast of luminaries in their own histories of rainfall.[92]

Symons reassured the British Association in 1865 that old observations would not be used to determine the absolute mean rainfall of a place or even the geographical distribution of rain. Rather, old observations would be restricted to the determination of the existence or otherwise of secular variation in rainfall: 'On these records turns the national (for such it really is) question of the secular variation of rainfall.'[93] In his evocation of notable early meteorologists, his promotion of rainfall histories, his own archival labour and self-conscious 'rescue' of observation from obscurity and destruction, Symons constructed an ordered and progressive history for rain studies.[94] Symons's archive of rain data was also positioned as a heritage site and space of atmospheric memory. In fact, in his assessment of the legacy of the British Rainfall Organisation, Sir William Napier

[88] Symons, 'On the Rainfall of the British Isles', 193.

[89] J. M. Craddock, 'Annual Rainfall in England since 1725', *Quarterly Journal of the Royal Meteorological Society*, 102 (1976), 823–40, 825.

[90] For a detailed study of the eighteenth-century 'height-catch problem', see W. Parker, 'Distinguishing Real Results from Instrumental Artifacts: The Case of the Missing Rain', in G. Hon, J. Schickore and F. Steinle (eds.) *Going Amiss in Experimental Research* (Dordrecht: Springer, 2009), pp. 161–77, p. 162.

[91] H. R. Mill, 'Introduction', in Anon, *Rainfall Atlas of the British Isles* (London: Royal Meteorological Society, 1926), pp. 5–12, p. 5.

[92] For instance, H. R. Mill, 'The Development of Rainfall Measurement in the Last Forty Years', *British Rainfall. 1900* (1901), 23–45.

[93] Symons, *British Rainfall. 1871*, 7.

[94] N. Michel and I. Smadja, 'Mathematics in the Archives: Deconstructive Historiography and the Shaping of Modern Geometry (1837–1852)', *British Journal for the History of Science*, 54 (2021), 423–41.

Shaw, director of the Meteorological Office and the first Professor of Meteorology at Imperial College London, viewed the organisation's archive, including its various publications, as 'part of the organised memory of British weather' and one that should be preserved.[95] The incorporation of rainfall returns into a rainfall archive also repackaged atmospheric events as data, which refocused attention away from individual events and onto the idea of averages and trends – from reporting facts to making explanations.[96] The amalgamation of returns into a single archive also promised to eradicate errors specific to certain localities. Symons expressed his confidence that with sufficient skill, time and thought, the returns of several thousand observers could be converted into an almost perfect statistical work, where the labours of observers acted as checks one on another and highlighted measurement errors and lack of care.[97]

Sites of Experiment

Operating an expanding rain observation network demanded the implementation of principles of uniformity and accuracy as they pertained to instruments and their placement. Meteorologists were well aware that the key factors to be taken account of were the gauge's exposure and its elevation above the ground. There was a suspicion that rain gauges themselves were having an effect on measurement and that more work was needed to ascertain the effects of gauges' construction materials, as well as their size and form. A dissatisfaction with the proliferation and limitations of rain gauges led to questions about their degree of accuracy and how that accuracy could be quantified.[98] To determine the effects of gauges and their exposure on measurement – referred to collectively by Symons as 'the rain gauge question'[99] – a series of experiments were implemented in various locations across England. The experiments ran from the early 1860s until 1890, with a view to ascertaining the best form of gauge and its ideal situation, as well as the corrections needed to adjust observations made at different elevations.

The first set of experiments was undertaken by Major (later Lieutenant-Colonel) Michael Foster Ward in the grounds of his home, Castle House, in Calne, Wiltshire. Having just retired from the Army,

[95] Letter from N. Shaw to H. R. Mill, 18 October 1915, National Meteorological Archive, RMS/2/5/1/1/8.
[96] Maas and Morgan, 'Timing History', 99.
[97] G. J. Symons, Preface, *British Rainfall. 1872* (1873), 5.
[98] Gooday, *The Morals of Measurement*, p. 80.
[99] G. J. Symons, 'Report', *British Rainfall. 1873* (1874), 8–9, 9.

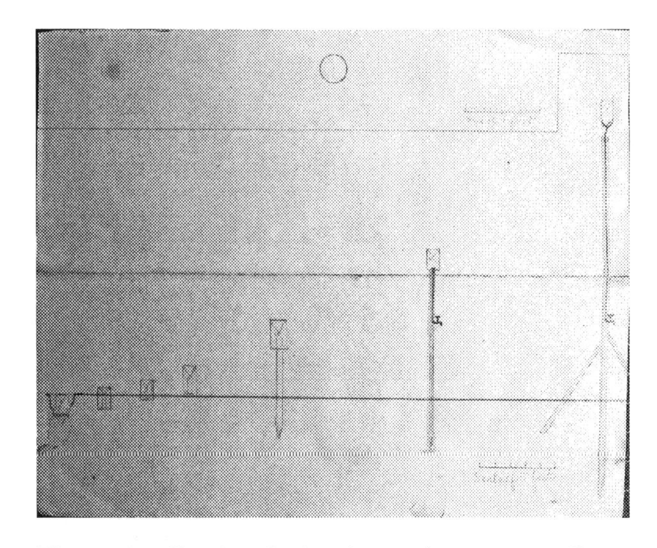

Figure 4.3 Sketch of elevation series gauges. (Source: National Meteorological Archive, RMS/2/3/1/1. Reproduced with permission of the Royal Meteorological Society.)

Ward offered his services to Symons in May 1863.[100] Symons was delighted, writing to Ward to 'express my conviction that nothing has yet been done so generally, practically useful to those who desire to render measurements of rainfall strictly comparable as the series you propose to conduct'.[101] Ward set out to investigate two factors that may affect rainfall collection: elevation and size of gauge. To investigate the effects of the height of the gauge above the ground on the amount of rain collected, Ward installed eight-inch-diameter Glaisher gauges with rims at ground level, at two inches, one foot, two feet, three feet, five feet, ten feet and twenty feet above the ground. An additional five-inch gauge was installed at twenty feet above the ground for comparison. The gauges were mounted on poles and at a distance from any trees or buildings (Figure 4.3). The use of poles rather than placement on the roofs of buildings made the experiment unique and was crucial given concerns about the possible effects buildings had on the amount of rain collected.[102] Water was drawn down from the highest gauges using

[100] Letter from M. F. Ward to G. J. Symons, 25 May 1863, National Meteorological Archive, RMS/1/5/13.

[101] Letter from G. J. Symons to M. F. Ward, 4 June 1863, National Meteorological Archive, RMS/1/5/13.

[102] G. J. Symons, no date, Calne rain gauge experiments, National Meteorological Archive, RMS/2/3/1/1.

composition gas tubing running through the centre of the pole.[103] To act as a check on Ward's experiments, another more limited trial was established in 1863, by the Rev. J. Chadwick Bates at St Martin's vicarage at Castleton Moor, near Rochdale. Five- and eight-inch gauges were placed at twenty feet, five feet and one foot above the ground. An additional three-inch gauge was placed one foot above the ground. Both sites were also equipped with self-recording anemometers. Ward's and Bates's social positions and the fact they volunteered their time and undertook to cover the costs of the experiments themselves were judged to ensure 'the fairest possible examination' of the rain gauge question.[104]

A second set of gauges was installed to measure the possible effects of gauge aperture on water collection. Gauges with 1, 2, 4, 5, 6, 8, 12 and 24 inch diameters, as well as square gauges of 25 and 100 inches in area, were used. These gauges were of various designs, chosen to represent the patterns most common for each aperture, although some attempt was made to standardise them.[105] An additional five-inch gauge was included with an in-sloping flange, along with pit and snow gauges. Symons pointed out that the Scottish Meteorological Society used 'odd little gauges of which I am rather suspicious' and wrote to the secretary of that Society to ask for the loan of one of their gauges to trial alongside Ward's.[106] All of these gauges sat one foot above the ground and were collectively referred to as the magnitude series.[107] Ward met the costs associated with the trials and employed a local watchmaker and chemist, Mr Rowdon, to construct some of his gauges, along with gauges supplied by Louis Casella.[108] It was later claimed that Symons first saw the Snowdon pattern of gauge in Ward's garden, devised by Rowdon and later adopted for a series of observations on the Welsh mountain from which the pattern received its name.[109] Ward committed to reading each of the gauges daily, along with the help of a domestic servant. He

[103] Symons, *British Rainfall. 1865*, 17.

[104] Symons, *British Rainfall. 1863*, 9. Bates also undertook a study of the temperature of rain, using amended gauges with thermometers fixed inside and out. Bates's 1865 report was included in Symons, *British Rainfall. 1865*, 10–12.

[105] Symons, *British Rainfall. 1865*, 17. Cans were fitted around those gauges – typically the smaller ones – that were previously formed by a simple funnel resting on a glass or earthenware bottle, so that they mirrored the construction of the larger gauges. C. H. Griffiths, 'Rain Gauge Experiments at Strathfield Turgiss', *British Rainfall. 1868* (1869), 12–30, 12.

[106] Letter from G. J. Symons to M. F. Ward, 13 June 1863, National Meteorological Archive, RMS/1/5/13.

[107] Symons, *British Rainfall. 1863*, 8.

[108] Letter from M. F. Ward to Symons, 5 June 1863, National Meteorological Archive, RMS/1/5/13; Anon, 'Colonel Michael Foster Ward', *Meteorological Magazine*, 50 (October 1915), 1.

[109] Anon, 'Obituary', *British Rainfall. 1915* (1916), 45–50, 48.

Figure 4.4 Experimental gauges at Castle House, Calne. (Source: *British Rainfall. 1865* (1866), frontispiece.)

estimated it took him almost ninety minutes to register the rain each morning at the twenty-nine gauges he had in operation.[110] The observations continued until 1868.

An image of the site of the trials featured as the frontispiece to the 1865 issue of *British Rainfall*, showing a fenced enclosure containing the gauges, anemometer and thermometer screen, set on level ground with trees in the mid- and background (Figure 4.4). Although photographs of the experimental sites never featured in the journal, engravings from photographs were included for all the rain gauge experimental sites. (One photograph survives in the National Meteorological Archive – see Figure 4.5.) The engravings helped readers to visualise the experimental work, 'almost as if they had absolutely seen the gauges'. They also illustrated best practice in the operation of a rainfall station and, most importantly, demonstrated the quality of the site's arrangement and its good exposure.[111] In the case of the Calne engraving Symons worried that the image's perspective foreshortened the distance between the gauges and the trees, giving the appearance that the trees overshadowed the gauges and undermining the crucial

[110] Letter from M. F. Ward to G. J. Symons, no date, National Meteorological Archive, RMS/1/5/13; Symons, 'On the Rainfall of the British Isles', 200.
[111] Symons, *British Rainfall. 1865*, 16.

Figure 4.5 Photograph of elevation series gauges. (Source: National Meteorological Archive, RMS/2/3/1/1. Reproduced with permission of the Royal Meteorological Society.)

principle that gauges were fully exposed to the atmosphere. A plan view of the enclosure was provided to assuage any worries on this score (Figure 4.6), along with the assurance that some of the trees had been cut down 'for fear they might be *thought* to affect the results'.[112]

Ward and Symons were initially cautious about releasing the results of the Calne experiments, for fear of inflaming the so-called battle of the gauges, particularly in relation to the best size and pattern of gauge. Different patterns had their own respective champions, Symons complained, 'ready to pounce on every hasty word'.[113] However, in the 1864 edition of *British Rainfall*, Symons published monthly abstracts for both the Calne elevation and magnitude series, along with a diagram that plotted depth of rain against elevation (Figure 4.7). (Symons continued

[112] Symons, *British Rainfall. 1865*, 16, original emphasis.
[113] Symons, *British Rainfall. 1864*, 12 and 13, respectively.

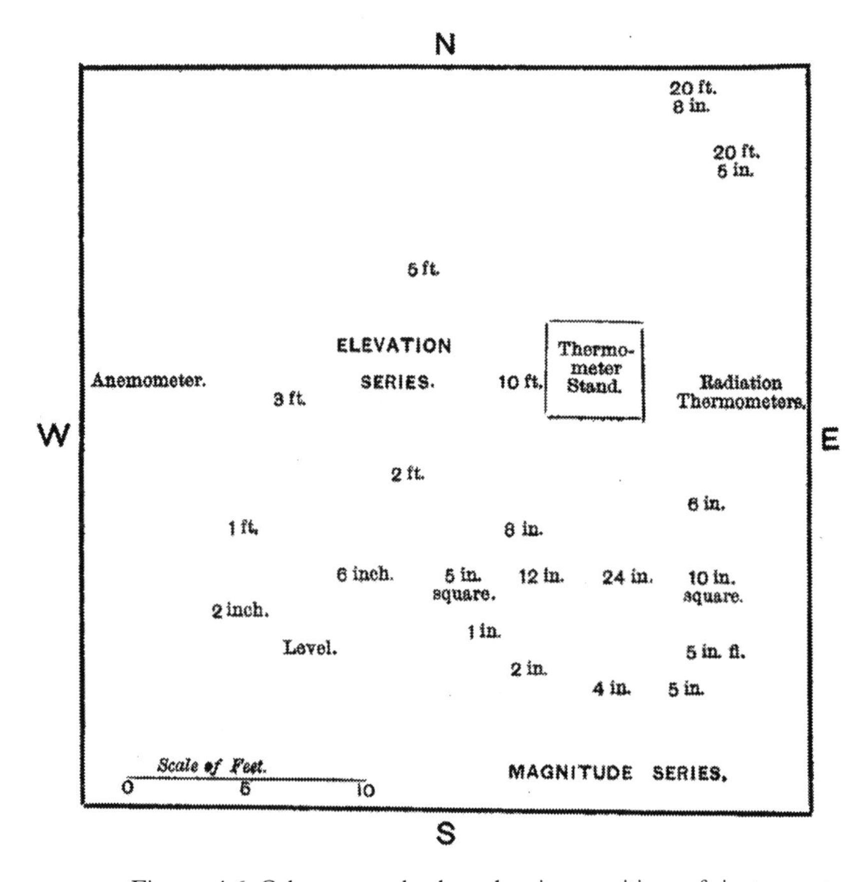

Figure 4.6 Calne ground plan showing position of instruments. (Source: *British Rainfall. 1865* (1866), 16.)

to publish monthly tables in subsequent editions of *British Rainfall* until 1868, alongside the monthly abstracts for Bates's observations at Castleton.) The curve was a good example of the increasing use of the graphical method in the mid-century to explore potential causal relations between two variables. Symons's plot so resembled a perfect theoretical curve that he could not resist drawing out some tentative explanations for the trend: that variation of rainfall with height might be amenable to a regular law; that in-splashing might explain the significant rise in rain collected below one foot; and that all gauges should be placed at a uniform height or would need corrections applied.[114] Symons's decision

[114] Symons, *British Rainfall. 1864*, 13.

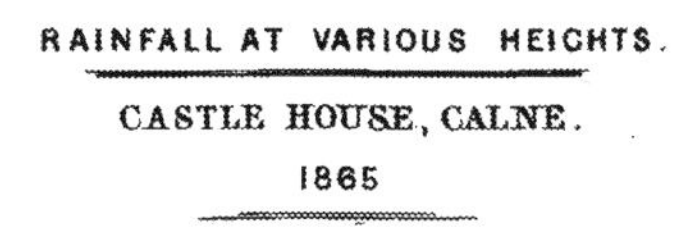

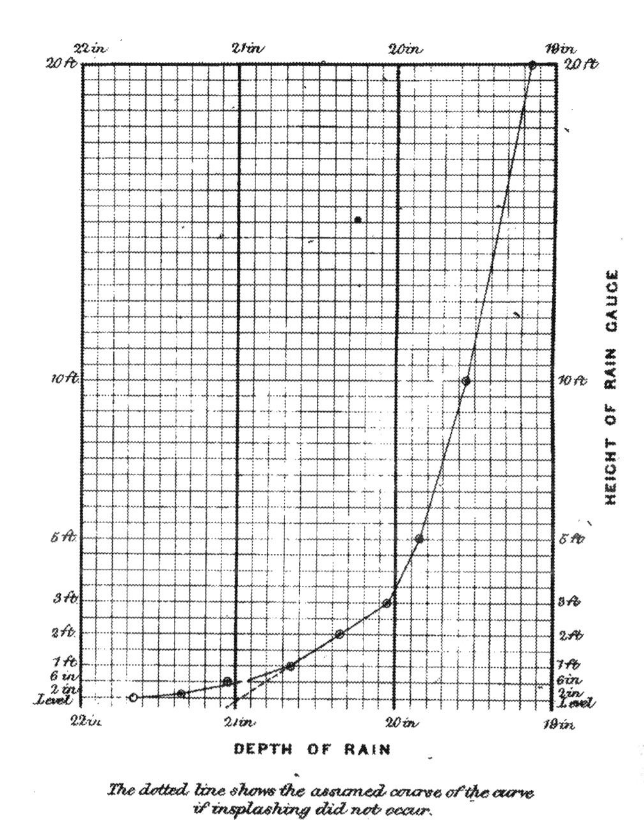

Figure 4.7 Graph of rainfall at various heights, observed at Castle House, Calne. (Source: *British Rainfall. 1864* (1865), facing 12.)

to discuss the results at Calne certainly did lead him to getting drawn into gauge battles. He was criticised in the *Mechanics' Magazine* in 1866 for supposedly advocating small rain gauges, which the author of the article called 'pipkins' and 'toy like things'.[115] Whilst complaining that he had expressed no view on the subject, Symons again presented data from the Calne experiments to show that small five-inch gauges registered the same

[115] Symons, *British Rainfall. 1866*, 27 and 28, respectively.

as those twenty times their size – a fact he enjoyed presenting to the 'advocates of washing tubs instead of pipkins'[116] – but he acknowledged that Ward's eight-inch gauges registered high compared to other apertures and was unable to explain away the anomaly.

Another aspect of the rain gauge question pertained to the materials with which gauges were constructed. Like observers, materials seemed to exhibit degrees of trustworthiness and untrustworthiness and were just as hard to regulate. Symons carried out experiments at his home in Camden, north London, in 1866, with gauges made of earthenware, glass and metal. His aim was to test the effects of material composition on gauge performance, particularly in relation to leakage, fluid retention and evaporation, but he quickly deemed the experiments 'desultory' and 'incomplete'.[117] Rather than pursue the experiments himself, Symons decided to pass the gauges on to another volunteer observer, T. E. Crallan. From January 1867 the Rev. Crallan – chaplain at the Sussex County Lunatic Asylum – began a set of gauge experiments at Framfield Lodge, Hurst Green, in Sussex. The gauges themselves were all of a similar pattern and with the same diameter. They all stood on a wooden shelf with their rims three feet above the ground. However, the funnels were made of different materials, including metal, glass, earthenware and ebonite. Some of the gauges were painted black and some white. One was insulated with wet sand, another with hair felt, with the purpose of arriving at 'some gauge which from a combination of form, size, construction, and material, should as nearly as possible give us a correct measure of the amount of rainfall upon the earth's surface, and to determine, if possible, the corrections required to reduce observations with the various forms of gauge hitherto used, to their true values'.[118] Crallan published the results of his experiment in *British Rainfall* in 1867. In terms of material, he found there was more capillary action on earthenware and glass funnels and the least on ebonite. He recommended the tubes on the ebonite funnels because they were longer and narrower, with a better conducting surface, so that they experienced lower evaporation than those with short, wide tubes. He also found vertical rims to be better at preventing wind eddies from causing water loss than inclined rims.[119]

[116] Symons, *British Rainfall. 1866*, 28.

[117] G. J. Symons, 'Incomplete Rain Gauge Experiments', *Meteorological Magazine*, 1 (December 1866), 96–7, 96.

[118] Crallan 'On Rain Gauge Experiments at Framfield Lodge', *British Rainfall. 1866*, 23.

[119] T. E. Crallan, 'Rain Gauge Experiments at Framfield Lodge, Hurst Green, Sussex', *British Rainfall. 1867* (1868), 45–9.

Due to work commitments, Crallan had to discontinue his experiments in late 1867 and his gauges were passed on to the Rev. Charles Higman Griffiths, who installed them at his rectory at Strathfield Turgiss, near Reading. Crallan's gauges were joined by another set – those passed on by Colonel Ward when he moved from Calne to Switzerland. Symons also passed on the gauges he had been using in London and Griffiths supplied a number of additional instruments himself. Griffith had become rector of All Saints Church in Strathfield Turgiss in 1862, when he began collecting rainfall observations. The church sat on the estate of the Duke of Wellington. Arthur Wellesley, the second Duke of Wellington, was Griffith's patron.[120] Griffith took advantage of his domestic situation to reconfigure the rectory grounds and neighbouring fields into a meteorological test site. Not only did he install forty-two rain gauges across the property, he also began a trial of thermometer screens for the British Meteorological Society. Running from 1868 to 1870 the thermometer housing trials involved testing a number of different patterns of screen, with a view to determining which screen provided the best protection for the thermometers without unduly affecting the performance of the instruments contained inside. Griffith's screen trials were crucial in the adoption of the double-louvred Stevenson screen by the Royal Meteorological Society and other bodies.[121]

Experiments in Exposure

Schaffer argues that the new Victorian physics laboratories were based in part on a 'tranquil fantasy and strenuously engineered reality of a place in the country', with a building boom on Britain's landed estates that provided spaces where country house physics could be pursued.[122] The rectory at Strathfield Turgiss did not represent the 'apparently effortless privilege' of nearby Strathfield Staye House, home of the dukes of Wellington and before them the Pitts, but it did share the more generic characteristics of the country house and its grounds, with its elite, 'secluded and improvised geography'.[123] The Strathfield Turgiss rectory and the country house also shared a highly deliberate regard to their natural setting and particularly to their exposures. Griffith described the place chosen for the installation of the gauges as

[120] S. Burt, 'An Unsung Hero in Meteorology: Charles Higman Griffith (1830–1896)', *Weather*, 68 (May 2013), 135–8.

[121] Naylor, 'Thermometer Screens and the Geographies of Uniformity', 208.

[122] S. Schaffer, 'Physics Laboratories and the Victorian Country House', in C. Smith and J. Agar (eds.), *Making Space for Science* (Macmillan: Basingstoke, 1998), pp. 149–80, p. 153.

[123] Schaffer, 'Physics Laboratories', 150 and 152, respectively.

Figure 4.8 Experimental gauges at Strathfield Turgiss Rectory, Hampshire. (Source: *British Rainfall. 1868* (1869), frontispiece.)

in a portion of the glebe of about 18 acres, perfectly free from trees or local influence of any kind, the large or timber trees being at a considerable distance, and the ornamental trees or shrubs being low and distant from the site selected. The land itself is perfectly level as far as the glebe extends . . . so that a more fair or more open exposure could hardly have been selected.[124]

An engraving of the rectory and its grounds further emphasised its open and exposed situation (Figure 4.8).

Griffiths divided the forty-two gauges into five groups. First, there was the magnitude series, based on Ward's collection of gauges along with several additions, numbering seventeen in total. These were set in a square plot of grass land, surrounded by an iron fence and wire game netting. Second, there was the elevation series, consisting of ten gauges, five of which were placed on roofs. The roof gauges, placed in various positions on the rectory barn and rectory house, were erected to show the effects of placing a gauge in a bad position. The third set of gauges was the material series, from Crallan, and the fourth comprised gauges with a variety of rims and flanges. Griffith added an evaporator and a storm gauge that measured

[124] Griffith, 'Rain Gauge Experiments at Strathfield Turgiss', *British Rainfall. 1868* (1869), 12–30, 15.

rate, rather than quantity, of fall. The material series was soon moved from the rectory garden to the open field and grouped with the elevation series. Griffith numbered the gauges and began his round of 9 am inspections with a different gauge each morning. One key challenge was the maintenance of a perfectly correct level for the receiving aperture of the gauges, which was crucial given the effect any variation had on the quantity of rainwater received. To help him mitigate this issue, Griffith made use of a tool designed by Symons and constructed by Casella, which was in effect a double spirit level at right angles to one another. A second issue was ensuring the readings of the rain were taken correctly. To do so Griffith made use of another improvised instrument – Ward's measuring table – on which the glass could be rested rather than trusting to the hand.[125] Writing in *British Rainfall* in 1869, Griffith reported on the performance of various patterns of gauge. He judged, like Symons, that aperture size made no difference to the efficacy of the gauge; advocated funnels with an upright sharp edge and no flange or shelving angle; pointed out the critical importance of the gauge being perfectly level; and supported Crallan's own findings that ebonite was an excellent construction material whilst also arguing for the use of copper funnels.[126]

The experiments at Strathfield Turgiss were terminated at the end of 1869. Griffith continued to monitor the elevation gauges but passed the magnitude gauges to the Rev. Fenwick Stow so that their performance could be tested in a 'rougher climate' – in 'an exposed situation on the Yorkshire upland, facing the sea'.[127] Despite the roads being blocked with snow, Stow was able to take ownership and erect the gauges on New Year's Day, 1870. The site had come to the attention of Britain's leading rain observers when Stow moved to Hawsker from Tunbridge Wells, writing to Glaisher shortly after his arrival in June 1869 to report that

> [t]he meteorology of this place seems interesting. When the wind is off the land it does not differ apparently from other places further inland, but when the wind is from the sea not only is the range of temperature much less, but the maximum generally occurs in the forenoon, with sometimes a second maximum between 5 and 6 pm. The humidity is also greatly increased, and one or two days last month when there was rain with an east wind, the clouds came so low as completely to envelope us at 3 pm, though we are only 340 ft above sea. . . . I hope to know more of the effect of the sea on the climate of the coast soon.[128]

[125] Griffith, 'Rain Gauge Experiments at Strathfield Turgiss', 14.

[126] C. H. Griffith, 'Rain Gauge Experiments at Strathfield Turgiss, Reading', *British Rainfall. 1869* (1870), 25–33.

[127] Mill, 'The Development of Rainfall Measurement', 24.

[128] Letter from F. Stow to J. Glaisher, 8 June 1869, National Meteorological Archive, RMS/1/5/13.

The gauges were originally placed on Ling Hill, about 100 feet above the cliff edge on which the local lighthouses stood and 315 feet above sea level. Some were left there under the charge of the light-keeper, while the rest were later moved to an experimental field near Stow's house, over 400 feet above the sea, 'almost level, but completely open and exposed on all sides'.[129] Stow proceeded with a set of experiments, called the material series, the magnitude series, the form series, the elevation series and the position series. The material series contained gauges made of tin, zinc, glass, copper and earthenware. The magnitude series contained gauges of varying diameter, from one inch to twenty-four inches. The form series contained ordinary gauges, those with flanges and upright rims, as well as some uncommon designs. While these and the elevation series mirrored those conducted at Strathfield Turgiss, Hurst Green, Camden and Calne, Stow's position series was the first set of experiments that attempted to measure the effects of wind and topography upon the amount of rain captured by the gauge.[130] Stow placed five gauges in the vicinity of Ling Hill: on top of the hill, on its northern slope, in a small valley, in the garden of the lighthouse and on the edge of the cliff (Figure 4.9). The engraving of the Hawsker gauges was oriented to emphasise the exposure of the site to the North Sea (Figure 4.10). The gauges' situation was further emphasised by a map of the coastline, which used vertical hachures to represent elevation.

Stow's observations over 1870 and 1871 supported the use of zinc as an ideal material to use in the construction of rain gauges. He also dismissed the use of flanges on the gauge rim and supported high-rimmed gauges, especially in mountainous terrain. His work with the position series showed that the highest gauge caught less than those at lower altitudes; that the sheltered gauges in the valley and by the lighthouse caught equal amounts; that the gauge on the slope caught more with southerly winds than those from the north; and that the clifftop gauge caught little rain with land winds and almost nothing in gales from the sea. On the basis of these observations Stow reasoned that higher winds on the hill top caused rain to fly at a greater angle to the vertical and so diminished the amount falling upon a horizontal surface. Contrary to this, falls of rain into sheltered areas fell with reduced force and at a smaller angle to the vertical. Upon the cliff edge, rain could travel almost horizontally and even upwards if caught by updrafts.[131] His work with the magnitude series showed a gradual but very small increase in the amount of rain

[129] F. W. Stow, 'Rain Gauge Experiments at Hawsker, near Whitby, Yorkshire', *British Rainfall. 1870* (1871) 9–22, 9.
[130] Stow, 'Rain Gauge Experiments', *British Rainfall. 1870*, 14.
[131] Stow, 'Rain Gauge Experiments', *British Rainfall. 1870*, 21–2.

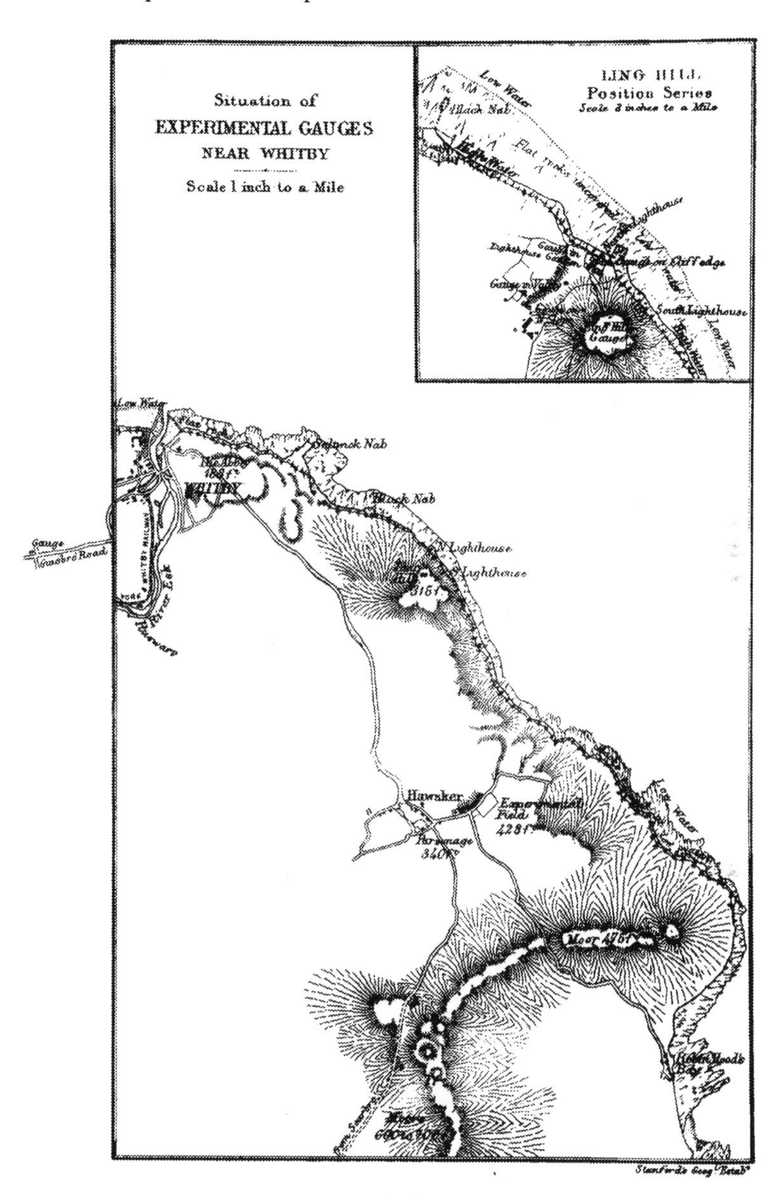

Figure 4.9 Situation of experimental gauges at Hawsker, near Whitby. (Source: *British Rainfall. 1870* (1871), facing 8.)

Figure 4.10 Experimental gauges at Hawsker, near Whitby, with the cliffs and sea in the background. (Source: *British Rainfall. 1870* (1871), frontispiece.)

caught with an increase in the size of the gauge. However, he supposed this was an artefact of excellent exposure. 'Had the exposure not been so good', he explained, 'I doubt whether so small a difference would have shown itself at all clearly'.[132] Lastly, Stow's work with the elevation series observed a diminution of the amount of rain caught in a gauge elevated above the ground, attributed to the angle of fall and the velocity and direction of the wind. Stow's own thoughts on the matter were informed by yet another set of rain gauge experiments, led by Richard Chrimes.

Richard Chrimes was one of the founders of Guest and Chrimes' Foundry and Brass Works in the Yorkshire town of Rotherham, employing 400 people by the 1870s. He had kept a record of the daily rain since May 1864 and began reporting his observations to Symons's network in 1865. Symons praised Chrimes's 'experience and skill in mechanical matters' and co-opted him as another experimentalist in late 1865.[133] Chrimes was initially hesitant to take on the role given his lack of

[132] F. W. Stow, 'Rain Gauge Experiments at Hawsker, near Whitby, Yorkshire', *British Rainfall. 1871* (1872), 16–32, 16.

[133] Letter from G. J. Symons to R. Chrimes, 6 March 1866, National Meteorological Archive, RMS/1/5/13.

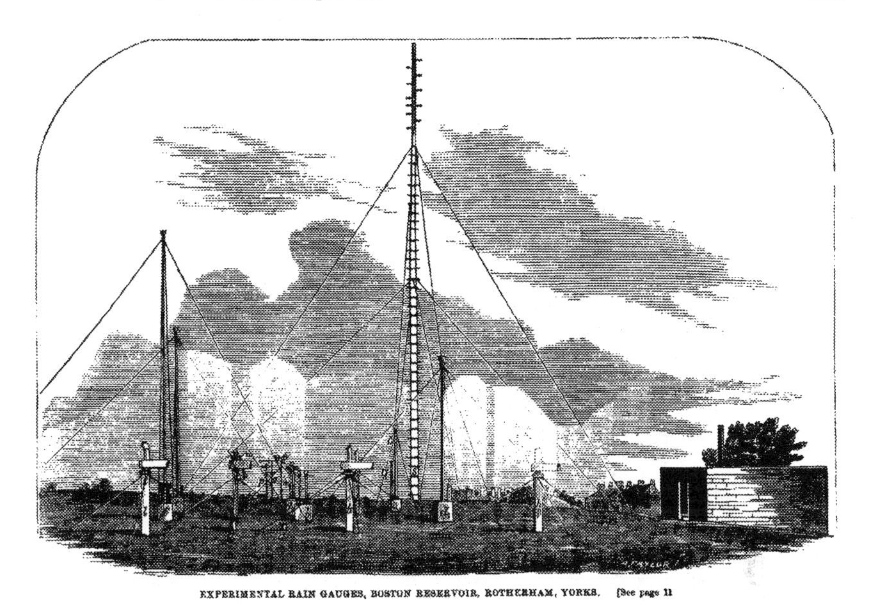

Figure 4.11 Experimental rain gauges at Boston Reservoir, Rotherham. (Source: *British Rainfall. 1869* (1870), frontispiece.)

facilities, lack of time and the poor exposure of the grounds of his house.[134] However, in late 1865, he found someone to help with the observations, and in early 1866 obtained permission to use the local water board's land at Rotherham Reservoir. Chrimes's intentions were to measure the effects of altitude above the ground on rain collected. He also established a set of unusual gauges that would investigate the effects of wind direction and the angle of the gauge mouth to the horizontal. As well as a dozen elevated gauges, Chrimes installed a five-mouthed gauge, with one vertical orifice and four facing horizontally in the cardinal directions, so as to allow calculation of the altitudinal angle at which the rain had fallen. Three inclined gauges were installed, inclined to the horizon at 22.5, 45, 67.5 and 90 degrees, respectively. A tipping-funnelled gauge was also included, designed to move in the wind so that the mouth would always stay at right angles to the path of the rain. Chrimes made use of his industrial wealth to install several anemometers and thermometers and to construct an office from which the observations could be administered (Figure 4.11). The Rotherham

[134] Letter from R. Chrimes to G. J. Symons, 11 December 1865, National Meteorological Archive, RMS/1/5/13.

experiments ran until 1872. The instruments were later presented to the Rotherham Corporation, which established a new gauge site on the bank of the Ulley Reservoir. A series of observations were continued there until 1890. Symons was effusive in his praise of the Rotherham site, describing the instruments as 'so varied, as to be not only self-checking, but almost speaking'.[135] Chrimes's gauges proved that gauges angled at around forty-five degrees and facing into the wind would collect the most rain possible.

Another observer in Symons's network experimented with tipping gauges. George Fox, from Kingsbridge in Devon, wrote to Symons in July 1871 to report on his own tipping gauge. Fox found that the stronger the wind and the greater the tip, the more rain was registered compared to a conventional gauge at ground level – sometimes almost 50 per cent more. Fox judged the quantity collected by the tipping gauge to be the 'actual amount'.[136] In a subsequent letter, he reported its good behaviour, even in very stormy weather: 'I take great pleasure in watching it.'[137] Symons had pointed out to his network that the purpose of their gauges was not to collect the most rain. Rather, his authorities were supposed 'to ascertain the accuracy with which the gauges in general use give the true fall on the ground – which obviously can neither be tilted to a varying angle, nor wheeled about in azimuth to meet the rain, and wherefore of course a horizontal-mouthed gauge is alone admissible for general use'.[138] Nonetheless, Fox submitted a description of his gauge in the *English Mechanic and World of Science* in 1871.[139] Another reader, John Thrustans, took issue with Fox's pattern, describing it as 'utterly valueless for any scientific purpose' and criticising Fox's claim that horizontal gauges do not register accurately. Thrustans accused Fox of a form of instrumental bias, which led to him adopting flawed measurements. Fox was not the first observer 'who has fallen into the error of supposing that because a tipped gauge collects more rain it more correctly represents the amount that falls'.[140]

[135] G. J. Symons, 'On the Results of the Various Sets of Experimental Rain Gauges', *British Rainfall. 1867* (1868), 29–31, 30.
[136] Letter from G. Fox to G. J. Symons, 14 July 1871, National Meteorological Archive, RMS/1/5/13.
[137] Letter from G. Fox to G.J. Symons, 1 August 1871, National Meteorological Archive, RMS/1/5/13.
[138] Symons, 'On the Results of the Various Sets of Experimental Rain Gauges', 31.
[139] G. Fox, 'Rain-Gauge', *English Mechanic and World of Science*, 13 (21 July 1871), 434.
[140] J. Thrustans, 'Rain-Gauge', *English Mechanic and World of Science*, 13 (21 July 1871), 492.

Rainfall Controversies

Thrustans's attack on Fox in the *English Mechanic* was not an isolated incident – it was part of a wider controversy that ran during 1871, largely in the pages of the *Meteorological Magazine*. As Thrustans's letter to the *English Mechanic* testified, what constituted the measurement of rain as a physical quantity was open to question. The controversy in the *Meteorological Magazine* also concerned the 'true interpretation or explanation of the results concerning what rain was, where it could be found and how it should be measured'.[141] It was ignited by a short commentary in 1870 on the experiments at Hawsker, Strathfield Turgiss and Rotherham, as well as a much more limited set of observations kept at Aldershot Camp by Colour-Sergeant J. Arnold from 1869 to 1874, where Arnold had also tested gauges at various heights and different sizes.[142] In the 1870 article Symons claimed that the various experiments, independently confirmed by those carried out at Aldershot, demonstrated that the angle of rain accounted for the decrease in the amount collected by horizontal gauges at height, and that if angled gauges were deployed, then any differences were removed. This, Symons claimed, solved a problem 'which has baffled observers for more than a century'.[143] However, while the experimental trials and the reports of their findings in *British Rainfall* were meant to demonstrate the trustworthiness of the practitioners and their methods, they ended up generating conflict and debate.

In a letter in the *Meteorological Magazine* in May 1871, George F. Burder took issue with Symons's conclusions. Dr Burder, a physician to the Bristol General Hospital and lecturer at the Bristol School of Medicine, complained that Symons had jumped too hastily to support the claim that the strength of the wind caused less rain to be caught in horizontal gauges at height. Although he supported the evidence from the various experiments regarding decrease of rainfall with elevation, Burder queried the acceptance of cause and effect between the deviation of rain due to wind and the amount of rain collected. Symons responded by questioning Burder's grasp of both the problem and the solution. He noted the long history of observations that showed gauges collected less rain if placed on the roofs of lofty buildings, making reference to the rain gauge observations carried out by William Heberden at various heights across Westminster Abbey in the

[141] G. J. Symons, 'Notes upon the Foregoing Papers', *British Rainfall. 1871* (1872), 56–7, 56.

[142] Mill, 'The Development of Rainfall Measurement', 32.

[143] G. J. Symons, 'Notes on the Preceding Papers', *British Rainfall. 1870* (1871), 43–5, 44.

1760s.[144] Symons went on to lay out the following elements of the problem as he saw them:

> Inference. – from the above fact it has been assumed that less rain falls the higher we rise above the ground.
>
> Real Problem. – To determine why an elevated horizontal gauge catches less than one on the ground.
>
> What We Think The Experiments Have Proved. – (1) That there is no sensible decrease in the amount which really falls within the first 20 or 30 ft. of the earth, although there is a decrease in the amount collected by horizontal-mouthed gauges. (2) That elevated horizontal rain gauges collect less because the rain falls at a greater angle with the vertical.[145]

Writing to the *Meteorological Magazine* the following month, Burder took issue with Symons's framing of the question at hand, particularly his inference that less rain fell at elevation. Burder argued that rainfall should be defined as what is measured at ground level: 'This, and this only, is "rainfall" in any intelligible or measurable sense, and the contested "inference" is, therefore, as it seems to me, only a varied expression of the "fact observed."'[146] This question of the exact relationship between the instrument, its placement and the phenomenon under observation was fundamental to the ensuing controversy. Burder took the position that rain should be treated not as a quantity found in nature but as the abstract outcome of a placed instrument, properly observed. Poor placement and poor measurement rendered observations error-laden and misleading. For the experimentalists, however, rainfall was a phenomenon observable throughout the atmosphere's profile, whether at the cloud base, at the top of a cathedral tower or one foot above the ground. Rain was a messy and complicated quantity that could be measured by instruments at different heights and at different attitudes. The rainfall controversy was in effect a battle between these two positions.

The June issue of the *Meteorological Magazine* contained pages of letters on the rain question. Fenwick Stow was the first to feature in the issue, with a four-page letter responding to what Stow saw as Burder's 'attack' on the results of the Hawsker and Rotherham experiments.[147] Stow explained that his work rested on the following assumptions: that the character of rain was not changed by its angle of fall; that raindrops did

144 W. E. K. Middleton, *A History of the Theories of Rain and Other Forms of Precipitation* (London: Oldbourne, 1965).

145 G. J. Symons, 'The Cause of the Decrease of Rainfall with Elevation', *Meteorological Magazine*, 6 (May 1871), 63–5, 64–5.

146 G. F. Burder, 'Letter to the Editor, 30 May 1871', *Meteorological Magazine*, 6 (June 1871), 75.

147 F. W. Stow, Letter to the Editor, no date, *Meteorological Magazine*, 6 (June 1871), 69–72, 69.

not approach or diverge from one another; nor did their size increase or diminish. With those factors ruled out, the explanation for reduced rainfall at a horizontal gauge had to be the fact that the mouth of a gauge presented a smaller area of aperture to rain falling obliquely. Any other explanation would, Stow warned, cause the Rotherham experiments to be cast aside 'as useless – a bold step to take in the face of very close agreement between calculation and observation exhibited by the results thus obtained'.[148] Stow set that explanation against the alternative – 'old-fashioned' – theory that raindrops increased in mass as they fell, arguing that any increase in the density of rain in a given space of air would affect both horizontal and vertical gauges in the same way, which was not the case with the experimental results. This old-fashioned theory was often attributed to Benjamin Franklin, who claimed that gauges at ground level collected more rain due to cold rain drops condensing dew upon themselves as they fell through warmer air at lower altitudes.[149]

The Norfolk vicar J. M. Du Port wrote in to support Dr Burder. Du Port had been sent the voluminous record sheets produced by Ward when observations at Calne had come to an end.[150] Du Port argued that the gauge itself was the cause of the problem and not the angle of fall, in that eddies would form around the gauge and deflect some of the water, and that the eddies would be greater at higher elevations. Du Port crucially aligned his own argument with that of William Stanley Jevons, the celebrated economist, logician and statistician, whose research on commercial crises encompassed weather, seasonality and agriculture.[151] Jevons had developed his own theories of rainfall, which he had presented at a meeting of the British Association and then published in the *Philosophical Magazine* in 1861. Jevons argued that the fall of rain was identical at all elevations and observed differences were due to imperfect collection by gauges. Gauges, houses, towers and other objects formed obstacles to the uniform passage of a rain-bearing air current, with a resultant reduction in the amount of rain recorded.[152]

Jevons was influenced by the earlier studies of Henry Meikle and Alexander Bache, whose work revealed the effects that gauges themselves had on the amount of rain collected, and the effects of elevation on gauge

148 F. W. Stow, Letter to the Editor, no date, *Meteorological Magazine*, 6 (June 1871), 69.

149 Middleton, *A History of the Theories of Rain*.

150 G. J. Symons, 'Report of the Rainfall Committee for the year 1868–69', 384.

151 P. J. FitzPatrick, 'Leading British Statisticians of the Nineteenth Century', *Journal of the American Statistical Association*, 55 (1960), 38–70. On Jevons's work on laboratory cloud studies and the average cirrus cloud, see Anderson, 'Looking at the Sky'.

152 W. S. Jevons, 'On the Deficiency of Rain in an Elevated Rain-Gauge, as Caused by Wind', *The London, Edinburgh, and Dublin Philosophical Magazine and Journal of Science*, 22 (1861), 421–33.

performance, respectively.[153] To support his own argument, Jevons made reference to other observational work with gauges earlier in the nineteenth century, including John Phillip's at York Minster, Luke Howard's in London, François Arago's at the Paris Observatory, Glaisher's at Greenwich and Dr Buist's at Bombay. All of these showed 'great irregularity and want of accordance' between gauges placed at different elevations. Jevons attributed these discordances to problems with gauges and to their positioning.[154] The returns from these various sites revealed to Jevons the unsatisfactory nature of the rain data, which in turn prevented the pursuit of meaningful statistical analysis of rain patterns and their causes. The production of average results would give the appearance of uniformity and law over sufficient spans of time and in accordance with the doctrine of probabilities, despite irregularities in the returns: 'the discrepancies of individual rain observations at different altitudes are such as can come under no law'.[155] Although Jevons presented his analysis of rainfall observations before Symons's experimental trials, his criticism of various observational regimens and rainfall trials in Britain, France and India fitted the position Du Port and Burder took against the rainfall experimentalists. Du Port used Jevons's arguments to support his own criticism of the experimental trials, which Du Port implied also produced faulty rain observations. Writing in the September issue, Burder claimed that in any investigation of the effects of obliquity on rain capture, 'mathematics are superfluous, and experiments are worse than useless'.[156]

John Thrustans contributed to the debate, writing to the *Meteorological Magazine* in July to disagree with Stow and his 'ardent supporters'. Thrustans took particular issue with Stow's assumption that raindrops were not brought closer together during their fall. Thrustans argued that drops would approach one another if they fell at an angle, which explained why tipped gauges collected more rain.[157] Du Port wrote in again to make the same point.[158] He explained his own argument with recourse to

[153] H. Meickle, 'On the Different Quantities of Rain Collected in Rain-Gauges at Different Heights', *Annals of Philosophy*, 14 (1819), 312–13; A. Bache, 'Note on the Effect of Deflected Currents on the Quantity of Rain Collected by a Rain Gauge', *Report of the Eighth Meeting of the British Association for the Advancement of Science, Transactions of Section 8* (1838), 25–27. See Parker, 'Distinguishing Real Results from Instrumental Artifacts', pp. 168–70.

[154] Jevons, 'On the Deficiency of Rain', 425.

[155] Jevons, 'On the Deficiency of Rain', 426.

[156] G. Burder, Letter to the Editor, 6 September 1871, *Meteorological Magazine*, 6 (September 1871), 139–40, 139.

[157] J. Thrustans, Letter to the Editor, 20 June 1871, *Meteorological Magazine*, 6 (July 1871), 92–3, 93.

[158] J. M. Du Port, Letter to the Editor, 3 July 1871, *Meteorological Magazine*, 6 (July 1871), 94.

a basic model made of four strings, each a foot long, tied at each end to knitting needles. When laid flat on a table and the needles pulled apart, the string represented the fall of rain, either vertically when at right angles to the needles, or obliquely if one needle was moved to the left or right. Du Port invoked the basic geometric principle that all parallelograms upon equal bases and between the same parallels are equal in area to show that the same volume of rain fell on the surface – whether following a vertical or oblique path – and that the paths of the raindrops did indeed draw nearer together as their fall's angle from the vertical increased. Writing in the July issue, Burder also proposed a geometrical exercise, this time compelling readers of the *Meteorological Magazine* to draw a set of equally spaced vertical parallel lines (representing raindrops), which then deflected at forty-five degrees. He pointed out that if a scale was placed across the vertical lines and then the deflected lines, more lines would be found contained within the scale where the lines were diagonal.[159]

Stow's letter to the August issue of the *Magazine* opened provocatively:

> Knitting needles, foot rules, strings, and a few parallelograms, as a mathematical sauce, are the potent weapons of common sense; and just as I was expecting to hear that the complicated problem admitted of no solution at all, owing to the want of facts to go upon, I find that neither facts nor mathematics are necessary, and that all is perfectly simple and obvious.[160]

Stow had taken issue with Burder's and Du Port's criticism of the instrumental practices and observational outcomes of the various English rainfall experiments, as well as with the two men's arguments that geometrical proofs offered a better solution to the rainfall question than experimental trials and a mass of observations. Stow also questioned his interlocutors' attempts to simplify the complexities of the atmosphere and the difficulties of the problem in front of meteorologists. Raindrops were unequal in size, he argued, were affected differently by the resistance of the air, fell with different velocities and were deflected by wind to a different extent. These 'great irregularities' made simple analogies like knitting needles and parallelograms inapplicable to the problem.[161] Stow attempted to undermine Du Port's and Burder's geometrical proofs by turning them on their side – 'as hard-hearted masters do with the figures of Euclid to puzzle boys'.[162] Their parallelograms should continue to work in the same way but did not, Stow claimed. That Du Port's proof only applied

[159] G. F. Burder, Letter to the Editor, 9 July 1871, *Meteorological Magazine*, 6 (July 1871), 95–6.
[160] F. W. Stow, Letter to the Editor, 19 July 1871, *Meteorological Magazine*, 6 (August 1871), 113–5, 113.
[161] F. W. Stow, Letter to the Editor, 19 July 1871, 113.
[162] F. W. Stow, Letter to the Editor, 19 July 1871, 113.

to horizontal gauges and not vertical gauges nullified the theory: 'I never heard that the properties of a parallelogram would tumble out of it if turned upside down; but I am open to conviction, and if convinced, I promise in the future to label my diagrams, "With care, this side up."'[163]

Stow's letter went on to ask what happened when raindrops quit their path, 'without the slightest basis either of fact, or even of thorough theory' that had been mapped out for them by Burder and Du Port?[164] The answer was a decrease received with elevation in horizontal gauges and an increase in vertical gauges. Stow admitted this was not an exact truth, but reasoned that if it was only roughly true, the calculations he presented would still be approximately correct. He claimed this was borne out in his experiments and he insisted that Burder and Du Port must agree that 'any clear result of experiments, when fairly ascertained, ought to outweigh any pre-existing theory unfounded upon experiment'. Richard Strachan, a Fellow of the Royal Meteorological Society, employee of the Meteorological Office and expert on marine meteorological instruments, wrote in to support Stow. Strachan called Burder's argument – that deflected raindrops preserved their parallelism – 'a pretty illustration' but deceptive and inconclusive, and that if the direction of raindrops changed at an oblique surface, different amounts of rain would be collected at a horizontal gauge.[165]

Another of Symons's experimentalists, T. E. Crallan, joined the debate to support Stow's call for an investigation of the path of the raindrop and a criticism of Burder's supposition in one of his earlier letters that a cloud only discharged rain from its lower horizontal surface. Crallan argued that a raindrop was a projectile that followed a parabolic trajectory. Drops discharged from a horizontal cloud surface would describe parallel curves; drops discharged from a vertical line through the cloud would converge as they reached the earth's surface. That clouds were volumes rather than horizontal surfaces therefore explained the increase of rain collected by horizontal gauges.[166] Burder replied in the November issue to complain about Crallan's caricature of his own theories of rain discharge from clouds, saying that he was proposing a 'hypothetical instance' from the point raindrops left the cloud and was not speculating on their origin point.[167] Others, like Charles Cator, tried to find a way through the

[163] F. W. Stow, Letter to the Editor, 19 July 1871, 114.
[164] F. W. Stow, Letter to the Editor, 19 July 1871, 113.
[165] R. Strachan, Letter to the Editor, 24 July 1871, *Meteorological Magazine*, 6 (August 1871), 115–7, 116.
[166] T. E. Crallan, Letter to the Editor, 5 September 1871, *Meteorological Magazine*, 6 (August 1871), 136–9.
[167] G. F. Burder, Letter to the Editor, 31 October 1871, *Meteorological Magazine*, 6 (August 1871), 180–1, 180.

opposing positions. In a long letter in the November issue, Cator suggested that both Stow and Burder were for the most part right, and that Crallan and Strachan also agreed as to the parabolic curve of falling rain.[168] In reply, Burder argued that Cator's calculations showed the truth of his own view and that 'both Mr Stow and Mr Cator are radically, absolutely, and fundamentally wrong'. Burder reiterated his argument that the quantity of rain appeared to increase when travelling at an angle, but this multiplication of drops was an artefact of the measurement itself and in no way represented a real increase in the quantity of rain:

[H]e who forgets that it is fictitious and for the moment regards it as real, will easily fall into the error of supposing that the horizontal gauge, because it catches less than the tilted one, catches therefore less than it ought; the truth being, not that the horizontal gauge catches too little, but that the tilted gauge catches too much.[169]

Both Stow and Burder were given the opportunity to express and summarise their arguments in articles in the 1871 issue of *British Rainfall*. Stow took the opportunity to rehearse various theories of rainfall, arguing that Franklin's theory that falling rain somehow gathered water to itself had been generally accepted for many years because it 'had the great advantage that it admitted of the mighty forces of heat and electricity being vaguely invoked'.[170] Stow dismissed the theory on the grounds that heat and electricity cannot generate rain out of nothing. Stow went on to review and discuss the horizontal and vertical flow of air; the impacts of obstacles such as hills, trees and buildings; and the cumulative effects of these flows and impediments on the movement of rain (Figure 4.12). Stow reiterated the value of his own position series experiments and made similarly positive reference to those at Strathfield Turgiss and Rotherham, stating that they were conclusive as to the effects of down- and up-currents of wind on rainfall – that down-currents caused an increase in rain collected and vice versa. If a rain shower was intercepted ten or twenty feet above the ground, Stow argued there would be as much rain at that height as actually fell to the ground within the whole area covered by the shower, but that the fall was unequally distributed due to the upward and downward movement of the air. The fact that gauges were usually placed in positions that experienced greater vertical downflow of air accounted for the excess measured in them. The non-horizontality of the wind's motion was the principal and

[168] C. Cator, Letter to the Editor, no date, *Meteorological Magazine*, 6 (November 1871), 183–8.

[169] G. F. Burder, Letter to the Editor, 9 December 1871, *Meteorological Magazine*, 6 (December 1871), 210–2, 210 and 212, respectively.

[170] F. W. Stow, 'Rain gauge experiments at Hawsker, near Whitby, Yorkshire, 1871', *British Rainfall. 1871* (1872), 16–31, 18.

most powerful 'disturbing influence' in determining the amount of rain caught at different elevations.[171]

Burder's article followed directly after Stow's. He continued to assert that Stow's theory that elevated gauges caught less rain because of their angle of travel was 'entirely erroneous'.[172] He similarly dismissed Crallan's argument regarding the confluence of raindrops as they approached the ground and criticised theories relating to the effects of eddies and out-splashing. His conclusion was that the old notion of rain increasing in quantity near the surface of the earth was in fact the correct one; that rain formed both in the clouds but also near the ground. Although he dismissed the hypothesis that this was due to condensation of vapour from warm air by cold raindrops, Burder supported the idea that rain particles of water existed suspended in the lower strata of air and that drops of rain 'lick up and incorporate these particles'.[173] This theory related back to a point Burder had made in a letter to the September issue of the *Meteorological Magazine*, where he presented the idea of 'spray' – a cloud of vapour caused by raindrops hitting terrestrial objects as well as condensation as air contacted cold ground and vegetation.[174] Burder concluded that the formation of rain was by no means a uniformly continuous process, coming mainly from the clouds but with a large addition from spray found right at the earth's surface, which became attached to falling raindrops.

The 1871 controversy ended with no clear resolution. The ongoing nature of the Rotherham observations as well as some new experiments continued to furnish data for commentators to analyse and discuss. George Dines, master builder, architectural advisor to Queen Victoria and Fellow of the Royal Meteorological Society, was motivated by the lack of consensus on the elevation question – calling it a 'standing disgrace to meteorologists' – and began his own investigations in the summer of 1876, making use of a large tower attached to his home in Surrey.[175] Dines placed a number of gauges both on the roof of the tower – along with a thermometer screen and wind vane – and upon the ground. Many of the gauges came from Charles Griffith, having previously served at Calne, Hawsker and Strathfield Turgiss. Dines claimed that the amount of rain caught at height was affected by winds, which caused eddies

[171] Stow, 'Rain Gauge Experiments at Hawsker, near Whitby, Yorkshire, 1871', 24.
[172] G. F. Burder, 'On the Cause of the Decrease of Rain with Elevation', *British Rainfall. 1871* (1872), 33–41.
[173] Burder, 'On the Cause of the Decrease of Rain', 35.
[174] G. F. Burder, Letter to the Editor, 6 September 1871, *Meteorological Magazine*, 6 (September 1871), 139–40, 140.
[175] G. Dines, 'Difference of Rainfall with Elevation', *British Rainfall. 1877* (1878), 15–25, 17. W. S. Pike, 'Master Builder Turned Meteorologist: George Dines 1812–1887', *Weather*, 42 (March 1987), 88–90.

Figure 4.12 Sketch of wind currents in distributing the fall of rain on the ground. (Source: *British Rainfall. 1871* (1872), facing 21.)

around the gauges, and was not due to any deficiency in the amount of rain at higher levels. Dines did not want to get drawn into the fraught discussions between Stow and Burder regarding the angle of fall or spray theory but did make positive reference to the drawing by Stow that showed the effects of obstructions on the fall of rain (Figure 4.12), which he said was entirely consistent with his own observations.

The conclusion drawn from Dines's work was that the ratio of the rainfall on the tower to that on the ground depended upon the force and direction of the wind; and that gauges in different positions on the tower would collect different quantities of rain, depending on wind direction (Figure 4.13). Speaking at the Jubilee Meeting of the British Association in 1881, Symons concluded that Dines's work meant that the 'terribly hard-fought battle' was nearly settled and there now existed sufficient proof that the apparent decrease in rainfall at elevation was due to the wind and imperfect collection by gauges placed there.[176]

[176] G. J. Symons, 'Different of Rainfall with Elevation', *British Rainfall. 1878* (1879), 24–30, 30; G. Dines, 'Difference of Rainfall with Elevation. Being the Results of Experiments Made at Woodside, Hersham, Surrey', *British Rainfall. 1880* (1881), 13–16.

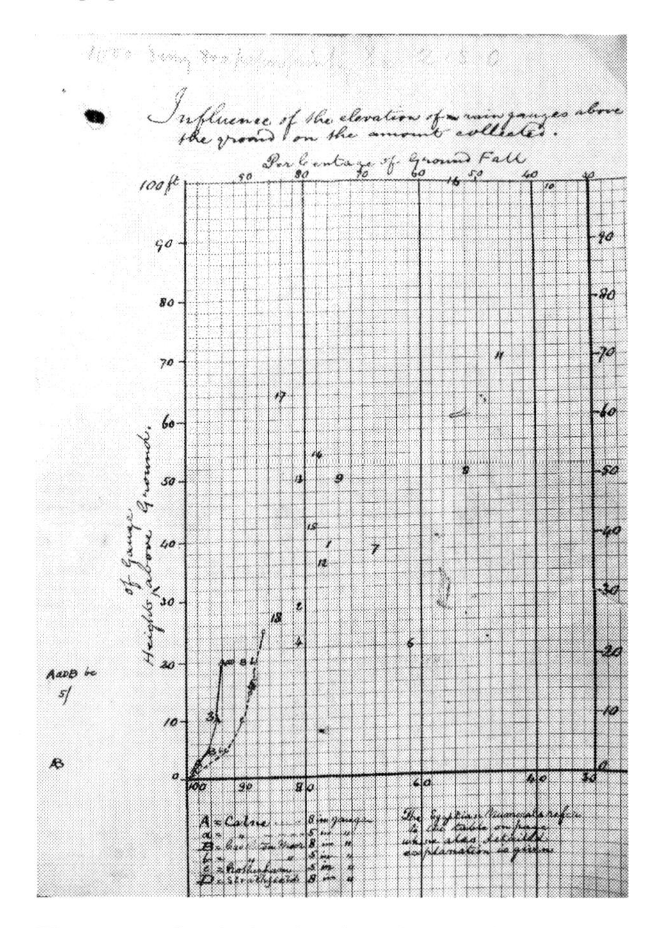

Figure 4.13 Graph showing the influence of elevation on the amount collected in rain gauges. (Source: Letter from George Dines, recipient unknown, 26 April 1869, National Meteorological Archive, RMS/1/5/13. Reproduced with permission of Royal Meteorological Society.)

Responding to the paper, the British Association president Sir William Thomson expressed his own satisfaction that the question was finally settled.[177] Despite his attempt to maintain a neutral position, Dines's own tower observations relied on the same instrumental infrastructure

[177] G. J. Symons, 'On the Rainfall Observations Made upon York Minster by Professor John Phillips', *British Rainfall. 1881* (1882), 41–5, 45. See Parker, 'Distinguishing Real Results from Instrumental Artifacts', p. 173.

of Symons's experimentalists; he gave implicit support to Stow's own arguments as to the effect of wind eddies on rain collection; and took a similar position as to the value of a programme of atmospheric observations. Subsequent commentators supported the value of the experiments and the mass of data they had produced. The noted Prussian meteorologist Gustav Hellmann conducted his own gauge experiments near Berlin, with the object of ascertaining how close together rain gauges ought to be set to fairly represent the land around them. As part of these trials he undertook roof experiments. In a 1906 article in *British Rainfall*, Bonacina argued that Hellmann's observations lent further support to the English experimentalists' conclusion that the loss of rain in elevated gauges was due to the effects of wind.[178] Despite attempts to query the work of Stow, Symons and others by Burder, Du Port and Thrustans, trustworthiness ended up being attributed to the methods and returns of the observational experiments, rather than to abstract geometrical proofs. The experiments' legacies for the establishment of principles of uniformity were later celebrated. Reflecting back on the experiments in *British Rainfall*, Mill reminded its readers that Symons had always urged the importance of uniformity in conditions and that the rainfall controversy had only confirmed the importance of that principle.[179] In his own summary of the history of rain gauge exposure, Salter said that the controversy had hastened the adoption of the rule that all gauges should be exposed with the top of the funnel one foot above the ground, claiming that the number of correctly placed gauges rose from 13 per cent in 1863 to 68 per cent in 1919.[180] The experimental trials were judged to have helped perpetuate a more uniform environment in terms of instrumental design and placement, such that 'fair corrections' could be calculated for observations taken under different conditions.[181] But questions remained about how to organise Symons's rain archive, how to analyse the data statistically and what to use the data for.

[178] L. C. W. Bonacina, 'The Effects of Exposure to Wind upon the Amount of Rain Caught by Rain-Gauges and the Methods of Protecting Rain-Gauges from Them', *British Rainfall. 1906* (1907), 27–45, 33. See Parker, 'Distinguishing Real Results from Instrumental Artifacts', p. 175.

[179] Mill, 'The Development of Rainfall Measurements', 39.

[180] Salter, *The Rainfall of the British Isles*, p. 59. This differed with practice in continental Europe, where it was common for the lip of the gauge to be placed one metre above the ground. This was partly due to the increased likelihood of heavy snowfall.

[181] F. Gaster, 'On the Monthly Percentage of Mean annual Rainfall', *British Rainfall. 1868* (1869), 50–4, 53.

Average Rain

In his discussion of secular variation of rainfall in England since the eighteenth century, Symons laid out what he saw as the main priorities for the organisation of rainfall observations into coherent and meaningful series: accuracy; temporal coverage; and geographical representation.[182] It was supposed that the work of Stow, Crallan and others had answered the first of these. The experimental trial sites promised unprecedented levels of accuracy and had already set new standards for other stations. With regard to the second priority, small numbers of rain stations at cathedrals, vicarages, country houses, centres of learning and industrial mill sites facilitated the production of long series of English and Scottish rain data. Work by Symons and the British Association's Rainfall Committee to secure and compile robust decennial groupings from these returns enabled the production of decadal summaries of British rainfall back to the first decades of the nineteenth century. Symons recognised that compiling long series of rain data often strained attempts to compile accurate series, arguing that, on balance, 'Permanence – definitiveness – and uniformity' should be prioritised over measurement standards.[183] Mill struck a cautionary note in his discussion of the application of the historical method to the study of rainfall, worrying that 'cycle hunts' in long series were 'difficult and dangerous', in that they could encourage some to project historical recurrences of wet or dry periods on to the future.[184]

If the study of rainfall variation over time in place was difficult and dangerous, Mill viewed the study of variation in different places at the same time to be 'safe and unadventurous'.[185] Developing long series and geographical series of rainfall had been opposed agendas until the 1850s. Although British rain observations stretched back to the eighteenth century, their spread was poor and their ability to represent regional, let alone national, variation was limited. One of the crucial aims of Symons's network was to address this lacuna. Poor geographical representation was not a uniquely British limitation. In his pamphlet 'Suggestions for a Uniform System of Meteorological Observations', Buys Ballot also argued for improved geographical coverage of meteorological stations across Europe, complaining that priority was instead often given to the publication in extenso of results, much of which was unnecessary for

[182] G. J. Symons, 'On the Secular Variation of Rainfall in England since 1725, Part II', *British Rainfall. 1871* (1872), 58–68.
[183] Symons, On the Secular Variation of Rainfall', 67.
[184] H. R. Mill, 'Map-Studies of Rainfall', *Quarterly Journal of the Royal Meteorological Society*, 34 (1908), 65–86, 67.
[185] H. R. Mill, 'Map-Studies of Rainfall', 67.

comparisons between observatories. Symons seized on Ballot's complaint to advocate for a 'thorough reform which shall sweep away individual crochets, and utilize the observing strength and pecuniary resources of British Meteorology'.[186]

Arguably the biggest challenge facing British rain observers was how to interpret the data archive they had compiled since the late 1850s. For many meteorologists, the most popular statistical tool to apply to rainfall returns was the arithmetical mean – the combination of direct observations of equal precision upon one and the same quantity – which could form the basis of a clear understanding of rainfall trends over time and across space.[187] This form of data treatment was both more difficult than it appeared and somewhat controversial, especially in light of Jevons's complaint about observational discrepancies getting hidden in the rush to produce rainfall averages – averages that exhibited a deceptively uniform and law-like appearance. The statistical concept of the mean had played an important role in meteorological analysis since at least the early nineteenth century. Humboldt had proposed the study of atmospheric phenomena through the mapping of mean values, most famously through the introduction of isotherms – contour lines of equal mean yearly temperature. As well as mean states, Humboldt encouraged other meteorologists to study deviations. Indeed, isolating local disturbances was one aspect of studying mean states. For instance, constant laws of the distribution of heat could only be discovered if constant variations produced by local causes were also understood. The emerging scientific discipline of climatology relied heavily on the statistical creation of average atmospheric conditions.[188]

Wladimir Köppen, the Russian meteorologist and climatologist, viewed the introduction of the arithmetical mean into meteorology as an important advance but also a potential source of confusion, in that the mean was an abstract quantity that buried the statistical diversity of the elements together and made causal connections in weather processes difficult to identify.[189] Buchan, who used data from the *Challenger* expedition to produce some of the first world maps of climate and had revised Elias Loomis's 1882 world map of precipitation, stated that the mapping of regional mean temperatures and pressures could be effected with comparatively few observational sites, but this was not the case for

[186] B. Ballot, *Suggestions on a Uniform System of Meteorological Observations* (Utrecht: Royal Dutch Meteorological Institute, 1872); G. J. Symons, 'Another Meteorological Congress', *Meteorological Magazine*, 7 (April 1872), 37–40, 39.

[187] M. Merriman, *Elements of the Method of Least Squares* (London: Macmillan, 1877), p. 24.

[188] Mahony, 'Climate and Climate Change'.

[189] O. B. Sheynin, 'On the History of the Statistical Method in Meteorology', *Archive for History of Exact Sciences*, 31 (1984), 53–95.

rain – numerous observing stations were needed to produce 'tolerable approximation to the average rainfall of a district'.[190] Even where long series of rain data were available, rainfall averages were unsuited to the analysis of differences in rainfall from year to year, or from site to site. In fact, Buchan argued that 'nothing could be more misleading' than the application of 'simple averages' to the study of the geographies of rainfall and complained about the dubious utilisation of such statistics in popular handbooks promoting British sanatoria.[191] In his own article on climate and health, Symons also acknowledged the limitations of using mean values in the determination of a place's climatic elements and urged others to pay attention to data ranges as much as to averages.

The Sunderland astronomer Thomas William Backhouse acknowledged in the Royal Meteorological Society's *Quarterly Journal* that the arithmetical mean was generally accepted as the best rule for combining direct observations of equal precision, on the basis that it would provide the most probable value – that errors or deviations of equal amounts in excess or defect were equally likely to occur.[192] Backhouse nonetheless cautioned that the arithmetical mean was unsuitable for the study of 'ideal' quantities like annual rainfall and put the case for the application of the geometrical mean to the statistical study of rainfall.[193] Backhouse, following the work of Francis Galton, argued that the geometrical mean provided a more precise treatment of deviations in annual rainfall observations and gave a more probable rainfall total for any given future year, especially where deviations could be large.[194]

Work by both Adrien-Marie Legendre and Carl Gauss had applied the calculus of probability to the study of geodesy and astronomy, whereby variations in observations were measured to allow the determination of errors in observations and the most accurate value of the object under study.[195] This theory of errors, or method of least squares, shifted

[190] A. Buchan, 'The Climate of the British Islands', *Journal of the Scottish Meteorological Society*, 7 (1884), 131–52, 131. J. Leighly, 'Climatology since the Year 1800', *Transactions of the American Geophysical Union*, 30 (1949), 658–72, 662.

[191] Buchan, 'The Climate of the British Islands', 132.

[192] D. McAlister, 'The Law of the Geometrical Mean', *Proceedings of the Royal Society of London*, 28 (1879) 367–76.

[193] T. W. Backhouse, 'The Problem of Probable Error as Applied to Meteorology', *Quarterly Journal of the Royal Meteorological Society*, 17 (April 1891), 87–92, 87, original emphasis, and 88, respectively.

[194] F. Galton, 'The Geometric Mean, in Vital and Social Statistics', *Proceedings of the Royal Society of London*, 28 (1879), 365–7. Jevons similarly promoted the superiority of the geometric mean. P. J. FitzPatrick, 'Leading British Statisticians of the Nineteenth Century', *Journal of the American Statistical Association*, 55 (1960), 38–70.

[195] J. W. L. Glaisher, 'On the Law of Facility of Errors of Observations, and on the Method of Least Squares', *Memoirs of the Royal Astronomical Society*, 39 (1872), part II, 75–124; Anderson, *Predicting the Weather*, p. 134.

probability from a measure of uncertainty to a measure of exactness – from a calculus of probability to a calculus of observations.[196] Herschel saw this as a crucial advancement in the sciences, in that it provided the means of measuring the degree of precision in all numerical investigation. Adolphe Quetelet turned the theory into a statistical tool for tracing regularities in groups of observations, where the mean was the 'best possible approximation of a quantity in nature, based on different measurements resulting from imperfect observations of a single object'.[197] Meanwhile, Francis Galton developed the statistical ideas of reversion and co-relation, later rephrased as regression and correlation, respectively, as a method of expressing numerically the relation between corresponding deviations of quantities from their respective mean values – since referred to as regression to the mean.[198] Galton used the phrase 'normal distribution', instead of astronomers' 'error distribution', to describe the tendency of measurements to arrange themselves symmetrically around the mean.

Astronomers and geodesists developed their statistical tools to deal with errors in the multiple measurements of single physical quantities that were supposed to be without variation, so as to determine which was the best value. Meteorologists borrowed from the science of probabilities but applied the methods to data sets such as variations in annual rainfall or temperature, on the assumption that the measurements included no errors. One significant consequence of meteorologists' use of the theory of probability was the need to substitute a normal curve of error for a curve of frequency of deviation from mean values.[199] Backhouse encouraged the application of the method of least squares to the probable error of the mean, while the more general tone of his paper was one of frustration at British meteorologists' reluctance to adopt modern numerical techniques, expressing surprise that Galton's views on the geometrical mean had not yet been applied to meteorological statistics. Although responses to Backhouse's paper were positive, other meteorologists, including Henry Blanford, continued to support the application of the arithmetic mean to rain data. Even with the use of the arithmetical mean, there seemed to be no statistical adjustment made for the possibility of observations of unequal precision, or weight, despite concerns about variable levels of accuracy. There was no acknowledgement that the arithmetical

[196] Stigler, *The History of Statistics*, p. 11.

[197] A. Desrosieres, *The Politics of Large Numbers: A History of Statistical Reasoning* (Cambridge, MA: Harvard University Press, 1998), p. 71.

[198] D. Curran-Everett, 'Explorations in Statistics: Correlation', *Advances in Physiology Education*, 34 (2010), 186–91; S. M. Stigler, 'Francis Galton's Account of the Invention of Correlation', *Statistical Science*, 4 (1989), 73–9.

[199] N. Shaw, *Manual of Meteorology. Volume I. Meteorology in History* (Cambridge: Cambridge University Press, 1926), p. 256.

mean was not always the most plausible value but itself created a new idea of mean value of a collection of meteorological observations.[200] Preference continued to be for the use of qualitative approaches for the estimation and elimination of errors and interferences, in similar manner to the determination of precision rainfall measurements and the establishment of what constituted a rainy day.[201]

In his analysis of Scotland's rainfall totals, Buchan noted the problems of calculating reliable rainfall averages, especially given the capriciousness of rain's occurrence.[202] Averages could only be calculated on long runs of annual observations. Although the production of decadal series had been prioritised by Symons, averages based on these periods could still exhibit significant differences from region to region due to the effects of significantly high or low rainfall in particular years. Mill supported the civil engineer Sir Alexander Binnie's argument for the use of thirty-year runs of rain data, which would be enough to produce satisfactory means and eliminated the effects of extreme local fluctuations on shorter runs of data, including decadal rain series.[203] A fifty-year-long series would have been better again, but the number of stations was too small. How then to produce trustworthy averages of rainfall at stations with short runs of observations; how to compare stations' averages that had been generated from different numbers of years of observations; and how to distinguish errant returns from those that were the result of unusually heavy rain, or no rain at all? Shaw, in his 1926 treatment of the history of meteorology, noted that '[w]hile astronomy and geodesy were concerned with the mathematical treatment of errors of observation on the basis of the theory of probability as applied to large numbers, meteorology was engaged in the graphic representation of its accumulating observations'. The answer, in other words, was the rainfall map.

Mapping the Mean

In his work on the climate of Britain, published in the *Journal of the Scottish Meteorological Society* in 1884, Buchan outlined his method for the management of rain data. Buchan's analysis utilised the returns from

[200] Shaw, *Manual of Meteorology*, p. 256.

[201] Gooday, *The Morals of Measurement*, p. 73; See Merriman, *Elements of the Method of Least Squares*, p. 30, for a contemporary explanation of the weights of observations and the computation of the probable error of the general mean.

[202] A. Buchan, 'The Monthly and Annual Rainfall of Scotland, 1866 to 1890', *Journal of the Scottish Meteorological Society*, 10 (1892), 3–24, 3.

[203] H. R. Mill, *On the Distribution of Mean and Extreme Annual Rainfall over the British Isles* (London: William Clowes, 1905), 4; Anon, 'Sir A.R. Binnie', obituary in *Nature*, 99 (1917), 267.

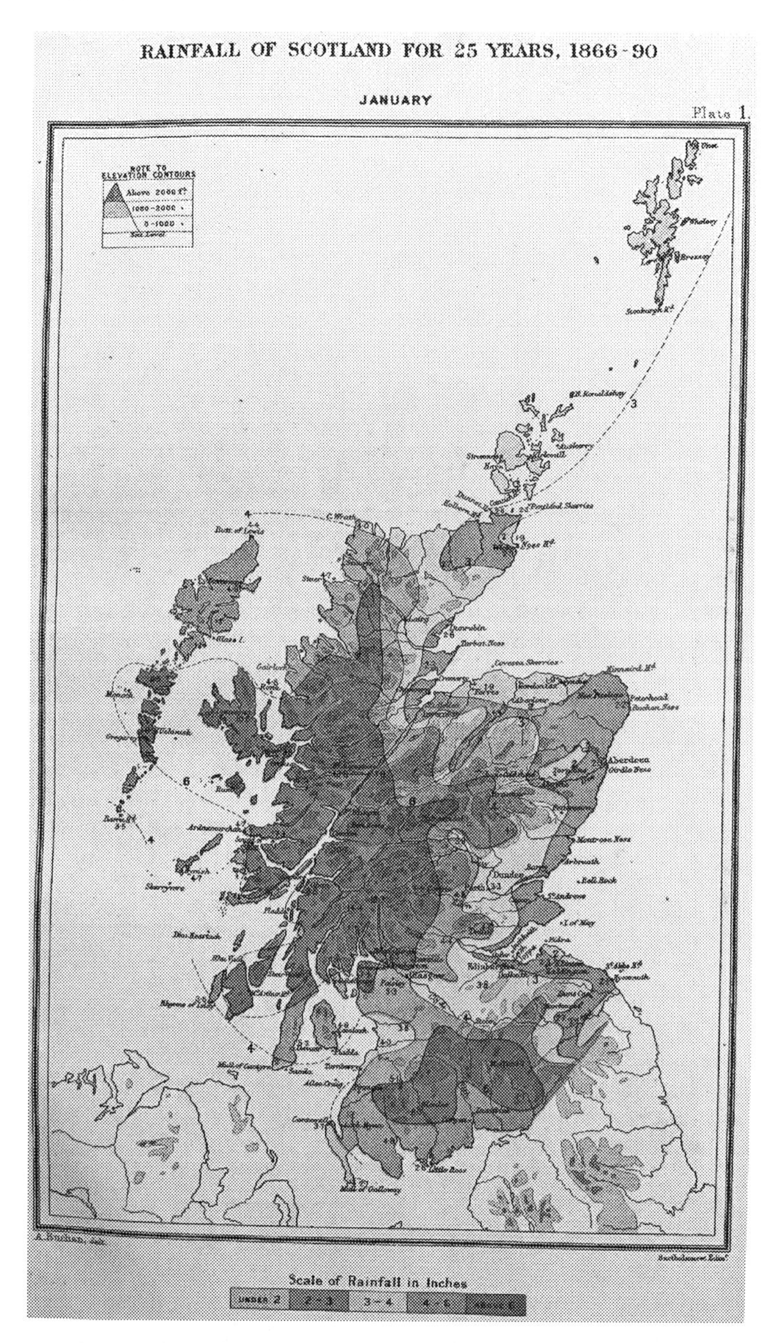

Figure 4.14 Average rainfall map of Scotland for the month of January, calculated from returns from 1866 to 1890. (Source: Buchan, 'The Monthly and Annual Rainfall of Scotland, 1866 to 1890', Plate 1.)

547 stations in Scotland, 1,080 in England and Wales and 213 in Ireland. Buchan grouped these stations into counties and then arranged them into geographical order, entering annual rain totals and arithmetical means under each station. This geographical arrangement allowed stations' means to be compared to their neighbours, and those that appeared to depart from their 'normals' were interrogated as being possible 'erroneous rainfalls' and 'doubtful or faulty readings'.[204] (Normals were means of observations spread over sufficient series of years, while 'average' – a 'very hard-worked word' according to Shaw – was often used for short or irregular series.[205]) Buchan then reduced the means of all stations to the same twenty-four-year period, from 1860 to 1883, the start date chosen because of the publication of the first volume of *British Rainfall* that year. For stations that did not have a full twenty-four-year run of data, their annual amounts and averages were calculated using neighbouring stations' returns – a technique Buchan referred to as 'differentiation' – starting with those with the longest runs and working down to those with the least. For some stations, data were so limited that no average was calculated, 'owing to the doubtful nature of the result'.[206] Isolated instances of exceptionally heavy rain were also omitted, as were the averages of gauges near exposed headlands, because they could not be regarded as representative of the rainfall over the district.[207] Buchan applied the same approach to the mapping of Scotland's monthly and annual rain averages for the period 1866–90, using data from 130 stations with full returns, along with 194 other stations where means were also calculated by differentiation (Figure 4.14).[208] This work was well received. In the *Edinburgh Medical Journal*, for instance, Buchan's maps of Scottish rain were described as one of the 'outstanding features' of the *Journal of the Scottish Meteorological Society*; the maps being 'at once comprehensible to the reader, and many useful lessons are learned by a mere glance at them'.[209] Buchan's approach to the analysis of regularities in groups of rain observations revealed outlying data points visually rather than statistically; their status as doubtful or faulty revealed to the cartographer's eye through a qualitative judgement rather than through a statistical calculation of the probable error from the mean.

[204] Buchan, 'The Climate of the British Islands', 132.
[205] Shaw, *Manual of Meteorology*, p. 256.
[206] Buchan, 'The Climate of the British Islands', 132.
[207] Buchan, 'The Climate of the British Islands', 133.
[208] A. Buchan, 'The Monthly and Annual Rainfall of Scotland, 1866 to 1890', *Journal of the Scottish Meteorological Society*, 10 (1892), 3–24.
[209] Anon, 'Reviews', *Edinburgh Medical Journal*, 40 (December 1894), 550–1.

Buchan transferred his observed and calculated averages onto maps of the British Isles and represented them using shadings that mapped mean annual rainfall. For his maps of the British Isles, Buchan used six colour tints, which ranged from light pinks for average rainfall under twenty-five inches to reds and then blues, with the darkest shade of blue representing mean annual rainfall over eighty inches. The same system of tinted colours was applied to the Scottish maps, although only five shades were used (the lowest tint was removed, which denoted average rainfall of less than twenty-five inches). However, the Scottish maps did incorporate shading to indicate elevation contours, making the maps more visually complex than the British maps but more useful for the exploration of relations between altitude and precipitation. The production of the Scottish rainfall maps was aided by John G. Bartholomew, who donated a large number of physical maps of Scotland to the Scottish Meteorological Society and upon which Buchan plotted the rainfall returns. Bartholomew was the owner of John Bartholomew & Company, a large Edinburgh map maker with a global reputation for the aesthetic quality and accuracy of its maps.[210] Bartholomew's map business had made use of layer-colouring and tinting to indicate altitude since the early 1880s, which likely influenced Buchan's own use of tint shading.[211] Bartholomew and Co. published the *Atlas of Meteorology* in 1899, described in one review as 'an epoch-making work in scientific geography'.[212] Buchan acted as editor and advisor on the *Atlas*. The volume represented graphically the results of the science of meteorology up to the end of the nineteenth century.[213] In his Introduction to the *Atlas*, Buchan acknowledged the much greater number of rainfall stations than for the other meteorological elements. He estimated that for the map of Britain's annual rainfall, 760 station returns and nearly 7 million observations were used. His commentary for the map of mean annual and monthly pressure over the British Isles noted its reliance on the work of Symons, as well as the Meteorological Office and Scottish Meteorological Society.[214] One of the authors of the *Atlas*, Andrew Herbertson – whose work at the Ben Nevis Observatory was discussed

[210] A. Feintuck, 'Producing Spatial Knowledge: Map Making in Edinburgh, c.1880–c.1920, *Imago Mundi*, 69 (2017), 269–71.
[211] T. Nicholson, 'Bartholomew and the Half-Inch Layer Coloured Map 1883–1903', *The Cartographic Journal*, 37 (2000), 123–45.
[212] J. Paul G., Review of *Bartholomew's Physical Atlas: An Atlas of Meteorology*, Volume III, *Journal of Geology*, 8 (1900), 573–7, 573.
[213] R. C. M., 'The New Atlas of Meteorology', Review of *Bartholomew's Physical Atlas: An Atlas of Meteorology*, Volume III, *Scottish Geographical Magazine*, 15 (1899), 646–8.
[214] J. G. Bartholomew and A. J. Herbertson (edited by A. Buchan), *Atlas of Meteorology: A Series of over 400 Maps* (London: Archibald Constable, 1899), p. 23.

in the previous chapter – also noted the use of coloured inks to show different periods of means.[215]

Like Buchan and indeed Herbertson, Hugh Robert Mill developed his scientific career in Edinburgh. Born in Caithness, Mill came south to study for a chemistry degree at the University of Edinburgh. He went on to run the Scottish Marine Station at Granton, near Leith, where one of his duties included operating the shoreline second-order weather station. Mill was supervised in the role by Buchan, whose position as secretary of the Scottish Meteorological Society required him to inspect Scotland's observatories. After several years working as a lecturer at Heriot-Watt College, Mill moved to London, where he took up the position of librarian at the Royal Geographical Society, and then, upon Symons's death, co-director of the British Rainfall Organisation. As an incoming director, Mill had a number of ambitions for the Organisation: to reorganise the administration of rainfall work at its Camden Square headquarters; to improve how data were discussed in *British Rainfall*; to give the Organisation a 'purely scientific outlook'; and to solve the question of the control of rainfall distribution by the land. Mill was dissuaded from some of his reforms by Herbert Sowerby Wallis, the other co-director, but he did take forward his scientific agenda. He did so by introducing mapping, 'not only as illustrations, but as devices for the computation of rainfall distribution more accurately than is possible by statistics alone'.[216] Mill viewed this as his most important contribution to the study of British rainfall. Upon Wallis's retirement in 1909, Mill transferred the lease of sixty-two Camden Square to a board of trustees, along with all the records and publications. This freed him up from some of the routine administrative work and gave him more research time, which he used to carry out a detailed mapping exercise. Mill took the Organisation's long-running rainfall records and plotted them on maps with a view to forming a new method of computation for a fifty-year average rainfall map. Mill later called this method 'cartometric hyetography'; Salter referred to it as Mill's 'cartometric method'.[217]

Mill preferred to use thirty-five-year series of rain records in the production of his annual rain maps, which was long enough to provide

[215] On the imperial underpinnings of Buchan's and Herbertson's climatological maps, see T. Simpson and M. Hulme, 'Climate, Cartography, and the Life and Death of the 'Natural Region' in British Geography', *Journal of Historical Geography*, 80 (2023), 44–57.

[216] H. R. Mill, 'Map-Studies of Rainfall', *Quarterly Journal of the Royal Meteorological Society*, 34 (1908), 65–86, 66; Salter, *The Rainfall of the British Isles*, p. 102; Mill, *Life Interests of a Geographer*, p. 95.

[217] Mill, *Life Interests of a Geographer*, p. 103.

'practical stability' to the value of the mean.[218] Where those long series did not exist he computed equivalent values for fifteen- and twenty-year records by comparing the station in question with two or more of its neighbours. He avoided using records of less than fifteen years except in places with few stations, such as the Scottish Highlands and parts of Ireland. Mill took the annual values of all stations whose returns were published in *British Rainfall*, placed their mean values on maps and looked for exceptional figures. This application of the cartometric method transformed a 'slow and tedious process' of inspecting printed forms into a 'simultaneous and almost automatic' process of visual inspection.[219] The method turned the office of *British Rainfall* into what Mill called a 'detective agency for the discovery of accidental errors'.[220] The method justified Mill's faith in the quality of the observations and in Symons's management of the data archive. The manner in which the data points merged into one another on the map and the uniformity of the values also 'inspired the greatest confidence in the substantial accuracy of the observations'.[221] Mill claimed that records less than thirty-five years long, when reduced to the average, proved to be reliable and even very short records gave excellent results.[222] For Mill, mapping rain was an exercise in mapping care and trustworthiness – attributes detectable both in the work carried out by rain authorities in their daily observations and in Symons's curatorial labour as the creator of the data archive.

Isohyetal lines were added in accordance with the available figures or drawn in hypothetically. Mill would often plot isohyets onto maps with no physical features or contours, to test whether rainfall corresponded with the configuration of the landscape and the nature of the vegetation. Mill described this process as 'laying down a foundation of absolutely neutral fact'.[223] The later reintroduction of physical features and contour lines onto the maps allowed theory to follow facts in relation to the effects of topography on rainfall. For regions where rainfall records were missing, it was still possible to add isohyetal lines, which could be estimated by observing the local configuration of the land or changes in vegetation type. Mill's method was based on the following principles: there were no abrupt transitions in nature; rainfall on the windward slope of a hill

[218] H. R. Mill, 'Introduction', *Rainfall Atlas of the British Isles*, p. 7. Computation of normal monthly rainfall required much longer rainfall series – 100 years at least – due to the much greater range found in rain totals at that temporal scale. Ibid., 10.
[219] Mill, 'Map-Studies of Rainfall', 70. [220] Mill, 'Map-Studies of Rainfall', 69.
[221] H. R. Mill, *On the Distribution of Mean and Extreme Annual Rainfall over the British Isles* (London: William Clowes, 1905), p. 8.
[222] H. R. Mill, 'The Rainfall of the Forth Valley and the Construction of a Rainfall Map', *British Rainfall. 1915* (1916), 1–12, 4.
[223] Mill, *On the Distribution of Mean and Extreme Annual Rainfall*, p. 6.

increased from the lower to higher level; rain on the leeward slope could have greater rain midway up than on the summit; and the distribution of rain would be similar on two different portions of land with similar configurations and exposure to prevailing wind. These principles could be applied in instances where returns were few or non-existent. For instance, one could draw a straight isohyetal line across a district with no data if the land was a flat plain, but where it was hilly 'experience shows that the distribution of rainfall bears a close relationship to the height and form of the land and to the prevailing direction of the wind'.[224] Lines would be drawn in accordingly. For this exercise, Mill followed Buchan in recommending Bartholomew's series of maps, particularly those on the scale of half an inch to a mile. He thought them a better basis for rainfall work than the Ordnance Survey maps on the same scale, due to the greater contrast presented between high and low land and thc absence of woodland colouring, which were distracting features on the Ordnance Survey sheets.[225] For county-level rainfall mapping, Mill found the Ordnance Survey quarter-inch scale the most convenient to use.

Mill tested his cartometric method in his paper on the rainfall of the Forth Valley in Scotland. The paper worked with data covering a thirty-five-year period from 1878. The table that accompanied the paper included the arithmetical mean for the observed data at each station; the ratio that shorter periods of rainfall bore to the average at adjacent stations where full records were available; and average rainfall for short records, computed by applying the ratio to the mean for the years observed. The map was prepared by Mortyn de Carle S. Salter using Ordnance Survey sheets and showed tint-shaded contours as well as the isohyets (Figure 4.15). Station annual averages were also marked on and those with short runs of data were marked with a circle. Twelve short records in the neighbourhood of Loch Katrine, none longer than seventeen months, were used to test the method, Mill claiming that when these new figures were added to the map, it was not necessary to alter rainfall lines by any appreciable amount.[226] For Mill, this both demonstrated the quality of available data, even where the series was short, and illustrated the veracity of his cartometric method.

Mountain stations that were exposed to high winds and storms posed another problem for map makers in that they presented lower rainfall figures compared to neighbouring sheltered stations. This risk of error from overexposure was particularly high in the winter months due to wind, storms and snow. Mill's cartometric method allowed for these exposed stations' means to be compared to their neighbours and

[224] Mill, 'The Rainfall of the Forth Valley, 5–6.
[225] Mill, 'Map-Studies of Rainfall', 68.
[226] Mill, 'The Rainfall of the Forth Valley', 4.

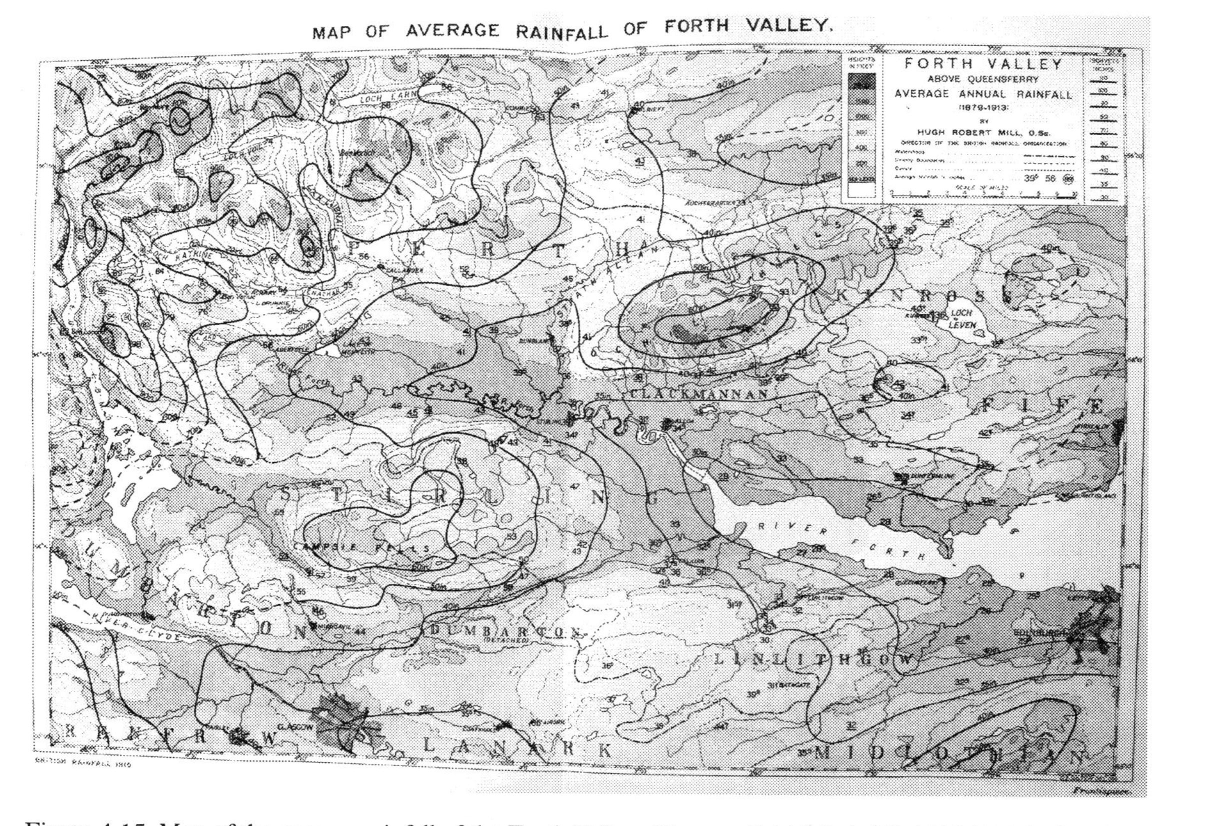

Figure 4.15 Map of the average rainfall of the Forth Valley. (Source: *British Rainfall. 1915* (1916), frontispiece.)

corrections applied where necessary. Salter, who went on to become director of the British Rainfall Organisation, cautioned that detecting errors due to overexposure remained a challenging task, in that for individual readings, the amount of error was usually smaller than the difference from the reading at a neighbouring station that arose naturally. Systematic errors – the product of an inappropriate measuring glass or a leaking gauge, for instance – became more apparent when the totals for a considerable period were compared, but even those could still be 'mistaken for a geographical variation'.[227] Salter nonetheless viewed the mapping of rainfall data as the 'surest safeguard against the employment of erroneous records from whatever cause the errors may arise'.[228] He also argued that the mapping of synchronous observations imparted vitality to the data, and that the full force of this reconstructive process was only increased when isohyetal lines were added and rainfall averages tinted. Mill suggested that the shading of isohyets turned a map into a picture, 'in which the eye is attracted instantly to the areas of highest and lowest values and can appreciate the range and rate of gradation between'.[229]

The mapping of rainfall visualised errors and brought numerical data to life. It also helped meteorologists to achieve other ends. The maps potentially allowed for the recognition of recurring types of distribution and their relationship to rainfall, such as other physical phenomena or seasonal variations. Identification of these correlations helped to find laws underlying rainfall, such as the effects of landforms on rain patterns, the movements of low-pressure centres, and general atmospheric circulation.[230] The imposition of isohyets onto physical maps also allowed for the estimation of a single general rainfall total for particular geographical areas – 'general rainfall' being Mill's term for the mean of a number of values over a geographical area. Working with the assumption that the total rainfall of an isohyetal zone was the mean between the two limiting values expressed by the lines, the general rain total of each zone could be calculated using a planimeter, or simply by dividing the area into small regular squares and counting them. Multiplying the general rain of each zone by its area produced its total volume of fall. Adding the volumes for all zones produced the volume for the whole map and for the region it represented.[231]

[227] Salter, *The Rainfall of the British Isles*, p. 96. Arithmetical mistakes or mistranscriptions were considered less of an issue because they would only affect the map on which the individual reading occurred, unless the error occurred repeatedly.
[228] Salter, *The Rainfall of the British Isles*, p. 97.
[229] Mill, 'Introduction', *Rainfall Atlas of the British Isles*, p. 6.
[230] Salter, *The Rainfall of the British Isles*, p. 100.
[231] Salter, *The Rainfall of the British Isles*, p. 102; Herbertson, *The Distribution of Rainfall over the Land*, p. 7.

The mapping of rainfall also had a practical importance. Rain maps provided information of use to farmers, civil engineers, city planners, local and national authorities and industrialists. Many of the maps produced by the British Rainfall Organisation had been produced as part of commissions for municipal or statutory authorities, for legal arbitrations, as evidence for Parliamentary Committees, for insurance disputes – investigation of floods and their impacts for instance – and to supply information to reservoir builders and water boards.[232] One of the earliest maps showing mean annual rainfall for the British Isles was produced as part of Symons's evidence to the 1868 River Pollution Commission. The Commission's remit was to assess the causes and extent of pollution of Britain's rivers in the context of increasing domestic, navigational and industrial demands for water. Symons's report to the Commission supplied information on the amount of rain falling on particular river basins and the effects of recent droughts on supply.[233] Although the map that accompanied the written evidence was based on only a five-year run of rainfall records (1860–65), Mill later claimed it was impossible to improve on it.[234] Many of Mill's own maps were produced for professional ends. His 1915 map of the Firth of Forth was part of a report on the amount of water available for the operation of locks on the Forth and Clyde Ship Canal. Others were produced as evidence for Parliamentary approval of large-scale waterworks such as reservoirs. For instance, the London Metropolitan Water Board signed an agreement with the British Rainfall Organisation in 1906 for monthly returns of rainfall for the Thames and Lee valleys, which had to be accompanied by maps, one of which was reproduced in the *Meteorological Magazine*.[235]

Conclusion

When Mill retired in 1919, he received a letter from Alfred Blackburn, the president of the Institution of Water Engineers, thanking him for the services he had provided to the profession and the nation and expressing his pleasure that the work of the British Rainfall Organisation was to be

[232] Mill, *Life Interests of a Geographer*, p. 95.
[233] Rivers Pollution Commission, *First Report of the Commissioners Appointed in 1868 to Inquire into the Best Means of Preventing the Pollution of Rivers. (Mersey and Ribble Basins.)* Vol. I Report and Plans (London: G.E. Eyre and William Spottiswoode, 1870).
[234] Mill, On the Distribution of Mean and Extreme Annual Rainfall, 1.
[235] Anon, Memorandum of Agreement between H. R. Mill and the Metropolitan Water Board, National Meteorological Archive, RMS/2/5/1/1/3; Anon, 'Thames Valley Rainfall – January 1911', *Meteorological Magazine*, 46 (1911), facing 10.

placed on a permanent footing.[236] Blackburn was referring to the amalgamation of the British Rainfall Organisation with the Meteorological Office. Conversations about incorporation had begun in 1915, when Napier Shaw, the director, admitted that it was embarrassing his office did little work pertaining to rainfall and had to refer inquiries on to the British Rainfall Organisation. He also recognised the value and public utility of the Organisation's body of rainfall statistics and thought it should become a public record.[237] Mill made the case for the value of the British Rainfall Organisation to Henry George Lyons, Shaw's replacement.[238] The Organisation now had a large number of daily records, well distributed across the country. He put the greatest emphasis on its cartographic methods of analysing daily, monthly and annual rainfall, which placed the work 'on a total different scientific plane' to earlier accomplishments.[239] The Organisation was eventually incorporated with the Meteorological Office in 1919. Mill's statement to the British Rainfall Organisation trustees in late 1918 cited his own ill health and the wider effects of the war on the rainfall network as reasons for the amalgamation.[240] With Mill's retirement in early 1919, Salter took over as the first superintendent of the British Rainfall Organisation under the Meteorological Office. Symons's leasehold house on Camden Square was given to the Royal Meteorological Society along with an endowment fund. The fund was used to research and prepare a rainfall atlas, published in 1926 as the *Rainfall Atlas of the British Isles*, with an Introduction by Mill.[241] Reflecting on the achievements of the British Rainfall Organisation, Salter praised Symons's work in England and Buchan's work in Scotland, saying that they 'laid the foundation for the systematic observation of the rainfall in all parts of the country on a scale unparalleled in any part of the world'. He went on to claim that their work had made possible the cartographical studies of Mill, from whose maps of rain distribution 'practically all our existing knowledge of the rainfall of the British Isles from the purely geographical side must be drawn'.[242]

[236] Letter from A. Blackburn to H. R. Mill, 7 November 1919, National Meteorological Archive, RMS/2/5/1/1/8.
[237] Letter from W. Napier Shaw to H. R. Mill, 18 October 1915, National Meteorological Archive, RMS/2/5/1/1/8.
[238] M. E. Crewe, 'The Met Office Grows Up: In War and Peace', *Occasional Papers on Meteorological History*, 8 (March 2009), 1–37.
[239] Letter from H. R. Mill to H. G. Lyons, 15 October 1918, National Meteorological Archive, RMS/1/5/12.
[240] H. R. Mill, Statement for Trustees, 10 December 1918, National Meteorological Archive, RMS/2/5/1/1/7.
[241] Anon, *Rainfall Atlas of the British Isles*.
[242] Salter, *The Rainfall of the British Isles*, p. 4.

In this chapter, I have shown that Symons's rainfall observation network was characterised by three key geographical projects: the establishment of a rainfall archive; the staging of rainfall experiments; and the production of rainfall maps. Developing a network of small-scale, distributed observatories that could produce a meaningful archive of rain data involved the imposition of principles of uniformity and accuracy as they pertained to instrument pattern, position and exposure. Gooday reminds us that '[w]hat counted as accuracy was what constituted a sufficient degree of accuracy for a particular purpose to be undertaken within existing contextual constraints of money and time to the satisfactions of relevant audiences'.[243] The same applied to uniformity. Both accuracy and uniformity were crucial factors in the establishment of an observational rain network, but the varied social terrain and the constraints imposed by the use of volunteer labour and domestic spaces challenged both and meant that uniformity was prioritised over accuracy. Certain aspects of uniformity were easy to resolve – such as timing of the daily gauge reading or data-entry protocols – but others required negotiation and experiment. In particular, securing good exposure was a crucial component in rainfall observation but relied on negotiation in contexts where observers were using a suburban garden or other domestic space. Degree of accuracy was less well defined, where the observer was often relied upon to demonstrate practices of care, and hence trustworthiness, in observation and inscription.

The establishment of experimental sites in various locations in England helped to develop other aspects of the network's uniformity, notably in relation to gauge pattern and construction materials (as a result of the experiments, by the late nineteenth century the five-inch copper Snowdon gauge was in widespread use across Britain). If Symons's network emphasised the horizontal extension of uniformity over geographical space, the experiments considered the integrity and trustworthiness of instruments and their placement vertically. In their investigation of the best situation for rain gauges above the ground, the experiments became caught up in a long-running debate about the nature of rain itself. While experiments were meant to create trust in gauges and rain study, they ended up engendering controversy. The trials opened to question the nature of measurement in rainfall work and the trustworthiness of gauges. Were they, like Chrimes's gauges, almost speaking, or were they, as Thrustans put it, utterly valueless for any scientific purpose? In the end, the consensus fell on the side of observations and the archive over geometrical proof and abstract reason.

[243] Gooday, *The Morals of Measurement*, p. 268.

Generating new rain series and combining them with historical observations turned localised rain events into data and discrete data sets into a coherent archive and organised memory of British rain. Combining rain data from family ledgers and putting them together into a single national data set promised the ability to conceptualise weather trends and investigate explanations for those trends. The archive also exerted its own form of care over its returns, whereby observations acted as checks on one another and eradicated errors specific to localities. While meteorologists continued to pursue interests in exceptional sites – those localities with very high rainfall for instance – increasing emphasis was placed on understanding average rain states and variations from those states. To this end, Symons and his observers relied on the arithmetical mean as the principal method of data summary and analysis, even if the meaning of the mean was contested. Detractors of the experimental trials mobilised the arguments of Jevons and Galton to undermine the quality and value of rain data's arithmetical mean. Even those who made use of it acknowledged that applying the mean to rain series ran the risk of projecting the impression of integrity and equivalence onto otherwise quite different and possibly flawed data sets. Rain maps took the statistical culture of the arithmetical mean and translated it into a visual form while offering a solution to the problem of data integrity. Mill's self-styled detective agency for accidental errors articulated cartography as a form of data care. Maps represented the rainfall archive at a glance and extended its terrain – from data points to planimetric quantities and isohyetal zones. Rainfall maps also emphasised comparative measurement, that is, differences from the mean rather than absolute values. The *Rainfall Atlas* was a case in point. It featured two main sets of maps. There were the small-scale maps of annual rainfall for each individual year from 1868, expressed as a percentage of the average for the period 1881–1915. These maps revealed years of excess and low rain relative to the mean. The second set featured maps of average monthly rainfall. Writing in *Nature*, Ernest Gold, the deputy director of the Meteorological Office, praised the average monthly maps for their practical utility but complained that 'they do not make quite the same appeal to the imagination: averages never do'.[244]

[244] E. Gold, 'An Atlas of Rainfall', *Nature*, 120 (27 August 1927), 291–2.

Conclusion: Historical Geographies of Future Weather

Just like the Introduction chapter to this book, the Conclusion chapter begins with an address by George Symons. In a talk to the Royal Meteorological Society he outlined plans for a meteorological exhibition which had been organised to celebrate the sixtieth anniversary of Queen Victoria's reign. It was to contain two sets of instruments, from 1837 and 1897, a comparison of which would commemorate the occasion of the Jubilee and show the progress made in meteorology since Victoria ascended to the throne.[1] Symons, the then secretary of the Society, provided a lecture detailing the aims of the exhibition and the activities of the Society over the period. However, reconstructing what Symons called the 'ordinary outfit' of a meteorological station in 1837 proved to be no mean feat. It was not enough to assume that instruments invented prior to 1837 would have been part of the usual equipment of a station. Symons turned instead to his brief employment at the Meteorological Office, his long involvement with the Royal Meteorological Society and his leadership of the British Rainfall Organisation to reconstruct a list of all the stations in operation across the British Isles in 1837, many of which he had visited and inspected himself. He also made use of the contemporary writings of notable meteorologists, including John Daniell, Luke Howard, John Lubbock and James Forbes. From these sources Symons was confident that he could describe the instruments that were in use by 'good, ordinary observers' in 1837 and were readily available in the 'shops of the opticians of that day – Newman of Regent St., Jones of Oxenden St., Cary of the Strand, Potter of the Poultry, Long of Martin's Lane, or Harris of Holborn'.[2] In the end, twenty-nine instruments from the period were exhibited.

Mapping out the essential instruments for an ordinary station in Britain in 1897 was obviously a much simpler task. In all, 116 instruments were on display at the exhibition, along with numerous diagrams and

[1] G. J. Symons, 'Meteorological Instruments in 1837 and 1897', *Quarterly Journal of the Royal Meteorological Society*, 23 (1897), 205–20, 205.

[2] Symons, 'Meteorological Instruments in 1837 and 1897', 210–1.

photographs.[3] One could expect to find a barometer (compared against the standard at Kew Observatory), a Stevenson screen containing dry and wet bulb hygrometer, maximum and minimum thermometers, a rain gauge and perhaps thermometers adapted for measuring earth temperature and solar and terrestrial radiation, an anemometer, a sunshine recorder and even self-recording instruments. By comparing the instruments and practices across the sixty-year span, Symons presented what he saw as two quite different communities of practice, with the '[v]ery nearly perfect uniformity' of 1890s meteorology compared to an ad hoc and antiquated regime, where 'every observer did what he thought best'.[4] Although Symons did strike some critical notes in his review of meteorological progress in Britain – particularly relating to experimental meteorology – he ended his presentation on a triumphal note with a discussion of the weather map, a self-consciously mundane feature of intellectual and public life in the 1890s and yet also an object that embodied the 'essence' of meteorological labour over the last sixty years:

> I have to show that which may in many respects be regarded as the essence of much that has gone before, and of the marvellous progress which has characterised the Victorian reign. Thanks to the help of Mr. Scott and Mr. Gaster I am able to show the state of the weather over Western Europe only three hours ago. We are so used to weather maps that they do not strike one as at all a marvel, but when we look into the details, they really are startling. We have here records for this evening from the sunny Riviera, eastwards as far as Berlin, and northwards, not merely to the top of the Gulf of Bothnia, but actually up to Bodo within the Arctic circle. If this does not prove the existence of wide-spread interest in meteorological work, excellent organisation, and the wonderful development of human intercourse, I do not know how otherwise it could be demonstrated.[5]

Symons's choice of temporal bookends for his exhibition also happen to frame this study, even if some of the examples in the book spill out into the years prior to Victoria's succession to the throne and those after her death in 1901. This book has also been concerned with many of the topics covered in Symons's anniversary exhibition and address: with the recruitment and management of 'good, ordinary observers'; the deployment of meteorological instruments in a range of observatory settings; the sustenance of regimes of uniformity across extended observatory networks; and the production of startling weather maps. In this study, meteorology has been treated as an observatory and a survey science; that is, a science defined and governed by a set of geographically constrained techniques

[3] Anon, 'Report of the Council', *Quarterly Journal of the Royal Meteorological Society*, 23 (1897), 83–5, 83.
[4] Symons, 'Meteorological Instruments in 1837 and 1897', 216 and 211, respectively.
[5] Symons, 'Meteorological Instruments in 1837 and 1897', 220.

and expectations on the one hand, while emphasising the value of mobile observation and the exploration of different environmental atmospheres on the other.

The book's most general argument, sustained across its four empirical chapters, is that meteorology's status as an observatory science was constantly on trial over the course of the nineteenth century. Meteorological observatories staged various experimental trials and new techniques, such as the rain gauge experiments discussed in Chapter 4, but ultimately meteorological observatories, and indeed the very idea of observatory meteorology, were the objects under scrutiny. Although some of the observatories considered in this book appeared more permanent than others, none were guaranteed longevity. Observatory techniques came and went and tested the limits and resolve of all the sites in question: Beaufort's sailing vessels were replaced by steam ships no longer solely reliant on the wind, the Colaba Observatory ceased to be part of the Indian Meteorological Department in 1971 when it became the Indian Institute of Geomagnetism, while the Ben Nevis Observatory was reduced to nothing more than a ruin and a commemorative plaque.[6] The observatory experiments discussed in this book interrogated the efficacy of four crucial conditions: the significance of geographical particularity in justifications of observatory operations; the sustainability of coordinated observatory networks at a distance; the ability to manage, manipulate and interpret large data sets; and the potential public value of meteorology as it was prosecuted in observatory settings. An observatory's life expectancy was determined by its response to these conditions.

Geographies of Operations and Networks

Similar to Symons's emphasis on the work done at Britain and Ireland's 161 weather stations in 1837, this book has considered the history of weather study from the perspective of the places where the weather was observed, as meteorology attempted to reinvent itself as a physical science on a par with terrestrial magnetism. This is not primarily a study of Victorian meteorology as it was re-fashioned at elite sites of science such as Kew or Greenwich, but rather a historical geography that concentrates on local, provincial, peripatetic and otherwise relatively marginal sites of scientific activity. We have seen how a wide variety of spaces were constituted and reconfigured as meteorological observatories –

[6] Aubin et al., 'Introduction: Observatory Techniques in Nineteenth-Century Science and Society'; J. C. Farman and A.S. Thow, 'Ben Nevis Observatory: A Plaque Unveiled', *Weather*, 25 (1970) 77–82.

suburban back gardens, naval vessels, mountain tops and colonial outposts. Observatory managers hoped to mimic the astronomical ideal of the observatory as a place apart, but in all the cases considered here this ideal proved very difficult to achieve. Instead, meteorological observatories were shown to be hybrid spaces where diverse social groups mingled and often made contributions to the science being done there – sailors on board surveying vessels, Army personnel and local computers at the Colaba observatory, mountaineers and tourists on Ben Nevis and family members in the back gardens of suburban villas. The inability of observatory managers to maintain the integrity of their observatory's physical and imagined boundaries was often interpreted as a threat to the successful prosecution of the observatory experiment and even evidence of their inadequacy.

Weather observatories were meant to act as sites that passively monitored the environments and atmospheres in which they were placed, whether at sea, in the tropics, high in the mountains or in – to northern Europeans at least – more mundane temperate settings. But observatories were not simply mirrors held up to the surrounding atmosphere. Rather, observatories made their weather and climate, adding to understandings of meteorological processes under different environmental conditions. Ships' weather logs helped to shape theories about storms at sea, work at India's observatories contributed to understandings of the monsoon, observations at Ben Nevis and Fort William promoted the idea of a vertical atmospheric ocean and rain gauges in suburban back gardens helped to establish the factors that determined rainfall distribution. If observatories shaped contemporary understandings of regional atmospheres, so too were they influenced by their local and environmental contexts. Different localities and natural environments actively shaped meteorological practices, knowledge and identities.[7] The fact that *where* the weather was observed often seemed to affect *how* it was observed presented an existential challenge to a science that established uniformity as its raison d'être. In fact, it often seemed that environmental ideals became internalised into observatory life. For instance, mountaintop observers promoted an Alpine aesthetic and heroic sublimity, while rainfall stations modelled themselves on the English country house garden and its ideal exposures. We have also seen that specific locations became disproportionately important in observatory culture and helped to further meteorological projects such as the Ben Nevis Observatory, Bermuda in the case of Reid's work on maritime meteorology and Strathfield Turgiss,

[7] On the regionalisation of scientific cultures, see S. Naylor, *Regionalising Science: Placing Knowledges in Victorian England* (Pittsburgh: University of Pittsburgh Press, 2010).

a crucial hub for Symons's rain gauge experiments. These charismatic sites were both a blessing and a curse. As the fate of the Ben Nevis Observatory demonstrated most explicitly, the inability of meteorological observatories to negotiate the treacherous boundary between located fieldsite and locationless laboratory often proved to be their undoing.

Keeping meteorological instruments and the personnel who operated them in good working order whilst under different environmental conditions was another challenge faced by observatory managers. Instruments, instrumental infrastructure and their observers often only operated well within narrow environmental envelopes and a lot of observatory work was devoted to coaxing them through their discomfort, to making ad hoc amendments or to the development of new designs. Discussion of the international maritime conferences in Chapter 1 showed that national boundaries also confounded the circulation of instruments and their units of measure. Symons's campaign to develop a rain network helped to establish the five-inch Snowdon gauge as the standard for British work but the use of the metre in France and elsewhere placed constraints on international circulation of rain data. A great deal of research went into the design and construction of new instruments, interpreted as markers of an increasingly sophisticated science. However, many basic meteorological instruments, including thermometers, barometers and rain gauges, could be owned and operated by both experts and enthusiasts. These instruments facilitated meteorological participation by different social groups and brought different communities of practice together.[8]

In his 1926 reflection on the history and present state of meteorology, Napier Shaw argued that contemporary instrumental accuracy far outstripped the capacity of meteorologists to coordinate existing observations.[9] Observatory networks promoted the ideal of voluntary participation and the control and subordination of individual components and effort into coherent and productive science. In doing so, meteorology was meant to demonstrate the value of collective effort.[10] We have seen that sustaining coordination was an abiding concern and constant challenge for observatory meteorologists. How to enforce uniformity of instrument use, exposure, recording practices, infrastructure and reduction across dense or long-range networks? Each of the observatory cases examined in this book involved the building or repurposing of networks along which advice and protocols were disseminated, personnel and instruments moved and data shared. For instance, the development of naval ships as weather observatories exploited extant military

[8] Anderson, *Predicting the Weather*. [9] Shaw, *Manual of Meteorology*.
[10] Anderson, *Predicting the Weather*, p. 287.

infrastructure but also benefited from a series of international congresses that encouraged the adoption of universal standards, observation regimes and the mutual exchange of information. The establishment of the observatory at Bombay also benefited from military and imperial infrastructures but later found itself part of a much larger national network of meteorological stations across the Indian subcontinent.

Atmospheric Data Empires

Due to discrete projects, national weather services, imperial infrastructures and international schemes, vast quantities of numerical weather data produced by observatories ended up accumulating in weather archives.[11] For some, the production of weather data over a long time frame and at a global scale was an end in itself. For others it was an important means to an end. Meteorologists working in the empiricist tradition cultivated a descriptive science, which assumed that laws of weather behaviour would emerge from the analysis of large data sets. As the previous chapters have shown, this approach dominated the field internationally over most of the nineteenth century. Although quantitative regularity was hard to find in the data, qualitative regularities were deployed to support theorisation, such as the law of storms discussed in Chapter 1 and the effects of sunspot activity mentioned in Chapter 3.[12] As we saw in Chapters 2 and 4, practitioners applied statistical techniques to meteorological data, although many studies were limited to the discovery of average weather. This work catalysed the study of climate in the later decades of the nineteenth century, defined by the celebrated Austrian meteorologist Julius von Hann as 'the totality of meteorological phenomena which characterize the average state of the atmosphere at some place on the earth's surface'.[13] In other words, climate came to be defined in statistical terms and climatology developed as a statistical science that relied on the data sets produced by nineteenth-century meteorologists, even if its main concerns were often limited to the computation of daily, monthly and yearly averages of the elements of the weather, demonstrated in the rainfall work discussed in the previous chapter.[14]

[11] J. R. Fleming, *Historical Perspectives on Climate Change* (Oxford: Oxford University Press, 1998), p. 9.
[12] P. Edwards, *A Vast Machine: Computer Models, Climate Data, and the Politics of Global Warming* (Cambridge, MA: MIT Press, 2010), p. 62.
[13] Quoted in Nebeker, *Calculating the Weather*, p. 22.
[14] Nebeker, *Calculating the Weather*, p. 22.

Across all the case studies discussed in this book, we have followed observatory meteorologists translating data into visual images – graphs, diagrams and, most of all, maps – so as to seek regularities visually as well as numerically. The visualisation of weather and climate data was promoted by many notable meteorologists throughout the nineteenth century. The growth in observing networks, combined with the use of telegraphy, enabled the production of daily weather maps to be compiled and published on the same day – often referred to as weather telegraphy. (Symons claimed that the first daily weather map had been produced for sale at the Great Exhibition in London in August 1851.[15]) Maps and atlases representing average weather also became more common in the final decades of the nineteenth century, as we learned in Chapter 4. Edwards argues that meteorologists aspired to an infrastructural globalism, which 'sought to establish permanent sociotechnical systems for monitoring the weather, modelling its processes, and preserving planetary data as scientific memory'.[16] An early example of this globalism involved the use of weather and climate data to envision the world as a single entity. The map was a crucial technology towards this end. Lehmann observes that climatologists like Wladimir Köppen incorporated data from around the world, including colonial observatory data, to facilitate the '"scaling up" of the climatological vision to encompass the entire earth'.[17] Discussing German weather and climate study in colonial East Africa, he notes that much of the numerical data produced were inaccurate, incomplete and unverifiable, but those traits were lost when data were subsumed into large-scale data repositories and into mapping projects (as we saw in relation to the Bombay Observatory data). Climate maps and atlases in particular gave an appearance of globality even if they were built on 'highly local and contested data, which in turn were based on highly local and contested practices of data gathering'.[18] More generally, imperial networks helped to rethink and support meteorology in the nineteenth and early twentieth centuries, not just by giving it a new spatial reach but also by translating data in the service of scaled-up, ultimately global visions of weather and climate.[19]

[15] G. J. Symons, 'The First Daily Weather Map', *Meteorological Magazine*, 31 (1896), 1. See also W. Marriott, 'The Earliest Telegraphic Daily Meteorological Reports and Weather Maps', *Quarterly Journal of the Royal Meteorological Society*, 28 (1903), 123–31; Anderson, 'Mapping Meteorology', p. 71.

[16] P. Edwards, *A Vast Machine*, p. 25.

[17] Lehmann, 'Average Rainfall and the Play of Colours', 46.

[18] Lehmann, 'Average Rainfall and the Play of Colours', 46.

[19] Mahony and Caglioti, 'Relocating meteorology', 3–4; M. Mahony, 'Meteorology and Empire', in A. Goss (ed.), *The Routledge Handbook of Science and Empire* (Abingdon: Routledge, 2021), pp. 47–58, p. 52; Naylor and Goodman, 'Atmospheric Empire'.

Others were less convinced by the benefits and uses of meteorology's global data archives. Some worried that observatories were needlessly duplicating efforts on the grounds of national pride.[20] As early as 1839 the German meteorologist Heinrich Dove complained that meteorology was held back not by the lack of data but due to 'inadequate utilization of the data already at hand'.[21] Later in the century Joseph Henry, first secretary of the Smithsonian Institution, claimed: 'There is, perhaps, no branch of science relative to which so many observations have been made and so many records accumulated, and yet from which so few general principles have been deduced.'[22] George Airy made the same point and was later paraphrased by Symons when he complained that in national weather services 'the observing is out of all proportion to the thinking strength in Meteorology', even if his claim that nineteen out of twenty new weather observers made no attempt to study the subject to which they were contributing could surely have been levelled at his own rainfall network.[23] Writing in *The Athanaeum*, meteorologist William Chappell praised the science for 'cultivating most zealously the powers of observation' in such a large and well-trained group of observers but lamented that 'there has been no corresponding development of the reflective powers; we have not produced even a small band of meteorological philosophers'.[24] This was not an argument against data gathering or an empirical meteorological agenda but it was a call for greater attention to generalisable facts currently buried in 'the vast accumulations of observations'.[25]

A cadre of applied physicists working at the start of the twentieth century proposed to develop understandings of atmospheric processes theoretically. Some worked with little recourse to data repositories or processes observable in the physical world, a movement often emblemised by Vilhelm Bjerknes's Bergen School of Meteorology.[26] Others, like Napier Shaw, also 'worked in a purely deductive way, deriving equations of meteorological import from physical laws or investigating the properties of a mathematical model while setting aside the question of the model's fidelity to the physical world'.[27] He, like Koppen, supplied

[20] Donnelly, 'Redeeming Belgian Science'.
[21] Quoted in Nebeker, *Calculating the Weather*, p. 15.
[22] Quoted in Fleming, *Meteorology in America*, p. 148.
[23] G. J. Symons, 'Meteorological Bibliography', *Meteorological Magazine*, 14 (1879), 17–22, 21.
[24] W. Chappell, 'Science', *The Athanaeum*, 9 March (1878), 317–9, 318.
[25] Chappell, 'Science', 319.
[26] R. M. Friedman, *Appropriating the Weather: Vilhelm Bjerknes and the Construction of a Modern Meteorology* (Ithaca: Cornell, 1989).
[27] Nebeker, *Calculating the Weather*, p. 29.

Bjerknes with data sets, expertise and infrastructure.[28] Shaw also defended the value of meteorological observations and the development of theory from data. The weather, he argued, was constituted by a continuous series of physical and chemical processes that could only be explained by laws established by experiments in the laboratory. However, these atmospheric processes could only be illustrated and not imitated by laboratory experiments – the scales on which these processes operated were simply too vast, horizontally and vertically. Meteorology was therefore the combination of the knowledge obtained from the 'laboratory and the synoptic weather-map.'[29] This was by no means a straightforward collaboration, due to the need to reconcile competing approaches, assumptions and ends. As historians of twentieth-century meteorology have noted, by the first years of the twentieth century, the study of the weather was being pursued in three key and largely discrete ways: as a theoretical science that modelled the atmosphere mathematically; as a data-driven science that produced average weather and climate maps; and as an arm of government that produced synoptic maps and forecasts.[30]

What were meteorological observatories ultimately for? And perhaps even more critically, who should fund these observatories' maintenance costs, and to what end? The answer to these questions often came back to forecasting. Forecasting became a crucial justification for significant investments of time and money but as both concept and practice it haunted many of the observatories considered in this book, and like all good ghosts it threatened to drive their inhabitants mad. Forecasting worked for the architects of many observatory schemes because it portrayed meteorology as first and foremost a useful science that justified such significant investments of labour and resources. A preoccupation with prediction also presented meteorology as a progressive science, whereby the future became something over which a measure of control could be exercised. This attitude conformed with liberal teleological ideas about how science could contribute to empire's civilising project.[31] However, the relationship between nineteenth-century observatory meteorology and forecasting was in large part a marriage of convenience. As the book shows, the accumulation of huge quantities of data about the weather at sea, on mountains and over towns and cities was not enough to produce a demonstrably useful science and often contributed to observatory meteorology's undoing. The development of dynamical meteorology and the processing power of the electronic computer in the twentieth

[28] Wille, 'Colonizing the Free Atmosphere'. [29] Shaw, *Manual of Meteorology*, p. 116.
[30] Nebeker, *Calculating the Weather*, p. 39. [31] Mahony, 'Meteorology and Empire', 51.

century were crucial in significantly improving forecasting but these features of the history of meteorology fall largely outside the parameters and time frame of this study.

The apparent gap between the mass of weather data generated and the utility of those data allowed many to conclude that the observatory experiment was a failure, or at least that the significant human and financial resources observatories demanded could no longer be justified. The accounts of the observatory experiments provided in this book have of necessity ended when physical sites closed their doors, or when independent scientific organisations were subsumed into larger bureaucratic entities. It is nonetheless worth noting the afterlives of many of the data sets discussed in previous chapters.

Over the last decade or so a number of studies have been published that document the use of historical observations to reconstruct and reanalyse past weather and climate. Pertinently, these include weather observations from Royal Navy logbooks, the Ben Nevis and Fort William Observatory data and the rainfall records of the British Rainfall Organisation.[32] Led by climate scientists, all of these studies have included the transcription of historical weather and climate data from paper to digital formats, routinely describing this act of transcription as a form of data rescue. That data needed to be rescued from scientific obscurity and obsolescence – that, in other words, data unreadable to modern computers was lost to science – was an endeavour to reduce data friction, to minimise the resistance paper-based data offered to electronic analysis.[33] This rescue involved the painstaking digitisation of numerical observations from paper archives into electronic databases. The labour of rescue was achieved by recruiting volunteer 'citizen scientists' through crowdsourced campaigns. There is a neat circularity to this arrangement, where an army of human computers in the nineteenth and early twentieth centuries produced a huge paper archive of weather observations, only for another cadre of human computers in the twenty-first century to ensure those paper archives could be read by another computer, this one electronic.

[32] R. García-Herrera *et al.*, 'Understanding Weather and Climate of the Last 300 Years from Ships' Logbooks', *WIREs Climate Change*, 9 (2018), 1–18; P. Brohan et al., 'Marine Observations of Old Weather', *Bulletin of the American Meteorological Society*, 90 (February 2009), 219–30; E. Hawkins et al., 'Hourly Weather Observations from the Scottish Highlands (1883–1904) Rescued by Volunteer Citizen Scientists', *Geoscience Data Journal*, 6 (2019), 160–73; E. Hawkins, et al., 'Millions of Historical Monthly Rainfall Observations Taken in the UK and Ireland Rescued by Citizen Scientists', *Geoscience Data Journal*, 10 (2023), 246–61; E. Hawkins et al., 'Rescuing Historical Weather Observations Improves Quantification of Severe Windstorm Risks', *Natural Hazards and Earth System Sciences*, 23 (2023), 1465–82.

[33] B. J. Strasser and P. N. Edwards, 'Big Data Is the Answer . . . But What Is the Question?', *Osiris*, 32 (2017), 328–44, 330.

Aronova et al. remind us that the project of translating the world into data has been under way for centuries and some of the challenges faced by managers of crowdsourced projects in the twenty-first century are stubbornly similar to those faced by managers of networks that produced the data in the first place.[34] Policing observer and computer error remains a pressing concern, for instance: in the work of digitising the Ben Nevis observations it was decided that the input of each column of data would have to be repeated three times as a pragmatic check on the accuracy of the keying by fallible volunteers.[35] George Symons would likely have approved.

In their essay on Big Data, Strasser and Edwards observe that data sets are often kept on the moral grounds that future communities will one day make epistemic use of them, even if that moral imperative to share data with an imagined future contributes to a data deluge and thus poses further moral questions about what data to keep and what to let go.[36] That moral relationship between practitioners separated by time but joined by data goes both ways, or so it is claimed. In a reanalysis of the Ben Nevis observations, climate scientists argue their work 'fulfils the ambitions of the meteorologists of over a century ago that their data be made available to aid weather forecasting and the study of mountain meteorology'.[37] The authors of a paper that reconstructs a storm which affected much of the British Isles in February 1903 – named Storm Ulysses because it featured in James Joyce's novel – similarly claim that the digitisation of data improves understandings of historical weather events.[38] Writing about their work with Britain's archive of rainfall data, the scientists claim that the transcription of over five million observations, spanning the period 1677–1960, 'will transform our understanding of historical weather variations'.[39]

While twenty-first-century climate scientists situate themselves as the intended recipients of past weather data – as digital curators, well placed to manage, read and reinterpret the weather's paper archive – they in turn imagine themselves in conversation with potential future actors. Writing about Storm Ulysses, they argue that the reconstruction of greater numbers of severe historical storms and other events, such as heat waves and

[34] E. Aronova, C. von Oertzen and D. Sepkoski, 'Introduction: Historicizing Big Data', *Osiris*, 32 (2017), 1–17, 8.

[35] Hawkins et al., 'Hourly Weather Observations from the Scottish Highlands', 163.

[36] Strasser and Edwards, 'Big Data Is the Answer . . . But What Is the Question?', 341.

[37] Hawkins et al., 'Hourly Weather Observations from the Scottish Highlands', 161–2.

[38] Hawkins et al., 'Rescuing Historical Weather Observations Improves Quantification of Severe Windstorm Risks', 1475.

[39] Hawkins et al., 'Millions of Historical Monthly Rainfall Observations Taken in the UK and Ireland Rescued by Citizen Scientists', 247.

floods, will improve our understanding of the risks from such events today and in the future. More generally they suggest that comprehensive 'rescue' of weather observations stored on paper in traditional archives 'would allow observed trends in extreme events to be put into a longer-term context and help identify where present-day and future risks have been underestimated because such extreme events may not have yet been observed during the modern period'.[40] The translation of weather data from paper to digital format is a first step on the path to imagining and responding to possible planetary futures.[41] Adamson et al. call this 'forecasting by analogy', warning that looking at our archived past to understand our weather and climate futures runs the risk of creating the illusion of linearity from past to future and allowing 'complex and intangible future changes to be presented as coherent narratives of demonstrable change'.[42] The intention here is not to cast doubt on the importance or efficacy of historical climatology and the reanalysis of historical weather events but to emphasise Adamson et al.'s point that we should not assume a direct and clear link between the intentions of historical data producers and the demands of data users today. Weather observatories were experimental sites. They operated on limited budgets and relied on staff willing to work for little or nothing in return for uncertain prospects. Recording the weather as data was certainly the principal motivation for most observers but that task was routinely questioned, as was the statistical manipulation of resulting data sets. The ends of data production were also unknown or only determined in retrospect, while debates about those data's use value, including in forecasting, were routinely contested. If, as climate scientists argue, there are futures to be gleaned from meteorological observatories' weather data, then it is vital that our histories show just how remarkable an achievement that is.

[40] Hawkins et al., 'Rescuing Historical Weather Observations Improves Quantification of Severe Windstorm Risks', 1466 and 1475, respectively.

[41] Strasser and Edwards, 'Big Data Is the Answer ... But What Is the Question?', 329.

[42] G. C. D. Adamson, M. J. Hannaford, E. J. Rohland, 'Re-thinking the Past: The Role of a Historical Focus in Climate Change Adaptation Research', *Global Environmental Change*, 48 (2018), 195–205, 198 and 199, respectively.

Bibliography

Archival Sources

Archives of the Royal Society, London

Hygrometrical Observations made on board His Majesty's Surveying Vessel *Ætna*, AP/19/1.
Meteorological observations made aboard HMS *Jackdaw*, AP/19/2.
Reports on the weather and on meteorological phenomena at Bermuda, AP/24/16.
Meteorological Register of HMS *Thunder*, AP/19/18.
Lord Glenelg, Memorandum respecting the Records to be kept of the state of the Weather, in the British Colonies, BM/3/118.
Minute Book 1, various Royal Society Committees, CMB/1.
Minutes of the Joint Committee of Physics and Meteorology 1838–1839, CMB/284.
Meteorological Committee Minutes and Letters 1830–1837, Domestic Manuscripts Volume III, DM/3.
Correspondence of John Frederick William Herschel, Volume 3, Babinet–Beaufort, HS/3.
Correspondence of John Frederick William Herschel, Volume 19, Miscellaneous, HS/19.
Meteorological Council Papers 1854–1901, MS/775.
Letter from Archibald Geikie to The Secretary, H.M. Treasury, 8 November 1904, NLB/23/2/557.

Lochaber Archive Centre, Fort William

Ben Nevis Observatory Visitors' Book 1, 1884–1888, L/D155.

National Library of Scotland, Edinburgh

Album of photographs, mainly of the meteorological station on Ben Nevis, and of views from the station, ca.1900, Phot.med.89.

National Meteorological Library and Archive, Meteorological Office, Exeter

Three albums of station photographs of Royal Meteorological Society which were taken at time of inspections, 1884–1906, RMS/4/2/1–3.
British Rainfall Organisation correspondence, RMS/1/5/12.
Royal Meteorological Society incoming correspondence, RMS/1/5/13.
Private weather diary of Francis Beaufort 1805–1807, MET/2/1/2/3/540.
Private weather diary of Francis Beaufort 1807–1812, MET/2/1/2/3/541a and 541b.
Calne rain gauge experiment, RMS/2/3/1/1.
Memorandum of agreement between Dr H. R. Mill and the Metropolitan Water Board, RMS/2/5/1/1/3.
British Rainfall Organisation partnership correspondence, RMS/2/5/1/1/6.
File of documents relating to Trusteeship of British Rainfall Organisation, RMS/2/5/1/1/7.
Letters re. Future of British Rainfall Organisation, RMS/2/5/1/1/8.

National Records of Scotland, Edinburgh

Climatological observations (daily) from Ben Nevis Observatory and intermediate stations in connection with the base or sea level Observatory at Achintore, Fort William, MET1/5/2/2/1.
Ben Nevis Observatory: Log Book, Volume 1, 1883–1886, MET1/5/2/3/1.
Operating and Closing Down of Observatories, MET1/5/3/2.
Operating of Observatories, MET1/5/3/3.
Finance and Possible Closures, MET1/5/3/4.
Treasury Committee Report and Possible Closure, 1901–1904, MET1/5/3/5.
Lecture notes: Alexander Drysdale, MET1/5/3/7.
Annual Reports to the British Association for the Advancement of Science, MET1/5/3/8.
Collection of published material relating to the Observatory, MET1/5/3/9.
Minute book of meetings of Directors of the Ben Nevis Observatory, 1883–1901, MET1/5/3/11.
Report to Charles Greaves on the Meteorological Systems sited on Ben Nevis, MET1/5/3/14.

The British Library, London

Proceedings and Consultations of the Government of India and of Its Presidencies and Provinces, IOR/P/346/16/1–3.

The National Archives, Kew

Admiralty Rough Minutes, January–March 1839, ADM3/245.
Letters from the Colaba Observatory, Bombay, BJ 1/2.

Correspondence between Kew Observatory and the India Office concerning the supply of instruments to the Colaba Observatory, Bombay, BJ 1/16.
International Polar Year 1932–1933: re-opening of Ben Nevis Observatory, BJ 5/25.
Proposed establishment of Meteorological Department: correspondence between Royal Society and Board of Trade, BJ 7/4.
Meteorological Council: Printed Minutes, 1877–1880, BJ 8/10.
Meteorological Council: Printed Minutes, 1880–1885, BJ 8/11.

UK Hydrographic Office Archives, Taunton

Letter book, 1836–1837, LB/7.
Letter book, 1837–1839, LB/8.
Letter Book, 1845–1846, LB/13.
Letter Book, 1846–1847, LB/14.
Letter Book, 1852–1853, LB/19.
Incoming Letters prior to 1857, Series C, LP1857/C.
Incoming Letters prior to 1857, Series H, LP1857/H.
Incoming Letters prior to 1857, Serices J, LP1857/J.
Incoming Letters prior to 1857, Series L, LP1857/L.
Incoming Letters prior to 1857, Series R, LP1857/R.
Minute Book 3, 1837–1842, MB/3.
Minute Book 7, MB/7.
Minute Book 8, MB/8.
Miscellaneous Files.

Published Primary Sources

Adams, G. *A Short Dissertation on the Barometer, Thermometer, and Other Meteorological Instruments: Together with an Account of the Prognostic Signs of the Weather* (London: R. Hindmarsh, 1790).
Adams, W. H. D. *Alpine Adventure; or, Narratives of Travel and Research in the Alps* (London: Thomas Nelson, 1878).
Anon, *Observations and Instructions for the Use of the Commissioned, the Junior and Other Officers of the Royal Navy* (London: C. Whittingham, 1804).
Anon, *Regulations and Instructions Relating to His Majesty's Service at Sea* (London: Stationery Office, 1808).
Anon, 'On Storms', *Littell's Spirit of the Magazine and Annuals*, 2 (1838), 856–8.
Anon, 'A Report Explaining the Progress Made towards Developing the Law of Storms', *The Athanaeum*, 25 (August 1838), 594–6.
Anon, 'Redfield's Law of Storms: Notice of Col. Reid's Work on Hurricanes', *The American Journal of Science and Arts*, 35 (1839), 182.
Anon, *Report of the President and Council of the Royal Society on the Instructions to be Prepared for the Scientific Expedition to the Antarctic Regions* (London: Richard and John E. Taylor, 1839).
Anon, 'Weather Almanacs and the Law of Storms', *The London Saturday Journal*, 1 (1839), 7.

Anon, 'Review of Reid's Law of Storms, 1838, along with Redfield's Articles on Atlantic Storms in Silliman's Journal, Blunt's American Coast Pilot and US Naval Magazine', *Edinburgh Review*, 68 (1839), 406–30.

Anon, *Report of the Committee of Physics, Including Meteorology, on the Objects of Scientific Inquiry in Those Sciences* (London: Richard and John E. Taylor, 1840).

Anon, *Admiralty Instructions for the Government of Her Majesty's Naval Services* (London: Stationery Office, 1844).

Anon, 'Proceedings of the Society', *Transactions of the Bombay Geographical Society*, 10 (1852), i–cxi.

Anon, *Maritime Conference Held at Brussels for Devising a Uniform System of Meteorological Observations at Sea*, MS, 1853, Exeter, National Meteorological Archive.

Anon, *Prospectus of an Association for Promoting the Observation and Classification of Meteorological Phenomena in Scotland*, 11 July 1855, Bound in Pamphlets, Edinburgh, National Library of Scotland.

Anon, *Report by the Council of the Meteorological Society of Scotland to the General Meeting Held on the 18th January 1859, with the Minutes of General Meeting, Laws and Regulations of the Society and List of Office-Bearers for the Year* (Edinburgh: Murray and Gibbons, 1859).

Anon, 'Recommendations Adopted by the General Committee at the Newcastle-Upon-Tyne Meeting in August and September 1863', *Report of the Thirty-Third Meeting of the British Association for the Advancement of Science* (1864), xxxix–xliii.

Anon, 'Report of the Council, Held on 7th March 1879', *Journal of the Scottish Meteorological Society*, 5, New Series (1879), 276–81.

Anon, 'Report of the Council held July 21st 1879', *Journal of the Scottish Meteorological Society*, 5, New Series (1879), 368–76.

Anon, 'The Ascent of the Matterhorn', *Chambers's Journal*, 17 (1880), 24–6.

Anon, 'Notes', *Nature*, 27 (1882), 18–20.

Anon, 'The Proposed Observatory on Ben Nevis', *Glasgow Herald* (1883), 75–6.

Anon, *Ben Nevis Meteorological Observatory: An Account of Its Foundation and Work* (Edinburgh: Directors of the Ben Nevis Observatory, 1885).

Anon, 'Subscriptions in aid of Ben Nevis Observatory', *Transactions of the Royal Society of Edinburgh*, 34 (1890), ix–xvi.

Anon, 'Meteorology of Ben Nevis', *Nature*, 43 (1891), 538–540.

Anon, *Guide to Ben Nevis: With an Account of the Foundation and Work of the Meteorological Observatory* (Edinburgh: Directors of the Ben Nevis Observatory, 1893).

Anon, 'Reviews', *Edinburgh Medical Journal*, 40 (December 1894), 550–1.

Anon, 'Report of the Council', *Quarterly Journal of the Royal Meteorological Society*, 23 (1897), 83–5.

Anon [initialed R. C. M.], 'The New Atlas of Meteorology', Review of *Bartholomew's Physical Atlas: An Atlas of Meteorology*, Volume III, *Scottish Geographical Magazine*, 15 (1899), 646–8.

Anon, 'Preface', *Transactions of the Royal Society of Edinburgh*, 42 (1902), viii–ix.

Anon, 'Meteorological Notes and Letters', *Meteorological Magazine*, 39 (1904), 206–10.

Anon, 'Colonel Michael Foster Ward', *Meteorological Magazine*, 50 (1915), 137.
Anon, 'Obituary', *British Rainfall. 1915* (1916) 45–50.
Anon, 'Sir A.R. Binnie', Obituary in *Nature*, 99 (1917), 267.
Anon, 'News and Views: Henry Francis Blanford, F.R.S. (1834–93)', *Nature*, 133 (1934), 824.
Anon, 'Sir John Eliot, K.C.I.E., F.R.S., 1839–1908', *Nature*, 143 (1939), 847.
Archibald, E. D. 'Barometric Pressure and Temperature in India', *Nature*, 20 (1879), 54–5.
Bache, A. 'Note on the Effect of Deflected Currents on the Quantity of Rain Collected By a Rain-Gauge', *British Association for the Advancement of Science, Transactions of Section* 8 (1838), 25–27.
T. W. Backhouse, 'The Problem of Probable Error as Applied to Meteorology', *Quarterly Journal of the Royal Meteorological Society*, 17 (April 1891), 87–92.
Ball, J. (ed.), *Peaks, Passes and Glaciers: A Series of Excursions by Members of the Alpine Club* (London: Longman, 1859).
Ball, J. 'Suggestions for Alpine Travellers', in J. Ball (ed.) *Peaks, Passes and Glaciers: A Series of Excursions by Members of the Alpine Club* (London: Longman, 1859), pp. 482–508.
Ballot, B. *Suggestions on a Uniform System of Meteorological Observations* (Utrecht: Royal Dutch Meteorological Institute, 1872).
Bartholomew J. G. and Herbertson, A. J. (edited by Buchan, A.), *Atlas of Meteorology: A Series of over 400 Maps* (London: Archibald Constable, 1899).
Bell, J. H. B. 'The Mountains and Hills of Scotland', in J. H. B. Bell, E. F. Bozman and J. Fairfax Blakeborough (eds.) *British Hills and Mountains* (London: B. T. Batsford, 1940), pp. 1–62.
Benson, C. E. *British Mountaineering* (New York: George Routledge, 1909).
Blanford, H. F. *The Indian Meteorologist's Vade-Mecum* (Calcutta: Thacker, Spink, 1877).
Blanford, H. F. *The Rainfall of India, Volume III of the Indian Meteorological Memoirs* (Calcutta: Superintendent Government Printing, 1886).
Blanford, H. F. *A Practical Guide to the Climates and Weather of India, Ceylon and Burmah and the Storms of Indian Seas Based Chiefly on the Publications of the Indian Meteorological Department* (London: Macmillan, 1889).
Bonacina, L. C. W. 'The Effects of Exposure to Wind upon the Amount of Rain Caught by Rain-Gauges and the Methods of Protecting Rain-Gauges from Them', *British Rainfall. 1906* (1907), 27–45.
Buchan, A. *A Handy Book of Meteorology* (Edinburgh: William Blackwood, 1867).
Buchan, A. 'Sun-Spots and Rainfall', *Nature*, 17 (1878), 505–6.
Buchan, A. 'Meteorology of Ben Nevis', *Nature*, 25 (1881), 11–13.
Buchan, A. 'The Climate of the British Islands', *Journal of the Scottish Meteorological Society*, 7 (1884), 131–52.
Buchan, A. (ed.), 'The Meteorology of Ben Nevis', Special Issue of *Transactions of the Royal Society of Edinburgh*, 34 (1890), i–lxiv and 1–406.
Buchan, A. 'The Monthly and Annual Rainfall of Scotland, 1866 to 1890', *Journal of the Scottish Meteorological Society*, 10 (1892), 3–24.

Buchan, A. and Omond, R. T. (eds.), 'The Ben Nevis Observations 1888–1892', Special Issue of *Transactions of the Royal Society of Edinburgh*, 42 (1902), i–xiv and 1–552.

Buchan, A. and Omond, R. T. (eds.). 'The Ben Nevis Observations 1893–1897', Special Issue of *Transactions of the Royal Society of Edinburgh*, 43 (1905), 1–564.

Buchan, A. and Omond R. T. (eds.). 'The Ben Nevis Observations 1898–1902', Special Issue of *Transactions of the Royal Society of Edinburgh*, 44, part I (1910), i–v and 1–463.

Buchan, A. and Omond, R. T. (eds.). 'The Ben Nevis Observations 1903, 1904 and Appendix', Special Issue of *Transactions of the Royal Society of Edinburgh*, 44, part II (1910), 464–714.

Burder, G. F. 'On the Cause of the Decrease of Rain with Elevation', *British Rainfall. 1871* (1872), 33–41.

Burder, G. F. Letter to the Editor, 30 May 1871, *Meteorological Magazine*, 6 (1871), 75.

Chambers, C. 'Report on the Instrumental and Other Requirements of the Government Observatory, Colaba, for Increasing Its Efficiency and Extending the Field of Its Operations', in *Report of the Superintendent of the Government Observatory, Colaba* (Bombay: Military Secretariat Press, 1866), pp. 9–50.

Chambers, C. *Report on the Condition and Proceedings of the Government Observatory, Colaba, for the Period from September 1865 to December 31st 1867* (Bombay: Military Secretariat Press, 1868).

Chambers, C. *Report on the Condition and Proceedings of the Government Observatory, Colaba, for the Year Which Ended with the 30th June 1875* (Bombay: Government Central Press, 1876).

Chambers, C. *Report on the Condition and Proceedings of the Government Observatory, Colaba, for the Year Which Ended with the 30th June 1876* (Bombay: Government Central Press, 1877).

Chambers, C. *The Meteorology of the Bombay Presidency* (London: Her Majesty's Stationery Office, 1878).

Chappell, W. 'Science', *The Athanaeum*, 9 March (1878), 317–9.

Clouston, C. *An Explanation of the Popular Weather Prognostics of Scotland on Scientific Principles* (Edinburgh: Adam & Charles Black, 1867).

Craddock, J. M. 'Annual Rainfall in England since 1725', *Quarterly Journal of the Royal Meteorological Society*, 102 (1976), 823–40.

Crallan, T. E. 'On Rain Gauge Experiments at Framfield Lodge, Hurst Green', *British Rainfall. 1866* (1867), 21–4.

Crallan, T. E. 'Rain Gauge Experiments at Framfield Lodge, Hurst Green, Sussex', *British Rainfall. 1867* (1868), 45–9.

Daniell, J. *Meteorological Essays and Observations* (London: Thomas and George Underwood, 1863 [1823]).

Darwin, C. *Voyage of the Beagle* (London: Penguin, 1989 [1839]).

Dines, G. 'Difference of Rainfall with Elevation', *British Rainfall. 1877* (1878), 15–25.

Dines, G. 'Difference of Rainfall with Elevation. Being the Results of Experiments Made at Woodside, Hersham, Surrey', *British Rainfall. 1880* (1881), 13–16.

Eliot, J. *Instructions to Observers of the India Meteorological Department* (Calcutta: Office of the Superintendent of Government Printing, 1902, 2nd edition).

Fletcher, I. 'Remarks on the Rainfall among the Cumberland Mountains in 1865', *British Rainfall. 1865* (1866), 9–10.

Forbes, J. D. 'Report of the First and Second Meetings of the British Association for the Advancement of Science', *British Association for the Advancement of Science Second Report* (1833), 196–258.

Forbes, J. D. 'Supplementary Report on Meteorology', *Report of the Tenth Meeting of the British Association for the Advancement of Science* (1841), 37–156.

Fox, G. 'Rain-Gauge', *English Mechanic and World of Science*, 13 (1871), 434.

Galton, F. 'The Geometric Mean, in Vital and Social Statistics', *Proceedings of the Royal Society of London*, 28 (1879), 365–7.

Gaster, F. 'On the Monthly Percentage of Mean Annual Rainfall', *British Rainfall. 1868*, (1869), 50–4.

Gastrell, J. E. and Blanford, H. F. *Report on the Calcutta Cyclone of the 5th October 1864* (Calcutta: O.T. Cutter, 1866).

Geikie, A. *A Long Life's Work: An Autobiography* (London: Macmillan, 1924).

Geikie, A. *The Scenery of Scotland Viewed in Connection with Its Physical Geology* (New York: Macmillan, 1901).

Glaisher, J. 'Rainfall at the Royal Observatory, Greenwich', *British Rainfall. 1870* (1871), 42.

Glaisher, J. W. L. 'On the Law of Facility of Errors of Observations, and on the Method of Least Squares', *Memoirs of the Royal Astronomical Society*, 39 (1872), part II, 75–124.

Gold, E. 'An Atlas of Rainfall', *Nature*, 120 (1927), 291–2.

Greenwood, J. *The Sailor's Sea-Book: Rudimentary Treatise on Navigation* (London: John Weale, 1850).

Griffiths, C. H. 'Rain Gauge Experiments at Strathfield Turgiss', *British Rainfall. 1868* (1869), 12–30.

Griffith, C. H. 'Rain Gauge Experiments at Strathfield Turgiss, Reading', *British Rainfall. 1869* (1870), 25–33.

Harris, W. S. *Remarkable Instances of the Protection of Certain Ships of Her Majesty's Navy from the Destructive Effects of Lightning* (London: Richard Clay, 1847).

Herschel J. (ed.). *A Manual of Scientific Enquiry; Prepared for the Use of Officers in Her Majesty's Navy; and Travellers in General* (London: John Murray, 1849).

Herschel, J. 'An Address to the British Association for the Advancement of Science at the Opening of their Meeting at Cambridge, June 19th, 1845', reproduced in J. Herschel (eds.) *Essays from the Edinburgh and Quarterly Reviews* (London: Longman, Brown, Green, Longmans and & Roberts, 1857), pp. 634–82.

Herschel, J. *Instructions for Making and Registering Meteorological Observations in Southern Africa, and Other Countries in the South Seas, and Also at Sea* (London: Bradbury and Evans, 1835).

Herschel, J. 'Terrestrial Magnetism', in J. Herschel (ed.) *Essays from the Edinburgh and Quarterly Reviews* (London: Longman, Brown, Green, Longmans and Roberts, 1857 [1840]), pp. 63–141.

Herschel, J. 'The Weather and Weather Prophets', *Good Words*, 5 (1864), 57–64.

Hodgkinson, G. C. 'Hypsometry and the Aneroid', in Edward Shirley Kennedy (ed.) *Peaks, Passes and Glaciers; Being Excursions by Members of the Alpine Club.* 2nd Series (London: Longman, 1862), pp. 461–500.

Jevons, W. S. 'On the Deficiency of Rain in an Elevated Rain-Gauge, as Caused by Wind', *The London, Edinburgh, and Dublin Philosophical Magazine and Journal of Science*, 22 (1861), 421–33.

Knott, C. G. *Collected Scientific Papers of John Aitken. LL.D. F.R.S.* (Cambridge: Cambridge University Press, 1923).

Lloyd, H. *Account of the Magnetical Observatory of Dublin and of the Instruments and Methods of Observation Employed There* (Dublin: University Press, 1842).

Lloyd, H. 'Notes on the Meteorology of Ireland, Deduced from the Observations Made in the Year 1851, under the Direction of the Royal Irish Academy', *Transactions of the Royal Irish Academy*, 22 (1849), 411–98.

Lords Commissioners of the Admiralty, *Remarks on Revolving Storms* (London: HMSO, 1851).

MacPhee, G. G. (ed.) *Ben Nevis* (Edinburgh: Scottish Mountaineering Club, 1936).

Markham, C. R. *A Memoir on the Indian Surveys* (Cambridge: Cambridge University Press, 2014 [1871]).

Marriott, W. 'The Earliest Telegraphic Daily Meteorological Reports and Weather Maps', *Quarterly Journal of the Royal Meteorological Society*, 28 (1903), 123–31.

McAlister, D. 'The Law of the Geometrical Mean', *Proceedings of the Royal Society of London*, 28 (1879), 367–76.

Meickle, H. 'On the Different Quantities of Rain Collected in Rain-Gauges at Different Heights', *Annals of Philosophy*, 14 (1819), 312–3.

Merriman, M. *Elements of the Method of Least Squares* (London: Macmillan, 1877).

Meteorological Committee, *Report of the Proceedings of the Conference on Maritime Meteorology held in London, 1874* (London: HM Stationery Office, 1875).

Methven, R. *Narratives Written by Sea Commanders, Illustrative of the Law of Storms, and of Its Practical Application to Navigation. No. 1. The Blenheim's Hurricane of 1851; with Some Observations of the Storms of the South-East Trade* (London: John Weale, 1851).

Mill, H. R. 'Introduction', in Anon, *Rainfall Atlas of the British Isles* (London: Royal Meteorological Society, 1926), pp. 5–12.

Mill, H. R. *Life Interests of a Geographer 1861–1944: An Experiment in Autobiography* (East Grinstead: Privately Issued, 1945).

Mill, H. R. 'Map-Studies of Rainfall', *Quarterly Journal of the Royal Meteorological Society*, 34 (1908), 65–86.

Mill, H. R. *On the Distribution of Mean and Extreme Annual Rainfall over the British Isles* (London: William Clowes, 1905).

Mill, H. R. *On the Distribution of Mean and Extreme Annual Rainfall over the British Isles* (London: William Clowes, 1905).

Mill, H. R. 'The Development of Rainfall Measurement in the Last Forty Years', *British Rainfall. 1900* (1901) 23–45.

Mill, H. R. 'The Rainfall of the Forth Valley and the Construction of a Rainfall Map', *British Rainfall. 1915* (1916), 1–12.

Milne-Home, D. 'Address by D. Milne Home, Chairman of the Council of the Scottish Meteorological Society, to the General Meeting of the Society, held on 26th July 1877', *Journal of the Scottish Meteorological Society*, 5, new series (1877), 110–6.

Milne-Home, G. *Biographical Sketch of David Milne-Home* (Edinburgh: David Douglas, 1891).

Mitchell, A., Sanderson, J. and Murray, J. 'Introductory Note by the Council of the Scottish Meteorological Society', *Transactions of the Royal Society of Edinburgh*, 34 (1890), iii–v.

Nash, W. C. 'One Hundred Years' Greenwich Rainfall, 1815–1914', *British Rainfall. 1915*, 55 (1916), 35–39.

Negretti, H. and Zambra, J. W. *A Treatise on Meteorological Instruments: Explanatory of the Scientific Principles, Methods of Construction, and Practical Utility* (London: Negretti & Zambra, 1864).

O'Byrne, W. R. *Naval Biographical Dictionary* (London: John Murray, 1849).

Omond, R. T. 'Abstract of Paper on a Comparison of Observations at the Observatory and at the Public School, Fort-William', *Transactions of the Royal Society of Edinburgh*, 42 (1892), 537–40.

Piddington, H. *The Sailor's Horn-Book for the Law of Storms: Being a Practical Exposition of the Theory of the Law of Storms, and Its Uses to Mariners of all Classes, in all Parts of the World, Shewn by Transparent Storm Cards and Useful Lessons* (London: Williams and Norgate, 1848 [1860]).

Redfield, W. C. 'Remarks on the Prevailing Storms of the Atlantic Coast, of the North American States', *American Journal of Science and Arts*, 20 (1831), 17–51.

Reid, W. *An Attempt to Develop the Law of Storms by Means of Facts, Arranged According to Time and Place, and Hence to Point Out a Cause for the Variable Winds, With the View to Practical Use in Navigation* (London: John Weale, 1838).

Reid, W. *The Progress of the Development of the Law of Storms and of the Variable Winds, with the Practical Application of the Subject to Navigation* (London: John Weale, 1849).

Rivers Pollution Commission, *First Report of the Commissioners Appointed in 1868 to Inquire into the Best Means of Preventing the Pollution of Rivers. (Mersey and Ribble Basins.)* Vol.I Report and Plans (London: G.E. Eyre and William Spottiswoode, 1870).

Ross, J. C. *A Voyage of Discovery and Research in the Southern and Antarctic Regions, during the Years 1839–43*, Volume 1 (London: John Murray, 1847).

Ruskin, J. 'Remarks on the Present State of Meteorological Science', *Transactions of the Meteorological Society*, 1 (1839), 56–9.

Sabine, E. 'On Some Points in the Meteorology of Bombay', *Report of the Fifteenth Meeting of the British Association for the Advancement of Science* (1846), 73–82.

Sabine, E. 'On What the Colonial Magnetic Observatories Have Accomplished', *Proceedings of the Royal Society of London*, 8 (1856–1857), 395–413.

M. de Salter, C. S. *The Rainfall of the British Isles* (London: University of London Press, 1921).

Scott, R. H. 'The History of the Kew Observatory', *Proceedings of the Royal Society of London*, 39 (1885), 37–86.

Shaw, N. *Manual of Meteorology. Volume I. Meteorology in History* (Cambridge: Cambridge University Press, 1926).

Stevenson, T. 'On Ascertaining the Intensity of Storms by the Calculation of Barometric Gradients', Extract from Paper Read at the General Meeting of the Scottish Meteorological Society, June 1867, National Library of Scotland, Edinburgh.

Stewart B. and Haldane Gee, W.W. *Lessons in Elementary Practical Physics* (London: Macmillan, 1885).

Stow, F. W. 'Rain Gauge Experiments at Hawsker, near Whitby, Yorkshire', *British Rainfall. 1870* (1871), 9–22.

Stow, F. W. 'Rain Gauge Experiments at Hawsker, near Whitby, Yorkshire', *British Rainfall. 1871* (1872), 16–32.

Stow, F. W. Letter to the Editor, no date, *Meteorological Magazine*, 6 (1871), 69–72.

Stupart, R. F. 'The Toronto Magnetic Observatory', *Journal of Geophysical Research*, 3 (1898), 145–8.

Symons, G. J. 'Another Meteorological Congress', *Meteorological Magazine*, 7 (1872), 37–40.

Symons, G. J. 'A Contribution to the History of Rain Gauges', *Quarterly Journal of the Royal Meteorological Society*, 17 (1891), 127–42.

Symons, G. J. 'British Rainfall', *British Rainfall. On the Distribution of Rain over the British Isles, during the Years 1860 and 1861* (1862), 3.

Symons, G. J. 'Different of Rainfall with Elevation', *British Rainfall. 1878* (1879), 24–30.

Symons, G. J. 'English Climatological Stations', *Quarterly Journal of the Royal Meteorological Society*, 7 (1881), 281–3.

Symons, G. J. 'Incomplete Rain Gauge Experiments', *Meteorological Magazine*, 1 (1866), 96–7.

Symons, G. J. 'Meteorological Bibliography', *Meteorological Magazine*, 14 (1879), 17–22.

Symons, G. J. 'Meteorological Instruments in 1837 and 1897', *Quarterly Journal of the Royal Meteorological Society*, 23 (1897), 205–20.

Symons, G. J. 'Notes on the Preceding Papers', *British Rainfall. 1870* (1871), 43–5.

Symons, G. J. 'Notes upon the Foregoing Papers', *British Rainfall. 1871* (1872), 56–57.

Symons, G. J. 'On the Fall of Rain in the British Isles during the Years 1862 and 1863', *Report of the Thirty-Fourth Meeting of the British Association for the Advancement of Science* (1865), 367–407.

Symons, G. J. 'On the Rainfall of the British Isles', *Report of the Thirty-Fifth Meeting of the British Association for the Advancement of Science* (1866), 192–242.

Symons, G. J. 'On the Rainfall Observations Made upon York Minster by Professor John Phillips', *British Rainfall. 1881* (1882), 41–5.

Symons, G. J. 'On the Results of the Various Sets of Experimental Rain Gauges', *British Rainfall. 1867* (1868), 29–31.

Symons, G. J. 'On the Secular Variation of Rainfall in England since 1725', *British Rainfall. 1870* (1871), 53–7.

Symons, G. J. 'On the Secular Variation of Rainfall in England since 1725, Part II', *British Rainfall. 1871* (1872), 58–68.

Symons, G. J. 'Preface', *British Rainfall. 1872* (1873), 5.

Symons, G. J. 'Rain Gauges, and Hints on Observing Them', *British Rainfall. 1864* (1865), 8–15.

Symons, G. J. *Rain: How, When, Where, Why Is It Measured: Being a Popular Account of Rainfall Investigations* (London: Edward Stanford, 1867).

Symons, G. J. 'Report', *British Rainfall. 1862* (1863), 4–12.

Symons, G. J. 'Report', *British Rainfall. 1863* (1864), 4–7.

Symons, G. J. 'Report', *British Rainfall. 1865* (1866), 4–19.

Symons, G. J. 'Report', *British Rainfall. 1866* (1867), 4–8.

Symons, G. J. 'Report', *British Rainfall. 1868* (1869), 6–11.

Symons, G. J. 'Report', *British Rainfall. 1871* (1872), 6–15.

Symons, G. J. 'Report', *British Rainfall. 1873* (1874), 8–9.

Symons, G. J. 'Report of the Rainfall Committee for the Year 1865–1866', *Report of the Thirty-Sixth Meeting of the British Association for the Advancement of Science* (1867), 281–351.

Symons, G. J. 'Report of the Rainfall Committee for the Year 1867–68', *Report of the Thirty-Eighth Meeting of the British Association for the Advancement of Science* (1869), 432–74.

Symons, G. J. 'Report of the Rainfall Committee for the Year 1868–69', *Report of the Thirty-Ninth Meeting of the British Association for the Advancement of Science* (1870), 383–404.

Symons, G. J. 'Report of the Rainfall Committee for the Year 1869–70', *Report of the Fortieth Meeting of the British Association for the Advancement of Science* (1871), 170–227.

Symons, G. J. 'Report on the Rainfall of the British Isles for the Years 1875–76', *Report of the Forty-Sixth Meeting of the British Association for the Advancement of Science* (1876), 172–203.

Symons, G. J. 'Rules for Rainfall Observers', *British Rainfall. 1868* (1869), 102–3.

Symons, G. J. 'Second Report of the Rainfall Committee', *Report of the Thirty-Seventh Meeting of the British Association for the Advancement of Science* (1868), 448–67.

Symons, G. J. 'The Cause of the Decrease of Rainfall with Elevation', *Meteorological Magazine*, 6 (1871), 63–5.

Symons, G. J. 'The First Daily Weather Map, Sold in the Great Exhibition of 1851', *Meteorological Magazine*, 31 (1896), 113.

Symons, G. J. 'The History of English Meteorological Societies, 1823 to 1880', *Quarterly Journal of the Meteorological Society*, 7 (1881), 65–98.

Thrustans, J. 'Rain-Gauge', *English Mechanic and World of Science*, 13 (1871), 492.

Waldeck, M. 'Natural Prognostics of the Weather', *Quarterly Journal of the Society for Literature and the Arts* (January–June 1827), 501–2.
Weld, C. R. *A History of the Royal Society* (London: John W. Parker, 1848).
Whymper, E. *A Guide to Chamonix and the Range of Mont Blanc* (London: John Murray, 1900, 5th edition).
Wragge, C. L. 'Ascending Ben Nevis in Winter', *Chambers's Journal of Popular Literature, Science and Art*, 19 (1882), 265–8.
Wragge, C. L. 'Resumption of the Ben Nevis Meteorological Observatory', *Meteorological Magazine*, 17 (1882), 81–4.
Wragge, C. L. 'Watching the Weather on Ben Nevis', *Good Words*, 23 (1882), 343–7 and 377–85.
Wragge, C. L. 'Ben Nevis Observatory', *Nature*, 27 (1883), 487–91.

Secondary Sources

Adamson, G. C. D. '"The Languor of the Hot Weather": Everyday Perspectives on Weather and Climate in Colonial Bombay, 1819–1828', *Journal of Historical Geography*, 38 (2012), 143–54.
Adamson, G. C. D., Hannaford, M.J. and Rohland, E.J. 'Re-thinking the Past: The Role of a Historical Focus in Climate Change Adaptation Research', *Global Environmental Change*, 48 (2018), 195–205.
Agnew, D. C. 'Robert FitzRoy and the Myth of the "Marsden Squares": Transatlantic Rivalries in Early Marine Meteorology', *Notes and Records of the Royal Society*, 58 (2004), 21–46.
Anderson, K. 'Looking at the Sky: The Visual Context of Victorian Meteorology', *British Journal for the History of Science*, 36 (2003), 301–32.
Anderson, K. 'Mapping Meteorology', in J. R. Fleming, V. Jankovic and D. R. Coen (eds.) *Intimate Universality: Local and Global Themes in the History of Weather and Climate* (Sagamore Beach: Science History, 2006), pp. 69–92.
Anderson, K. *Predicting the Weather: Victorians and the Science of Meteorology* (Chicago: University of Chicago Press, 2005).
Arnold, D. *Science, Technology and Medicine in Colonial India* (Cambridge: Cambridge University Press, 2000).
Arnold, D. *The Tropics and the Travelling Gaze: India, Landscape, and Science, 1800–1856* (Seattle: University of Washington Press, 2006).
Aronova, E., von Oertzen C. and Sepkoski, D. 'Introduction: Historicizing Big Data', *Osiris*, 32 (2017), 1–17.
Ashworth, W. J. 'John Herschel, George Airy, and the Roaming Eye of the State', *History of Science*, 36 (1998), 151–78.
Ashworth, W. J. 'The Calculating Eye: Baily, Herschel, Babbage and the Business of Astronomy', *British Journal for the History of Science*, 27 (1994), 401–44.
Aubin, D. 'A History of Observatory Sciences and Techniques', in J.-P. Lasota (ed.) *Astronomy at the Frontiers of Science* (Heidelberg: Springer, 2011), pp. 109–21.
Aubin, D. 'The Fading Star of the Paris Observatory in the Nineteenth Century: Astronomers' Urban Culture of Circulation and Observation', *Osiris*, 18 (2003), 79–100.

Aubin, D. 'The Hotel that Became an Observatory: Mount Faulhorn as Singularity, Microcosm, and Macro-Tool', *Science in Context*, 22 (2009), 365–86.

Aubin, D., Bigg, C. and Sibum, H. O. 'Introduction: Observatory Techniques in Nineteenth-Century Science and Society', in D. Aubin, C. Bigg and H. O. Sibum (eds.) *The Heavens on Earth: Observatories and Astronomy in Nineteenth-Century Science and Culture* (Durham: Duke University Press, 2010), pp. 1–32.

Axelby, R. and Nair, S. P. *Science and the Changing Environment in India, 1780–1920* (London: British Library, 2010).

Barford, M. 'Fugitive Hydrography: The Nautical Magazine and the Hydrographic Office of the Admiralty, c.1832–1850', *International Journal of Maritime History*, 27 (2015), 208–26.

Bayly, C. A. *Empire and Information: Intelligence Gathering and Social Communication in India, 1780–1870* (Cambridge: Cambridge University Press, 1999).

Behrisch, E. *Discovery, Innovation, and the Victorian Admiralty: Paper Navigators* (Switzerland: Palgrave Macmillan, 2022).

Belteki, D. ''The Grand Strategy of an Observatory': George Airy's Vision for the Division of Astronomical Labour among Observatories during the Nineteenth Century', *Notes and Records of the Royal Society*, 77 (2023), 135–51.

Belteki, D. 'The Spring of Order: Robert Main's Management of Astronomical Labor at the Royal Observatory, Greenwich', *History of Science*, 60 (2022), 575–93.

Belteki, D. 'The Winter of Raw Computers: The History of the Lunar and Planetary Reductions of the Royal Observatory, Greenwich', *British Journal for the History of Science*, 56 (2023), 65–82.

Bigg, C., Aubin, D. and Felsch, P. 'Introduction: The Laboratory of Nature – Science in the Mountains', *Science in Context*, 22 (2009), 311–21.

Blouet, O. M. 'Sir William Reid, F.R.S., 1791–1858: Governor of Bermuda, Barbados and Malta', *Notes and Records of the Royal Society*, 40 (1986), 169–91.

Bravo, M. 'Geographies of Exploration and Improvement: William Scoresby and Arctic Whaling (1722–1822)', *Journal of Historical Geography*, 32 (2006), 512–38.

Brohan, P., Allen, R., Freeman J. E. et.al., 'Marine Observations of Old Weather', *Bulletin of the American Meteorological Society*, 90 (2009), 219–30.

Burnett, D. G. 'Matthew Fontaine Maury's "Sea of Fire": Hydrography, Biogeography, and Providence in the Tropics', in F. Driver and L. Martins (eds.) *Tropical Visions in an Age of Empire* (Chicago: Chicago University Press, 2005), pp. 113–36.

Burton, J. 'Robert FitzRoy and the Early History of the Meteorological Office', *British Journal for the History of Science*, 19 (1986), 147–76.

Caputo, S. 'Exploration and Mortification: Fragile Infrastructures, Imperial Narratives, and the Self-Sufficiency of British Naval "Discovery" Vessels, 1760–1815', *History of Science*, 61 (2023), 40–59.

Cardoso, S. S. S., Cartwright, J. H. E. and Huppert, H. E. 'Stokes, Tyndall, Ruskin and the Nineteenth-Century Beginnings of Climate Science', *Philosophical Transactions of the Royal Society A*, 378 (2020), 1–15.

Carter, C. 'Magnetic Fever: Global Imperialism and Empiricism in the Nineteenth Century', *Transactions of the American Philosophical Society*, 99 (2009), i–xxvi and 1–168.

Cawood, J. 'Terrestrial Magnetism and the Development of International Collaboration in the Early Nineteenth Century', *Annals of Science*, 34 (1977), 551–87.

Cawood, J. 'The Magnetic Crusade: Science and Politics in Early Victorian Britain', *Isis*, 70 (1979), 492–518.

Chakrabarti, P. *Western Science in Modern India: Metropolitan Methods, Colonial Practices* (Delhi: Permanent Black, 2004).

Cock, R. 'Scientific Servicemen in the Royal Navy and the Professionalisation of Science, 1816–55', in D. M. Knight and M. D. Eddy (eds.) *Science and Beliefs: From Natural Philosophy to Natural Science, 1700–1900* (Aldershot: Ashgate, 2005), pp. 95–112.

Coen, D. R. 'The Storm Lab: Meteorology in the Austrian Alps', *Science in Context*, 22 (2009), 463–86.

Courtney, N. *Gale Force 10: The Life and Legacy of Admiral Beaufort* (London: Review, 2002).

Crewe, M. E. 'The Met Office Grows Up: In War and Peace', *Occasional Papers on Meteorological History*, 8 (2009), 1–37.

Curran-Everett, D. 'Explorations in Statistics: Correlation', *Advances in Physiology Education*, 34 (2010), 186–91.

Damodaran, V. 'The East India Company, Famine and Ecological Conditions in Eighteenth-Century Bengal' in V. Damodaran, A. Winterbottom and A. Lester (eds.) *The East India Company and the Natural World* (Basingstoke: Palgrave Macmillan, 2015), pp. 80–101.

Damodaran, V., Winterbottom, A. and Lester, A. (eds.). *The East India Company and the Natural World* (Basingstoke: Palgrave Macmillan, 2015).

Daston, L. and Galison, P. *Objectivity* (New York: Zone Books, 2007).

Davis, M. *Late Victorian Holocausts: El Niño Famines and the Making of the Third World* (London: Verso, 2002).

Day, A. *The Admiralty Hydrographic Service 1795–1919* (London: Stationery Office, 1967).

Deacon, M. *Scientists and the Sea, 1650–1900: A Study of Marine Science* (London: Academic Press, 1971).

Desrosieres, A. *The Politics of Large Numbers: A History of Statistical Reasoning* (Cambridge, MA.: Harvard University Press, 1998).

Dettelbach, M. 'Humboldtian Science', in N. Jardine, J. A. Secord and E. C. Spary (eds.) *Cultures of Natural History* (Cambridge: Cambridge University Press, 1996), pp. 287–304.

Donnelly, K. *Adolphe Quetelet, Social Physics and the Average Man of Science 1796–1874* (London: Pickering and Chatto, 2015).

Donnelly, K. 'Redeeming Belgian Science: Periodic Phenomena and Global Physics in Brussels, 1825–1870', *History of Meteorology*, 8 (2017), 54–73.

Driver, F. and Martins, L. 'Shipwreck and Salvage in the Topics: The Case of HMS *Thetis*, 1830–1854', *Journal of Historical Geography*, 32 (2006), 539–62.

Driver, F. and Martins, L. (eds.). *Tropical Visions in an Age of Empire* (Chicago: University of Chicago Press, 2005).

Dry, S. 'Safety Networks: Fishery Barometers and the Outsourcing Judgement at the Early Meteorological Department', *British Journal for the History of Science*, 42 (2009), 35–56.

Dunn, R. '"Their Brains Over-Taxed": Ships, Instruments and Users', in D. Leggett and R. Dunn (eds.) *Re-inventing the Ship: Science, Technology and the Maritime World* (Aldershot: Ashgate, 2012), pp. 131–55.

Edney, M. *Mapping an Empire: The Geographical Construction of British India, 1765–1843* (Chicago: Chicago University Press, 1997).

Edwards, P. *A Vast Machine: Computer Models, Climate Data, and the Politics of Global Warming* (Cambridge, MA: MIT Press, 2010).

Endersby, J. *Imperial Nature: Joseph Hooker and the Practices of Victorian Science* (Chicago: University of Chicago Press, 2008).

Feintuck, A. 'Producing Spatial Knowledge: Map Making in Edinburgh, c.1880–c.1920', *Imago Mundi*, 69 (2017), 269–71.

Felsch, P. 'Mountains of Sublimity, Mountains of Fatigue: Towards a History of Speechlessness in the Alps', *Science in Context*, 22 (2009), 341–64.

FitzPatrick, P. J. 'Leading British Statisticians of the Nineteenth Century', *Journal of the American Statistical Association*, 55 (1960), 38–70.

Fleetwood, L. *Science on the Roof of the World: Empire and the Remaking of the Himalayas* (Cambridge: Cambridge University Press, 2022).

Fleming, F. *Barrow's Boys: A Stirring Story of Daring, Fortitude and Outright Lunacy* (London: Granta, 1999).

Fleming, J. R. *Historical Perspectives on Climate Change* (Oxford: Oxford University Press, 1998).

Fleming, J. R. *Meteorology in America 1800–1870* (Baltimore: Johns Hopkins University Press, 1990).

Friedman, R. M. *Appropriating the Weather: Vilhelm Bjerknes and the Construction of a Modern Meteorology* (Ithaca: Cornell, 1989).

Friendly, A. *Beaufort of the Admiralty: The Life of Sir Francis Beaufort 1774–1857* (New York: Random House, 1977).

García-Herrera, R., Barriopedro, D., Gallego, D. et al., 'Understanding Weather and Climate of the Last 300 Years from Ships' Logbooks', *WIREs Climate Change*, 9 (2018), 1–18.

Geyer, M. H. 'One Language for the World: The Metric System, International Coinage, Gold Standard, and the Rise of Internationalism, 1850–1900', in M. H. Geyer and J. Paulmann (eds.) *The Mechanics of Internationalism: Culture, Society, and Politics from the 1840s to the First World War* (Oxford: Oxford University Press, 2001), pp. 55–92.

Gillham, N. W. *A Life of St Francis Galton: From African Exploration to the Birth of Eugenics* (Oxford: Oxford University Press, 2001).

Gillin, E. 'The Instruments of Expeditionary Science and the Reworking of Nineteenth-Century Magnetic Experiment', *Notes and Records of the Royal Society*, 76 (2022), 565–92.

Golinski, J. *British Weather and the Climate of Enlightenment* (Chicago: University of Chicago Press, 2007).

Good, G. A. 'A Shift of View: Meteorology in John Herschel's Terrestrial Physics', in J. R. Fleming, V. Jankovic and D. R. Coen (eds.) *Intimate Universality: Local and Global Themes in the History of Weather and Climate* (Sagamore Beach: Science History, 2006), pp. 35–68.

Good, G. A. 'Between Data, Mathematical Analysis and Physical Theory: Research on Earth's Magnetism in the 19th century', *Centaurus*, 50 (2008), 290–304.

Gooday, G. J. N. *The Morals of Measurement: Accuracy, Irony, and Trust in Late Victorian Electrical Practice* (Cambridge: Cambridge University Press, 2010 [2004]).

Gooding, D., Pinch, T. and Schaffer, S. *The Uses of Experiment: Studies in the Natural Sciences* (Cambridge: Cambridge University Press, 1989).

Greer, K. 'Zoogeography and Imperial Defence: Tracing the Contours of the Nearctic Region in the Temperate North Atlantic, 1838–1880s', *Geoforum*, 65 (2015), 454–64.

Grove, R. 'The East India Company, the Raj and the El Niño: The Critical Role Played by Colonial Scientists in Establishing the Mechanisms of Global Climate Teleconnections 1770-1930,' in R. Grove, V. Damodaran and S. Sangwan (eds.) *Nature and the Orient: The Environmental History of South and Southeast Asia* (Oxford: Oxford University Press, 1998), pp. 301–23.

Hacking, I. *The Taming of Chance* (Cambridge: Cambridge University Press, 1990).

Hall, M. B. 'Public Science in Britain: The Role of the Royal Society', *Isis*, 72 (1981), 627–9.

Hankins, T. L. 'A "Large and Graceful Sinuosity": John Herschel's Graphical Method', *Isis*, 97 (2006), 605–33.

Hansen, P. H. 'Albert Smith, the Alpine Club, and the Invention of Mountaineering in Mid-Victorian Britain', *Journal of British Studies*, 34 (1995), 300–24.

Hawkins, E., Burt, S., Brohan, P. et al., 'Hourly Weather Observations from the Scottish Highlands (1883–1904) Rescued by Volunteer Citizen Scientists', *Geoscience Data Journal*, 6 (2019), 160–73.

Hawkins, E., Burt, S., McCarthy, M. et al., 'Millions of Historical Monthly Rainfall Observations Taken in the UK and Ireland Rescued by Citizen Scientists', *Geoscience Data Journal*, 10 (2023), 246–61.

Hawkins, E., Brohan, P., Burgess, S. N. et al., 'Rescuing Historical Weather Observations Improves Quantification of Severe Windstorm Risks', *Natural Hazards and Earth System Sciences*, 23 (2023), 1465–82.

Hevly, B. 'The Heroic Science of Glacier Motion', *Osiris*, 11 (1996), 66–86.

Heymann, M. 'The Evolution of Climate Ideas and Knowledge', *WIREs Climate Change*, 1 (2010), 581–97.

Higgitt, R. 'A British National Observatory: The Building of the New Physical Observatory at Greenwich, 1889–1898', *British Journal for the History of Science*, 47 (2014), 609–35.

Inkpen, D. 'The Scientific Life in the Alpine: Recreation and Moral Life in the Field', *Isis*, 109 (2018), 515–37.

Jankovic, V. 'Ideological Crests versus Empirical Troughs: John Herschel's and William Radcliffe Birt's Research on Atmospheric Waves, 1843–50', *British Journal for the History of Science*, 31 (1998), 21–40.

Jankovic, V. *Reading the Skies: A Cultural History of English Weather, 1650–1820* (Manchester: Manchester University Press, 2000).

Josefowicz, D. G. 'Experience, Pedagogy, and the Study of Terrestrial Magnetism', *Perspectives on Science*, 13 (2005), 452–94.

Keighren, I. and Withers, C. W. J. 'The Spectacular and the Sacred: Narrating Landscape in Works of Travel', *Cultural Geographies*, 19 (2012), 11–30.

Knowles Middleton, W. E. *A History of the Theories of Rain and Other Forms of Precipitation* (London: Oldbourne, 1965).

Larson, E. J. 'Public Science for a Global Empire: The British Quest for the South Magnetic Pole', *Isis*, 102 (2011), 34–59.

Lehmann, P. 'Average Rainfall and the Play of Colours: Colonial Experience and Global Climate Data', *Studies in History and Philosophy of Science*, 70 (2018), 38–49.

Leighly, J. 'Climatology since the Year 1800', *Transactions of the American Geophysical Union*, 30 (1949), 658–72.

Levere, T. *Science and the Canadian Arctic: A Century of Exploration, 1818–1918* (Cambridge: Cambridge University Press, 1993).

Livingstone, D. N. 'Tropical Climate and Moral Hygiene: The Anatomy of a Victorian Debate', *British Journal for the History of Science*, 32 (1999), 93–110.

Lloyd, C. *Mr. Barrow of the Admiralty: A Life of Sir John Barrow* (London: Collins, 1970).

Locher, F. 'The Observatory, the Land-Based Ship and the Crusades: Earth Sciences in European Context, 1830–50', *British Journal for the History of Science*, 40 (2007), 491–504.

Maas, H. and Morgan, M. S. 'Timing History: The Introduction of Graphical Analysis in 19th Century British Economics', *Revue d'Histoire des Sciences Humaines*, 7 (2002), 97–127.

Macdonald, L. T. 'Making Kew Observatory: The Royal Society, the British Association and the Politics of Early Victorian Science', *British Journal for the History of Science*, 48 (2015), 409–33.

Macdonald, L. T. *Kew Observatory and the Evolution of Victorian Science* (Pittsburgh: University of Pittsburgh Press, 2018).

MacLeod, R. 'Science and the Treasury: Principles, Personalities and Policies, 1870–85', in G. L'E Turner (ed.) *The Patronage of Science in the Nineteenth Century* (Leiden: Noordhoff International Publishing, 1976), pp. 115–72.

MacLeod, R. 'The Royal Society and the Government Grant: Notes on the Administration of Scientific Research, 1849–1914', *Historical Journal*, 14 (1971), 323–58.

MacLeod, R. 'Whigs and Savants: Reflections on the Reform Movement in the Royal Society, 1830–48', in I. Inkster and J. Morrell (eds.) *Metropolis and Province: Science in British Culture, 1780–1850* (London: Hutchinson, 1983), pp. 55–90.

MacLeod, R. *Public Science and Public Policy in Victorian England* (Ashgate: Aldershot, 1996).

Mahony, M. 'Climate and Climate Change', in M. Domosh, M. Heffernan and C. W. J. Withers (eds.) *The SAGE Handbook of Historical Geography* (London: Sage, 2020), pp. 579–601.

Mahony, M. 'Meteorology and Empire', in A. Goss (ed.) *The Routledge Handbook of Science and Empire* (Abingdon, Routledge, 2021), pp. 47–58.

Mahony, M. 'The "Genie of the Storm": Cyclonic Reasoning and the Spaces of Weather Observation in the Southern Indian Ocean, 1851–1925', *British Journal for the History of Science*, 51 (2018), 607–33.

Mahony, M. and Caglioti, A. M. 'Relocating Meteorology', *History of Meteorology*, 8 (2017), 1–14.

Mawer, G. A. *South by Northwest: The Magnetic Crusade and the Contest for Antarctica* (Edinburgh: Birlinn, 2006).

McAleer, J. '"Stargazers at the World's End": Telescopes, Observatories and "Views" of Empire in the Nineteenth-Century British Empire', *British Journal for the History of Science*, 46 (2013), 389–414.

McNee, A. *The New Mountaineer in Late Victorian Britain: Materiality, Modernity, and the Haptic Sublime* (London: Palgrave Macmillan, 2016).

Michel N. and Smadja, I. 'Mathematics in the Archives: Deconstructive Historiography and the Shaping of Modern Geometry (1837–1852)', *British Journal for the History of Science*, 54 (2021), 423–41.

Millar, S. 'Science at Sea: Soundings and Instrumental Knowledge in British Polar Expedition Narratives, c.1818–1848', *Journal of Historical Geography*, 42 (2013), 77–87.

Miller, D. P. 'Between Hostile Camps: Sir Humphrey Davy's Presidency of the Royal Society of London, 1820–1827', *British Journal for the History of Science*, 16 (1983), 1–47.

Miller, D. P. 'The Revival of the Physical Sciences in Britain, 1815–1840', *Osiris*, 2 (1986), 107–34.

Moore, P. *The Weather Experiment: The Pioneers Who Sought to See the Future* (London: Chatto & Windus, 2015).

Morgan, B. 'After the Arctic Sublime', *New Literary History*, 47 (2016), 1–26.

Morrell J. and Thackray, A. *Gentlemen of Science: Early Years of the British Association of the Advancement of Science* (Oxford: Clarendon Press, 1981).

Musselman, E. G. 'Swords into Ploughshares: John Herschel's Progressive View of Astronomical and Imperial Governance', *British Journal for the History of Science*, 31 (1998), 419–35.

Musselman, E. G. 'Worlds Displaced: Projecting the Celestial Environment from the Cape Colony', *Environmental History*, 3 (2003), 64–85.

Naylor, S. 'Thermometer Screens and the Geographies of Uniformity in Nineteenth-Century Meteorology', *Notes and Records of the Royal Society*, 73 (2019), 203–21.

Naylor, S. *Regionalising Science: Placing Knowledges in Victorian England* (Pittsburgh: University of Pittsburgh Press, 2010).

Naylor, S. 'Nationalising Provincial Weather: Meteorology in Nineteenth-Century Cornwall', *British Journal for the History of Science*, 39 (2006), 407–33.

Naylor, S. and Goodman, M. 'Atmospheric Empire: Historical Geographies of Meteorology at the Colonial Observatories', in M. Mahony and S. Randalls (eds.) *Weather, Climate and the Geographical Imagination: Placing Atmospheric Knowledges* (Pittsburgh: University of Pittsburgh Press, 2020), pp. 25–42.

Naylor, S. and Schaffer, S. 'Nineteenth-Century Survey Sciences: Enterprises, Expeditions and Exhibitions', *Notes and Records of the Royal Society*, 73 (2019), 135–47.

Nebeker, F. *Calculating the Weather: Meteorology in the 20th Century* (New York: Academic Press, 1995).

Nicholson, T. 'Bartholomew and the Half-Inch Layer Coloured Map 1883–1903', *The Cartographic Journal*, 37 (2000), 123–45.

von Oertzen, C. 'Machineries of Data Power: Manual versus Mechanical Census Compilation in Nineteenth-Century Europe', *Osiris*, 32 (2017), 129–50.

Parker, W. 'Distinguishing Real Results from Instrumental Artifacts: The Case of the Missing Rain', in G. Hon, J. Schickore and F. Steinle (eds.) *Going Amiss in Experimental Research* (Dordrecht: Springer, 2009), pp. 161–77.

Pedgley, D. E. 'A Short History of the British Rainfall Organization', *Occasional Papers on Meteorological History*, 5 (2002), 1–19.

Pratt, M. L. *Imperial Eyes: Travel Writing and Transculturation* (London: Routledge, 1992).

Raj, K. 'Beyond Postcolonialism ... and Postpositivism: Circulation and the Global History of Science', *Isis*, 104 (2013), 337–47.

Raj, K. *Relocating Modern Science: Circulation and the Construction of Knowledge in South Asia and Europe, 1650–1900* (Basingstoke: Palgrave Macmillan, 2007).

Raposo, P. M. P. 'Time, Weather and Empires: The Campos Rodrigues Observatory in Lourenço Marques, Mozambique (1905–1930)', *Annals of Science*, 72 (2015), 279–305.

Ratcliff, J. 'The East India Company, the Company's Museum, and the Political Economy of Natural History in the Early Nineteenth Century', *Isis*, 107 (2016), 459–517.

Ratcliff, J. 'Travancore's Magnetic Crusade: Geomagnetism and the Geography of Scientific Production in a Princely State', *British Journal for the History of Science*, 49 (2016), 325–52.

Ratcliff, J. *The Transit of Venus Enterprise in Victorian Britain* (London: Pickering & Chatto, 2008).

Reidy, M. S. 'Mountaineering, Masculinity, and the Male Body in Mid-Victorian Britain', *Osiris*, 30 (2015), 158–81.

Reidy, M. S. *Tides of History: Ocean Science and Her Majesty's Navy* (Chicago: Chicago University Press, 2008).

Risio, S.D. *Making Space in the Capital: Scientific Knowledge in Edinburgh's Calton Hill Observatory c.1811–1888*, Unpublished PhD thesis, University of Edinburgh, 2024.

Ritchie, G. S. *The Admiralty Chart: British Naval Hydrography in the Nineteenth Century* (London: Hollis & Carter, 1967).

Roberts, G. W. 'Magnetism and Chronometers: The Research of the Reverend George Fisher', *British Journal for the History of Science*, 42 (2009), 57–72.

Roberts, L. 'Accumulation and Management in Global Historical Perspective: An Introduction', *History of Science*, 52 (2014), 227–46.

Rodger, N. A. M. 'From the "Military Revolution" to the "Fiscal-Naval" State', *Journal of Maritime Research*, 13 (2011), 119–28.

Rozwadowski, H. *Fathoming the Ocean: The Discovery and Exploration of the Deep Sea* (Cambridge: The Belknap Press of Harvard University Press, 2005).

Sangwan, S. 'From Gentlemen Amateurs to Professionals: Reassessing the Natural Science Tradition in Colonial India 1780–1840' in R. Grove, V. Damodaran and S. Sangwan (eds.) *Nature and the Orient: The Environmental History of South and Southeast Asia* (Oxford: Oxford University Press, 1998), pp. 210–36.

Sanhueza-Cerda, C. 'Stabilizing Local Knowledge: The Installation of a Meridian Circle at the National Astronomical Observatory of Chile (1908–1913)', *Isis*, 113 (2022), 710–27.

Schaffer, S. 'Astronomers Mark Time: Discipline and the Personal Equation', *Science in Context*, 2 (1988), 115–45.

Schaffer, S. 'Astronomy at the Imperial Meridian: The Colonial Production of Hybrid Spaces', Plenary Lecture at the International Conference of Historical Geographers, London, 2015.

Schaffer, S. 'Babbage's Calculating Engines and the Factory System', *Réseaux: The French Journal of Communication*, 4 (1996), 271–98.

Schaffer, S. 'Keeping the Books at Paramatta Observatory', in D. Aubin, C. Bigg and H. Otto Sibum (eds.) *The Heavens on Earth: Observatories and Astronomy in Nineteenth-Century Science and Culture* (Durham: Duke University Press, 2010), pp. 118–47.

Schaffer, S. '"On Seeing Me Write": Inscription Devices in the South Seas', *Representations*, 97 (2007), 90–122.

Schaffer, S. 'Physics Laboratories and the Victorian Country House', in C. Smith and J. Agar (eds.) *Making Space for Science* (Macmillan: Basingstoke, 1998), pp. 149–80.

Schaffer, S. 'The Bombay Case: Astronomers, Instrument Makers and the East India Company', *Journal of the History of Astronomy*, 43 (2012), 151–80.

Schaffer, S. 'Oriental Meteorology and the Politics of Antiquity in Nineteenth-Century Survey Sciences', *Science in Context*, 30 (2017), 173–212.

Schiebinger, L. and Swan, C. (eds.) *Colonial Botany: Science, Commerce, and Politics in the Early Modern World* (Philadelphia: University of Pennsylvania Press, 2005).

Sen, J. *Astronomy in India, 1784–1876* (London: Pickering & Chatto, 2014).

Sheynin, O. B. 'On the History of the Statistical Method in Meteorology', *Archive for History of Exact Sciences*, 31 (1984), 53–95.

Sikka, D. R. 'The Role of the India Meteorological Department, 1875–1947', in U. D. Gupta (ed.) *Science and Modern India: An Institutional History, c.1784–1947* (Delhi: Pearson, 2011), pp. 381–426.

Simpson, T. and Hulme, M. 'Climate, Cartography, and the Life and Death of the "Natural Region" in British Geography', *Journal of Historical Geography*, 80 (2023), 44–57.

Smith, I. *Shadow of the Matterhorn: The Life of Edward Whymper* (Herefordshire: Carreg, 2011).

Smith, P. *Weather Pioneers: The Signals Corps Station at Pikes Peak* (Athens: Ohio University Press, 1993).

Smyth, F. S. *Edward Whymper* (London: Hodder and Stoughton, 1940).
Sorrenson, R. 'The Ship as a Scientific Instrument in the Eighteenth Century', in H. Kuklick and R. E. Kohler (eds.) *Science in the Field* (Chicago: Chicago University Press, 1996), pp. 221–36.
Stigler, S. *The History of Statistics: The Measurement of Uncertainty before 1900* (London: Belknap Press of Harvard University Press, 1986).
Stigler, S. M. 'Francis Galton's Account of the Invention of Correlation', *Statistical Science*, 4 (1989), 73–9.
Strasser, B. J. and Edwards, P. N. 'Big Data Is the Answer ... But What Is the Question?', *Osiris*, 32 (2017), 328–44.
Sutton, J. *Lords of the East: The East India Company and Its Ships (1600–1874)* (London: Conway Maritime Press, 2000).
Swinney, G. N. 'William Speirs Bruce, the Ben Nevis Observatory, and Antarctic Meteorology', *Scottish Geographical Journal*, 118 (2002), 263–82.
Tucker, J. 'Objectivity, Collective Sight, and Scientific Personae', *Victorian Studies*, 50 (2008), 648–57.
Vetter, J. *Field Life: Science in the American West during the Railroad Era* (Pittsburgh: University of Pittsburgh Press, 2016).
Walker, M. *History of the Meteorological Office* (Cambridge: Cambridge University Press, 2012).
Ward, C. and Dowdeswell, J. 'On the Meteorological Instruments and Observations Made during the 19th Century Exploration of the Canadian Northwest Passage', *Arctic, Antarctic, and Alpine Research*, 38 (2006), 454–64.
Waring, S. 'The Board of Longitude and the Funding of Scientific Work: Negotiating Authority and Expertise in the Early Nineteenth Century', *Journal of Maritime Research*, 16 (2014), 55–71.
Webb, A. 'More Than Just Charts: Hydrographic Expertise within the Admiralty, 1795–1829', *Journal for Maritime Research*, 16 (2014), 43–54.
Wille, R.-J. 'Colonizing the Free Atmosphere: Wladimir Köppen's "Aerology", the German Maritime Observatory, and the Emergence of a Trans-Imperial Network of Weather Balloons and Kites, 1873–1906', *History of Meteorology*, 8 (2017), 95–123.
Williamson, F. 'Weathering the Empire: Meteorological Research in the Early British Straits Settlements', *British Journal for the History of Science*, 48 (2015), 475–49.
Winter, A. '"Compasses All Awry": The Iron Ship and the Ambiguities of Cultural Authority in Victorian Britain', *Victorian Studies*, 38 (1994), 69–98.
Wise, M. N. and Smith, C. 'Work and Industry in Lord Kelvin's Britain', *Historical Studies in the Physical and Biological Sciences*, 17 (1986), 147–73.
Withers, C. W. J. *Geography, Science and National Identity: Scotland Since 1520* (Cambridge: Cambridge University Press, 2001).
Withers, C. W. J. 'Science, Scientific Instruments and Questions of Method in Nineteenth-Century British Geography', *Transactions of the Institute of British Geographers*, 38 (2012), 167–79.
Zeller, S. *Inventing Canada: Early Victorian Science and the Idea of a Transcontinental Nation* (Montreal: McGill-Queen's University Press, 2009).

Index

Printed in the United States
by Baker & Taylor Publisher Services